Numbers and Computers

Ronald T. Kneusel

Numbers and Computers

Second Edition

Springer

Ronald T. Kneusel
Broomfield, Colorado, USA

ISBN 978-3-319-50507-7 ISBN 978-3-319-50508-4 (eBook)
DOI 10.1007/978-3-319-50508-4

Library of Congress Control Number: 2016960676

Printed on acid-free paper

This Springer imprint is published by Springer Nature
The registered company is Springer International Publishing AG
The registered company address is: Gewerbestrasse 11, 6330 Cham, Switzerland

To my parents, Janet and Tom, for fostering my love of science.

Preface to the Second Edition

This book is about numbers and how they are represented and manipulated by computers. In this second edition, we expand on the coverage of the first edition by adding three new chapters and new material to several sections.

Broomfield, CO, USA
November 2016

Ronald T. Kneusel

Preface

This is a book about numbers and how those numbers are represented in and operated on by computers.

Of course, numbers are fundamental to how computers operate because, in the end, everything a computer works with is a number. It is crucial that people who develop a code understand this area because the numerical operations allowed by computers and the limitations of those operations, especially in the area of floating point math, affect virtually everything people try to do with computers. This book aims to help by exploring, in sufficient, but not overwhelming, detail, just what it is that computers do with numbers.

Who Should Read This Book

This book is for anyone who develops software including software engineers, scientists, computer science students, engineering students, and anyone who programs for fun.

If you are a software engineer, you should already be familiar with many of the topics in this book, especially if you have been in the field for any length of time. Still, I urge you to press on, for perhaps you will find a gem or two which are new to you.

If you are old enough, you will remember the days of Fortran and mainframes. If so, like the software engineers above, you are probably also familiar with the basics of how computers represent and operate on numbers, but, also like the software engineers above, you will likely find a gem or two of your own. Scientists in particular should be aware of the limitations and pitfalls associated with using floating point numbers since few things in science are restricted to integers.

Students need this book because it is essential to know what the computer is doing under the hood. After all, if you are going to make a career of computers, why would you not want to know how the machine works?

How to Use This Book

This book consists of two main parts. The first deals with standard representations of integers and floating point numbers, while the second details several other number representations which are nice to know about and handy from time to time. Either part is a good place to start, though it is probably best if the parts themselves are read from start to end. Later, after the book has been read, you can use it as a reference.

There are exercises at the end of each chapter. Most of these are of the straightforward pencil and paper kind, just to test your understanding, while others are small programming projects meant to increase your familiarity with the material. Exercises that are (subjectively) more difficult will be marked with either one or two stars (* or **) to indicate the level of difficulty.

Example code is in C and/or Python version 2.7 though earlier 2.x versions should work just as well. Intimate knowledge of these programming languages is not necessary in order to understand the concepts being discussed. If something is not easy to see in the code, it will be described in the text. Why C? Because C is a low-level language, close to the numbers we will be working with, and because C is the grandfather of most common programming languages in current use including Python. In general, code will be offset from text and in a `monospace` font. For readers not familiar with C and/or Python, there are a plethora of tutorials on the web and reference books by the bookcase. Two examples, geared toward people less familiar with programming, are *Beginning C* by Ivor Horton and *Python Programming Fundamentals* by Kent Lee. Both of these texts are available from Springer in print or e-book format.

At the end of each chapter are references for material presented in the chapter. Much can be learned by looking at these references. Almost by instinct we tend to ignore sections like this as we are now programmed to ignore advertisements on web pages. In this former case, resist temptation; in the latter case, keep calm and carry on.

Acknowledgments

This book was not written in a vacuum. Here, I want to acknowledge those who helped make it a reality. First the reviewers, who gave of their time and talent to give me extremely valuable comments and friendly criticism: Robert Kneusel, M.S.; Jim Pendleton; Ed Scott, Ph.D.; and Michael Galloy, Ph.D. Gentlemen, thank you. Second, thank you to Springer, especially my editor, Courtney Clark, for moving ahead with this book. Lastly, and most importantly, thank you to my wife, Maria, and our children: David, Peter, Paul, Monica, Joseph, and Francis. Without your patience and encouragement, none of this would have been written.

Broomfield, CO, USA — Ronald T. Kneusel
December 2016 — AM+DG

Contents

Part I Standard Representations

1 Number Systems ... 3
- 1.1 Representing Numbers ... 3
- 1.2 The Big Three (and One Old Guy) ... 8
- 1.3 Converting Between Number Bases ... 10
- 1.4 Chapter Summary ... 16
- Exercises ... 16
- References ... 17

2 Integers ... 19
- 2.1 Bits, Nibbles, Bytes, and Words ... 19
- 2.2 Unsigned Integers ... 21
  - 2.2.1 Representation ... 21
  - 2.2.2 Storage in Memory: Endianness ... 22
- 2.3 Operations on Unsigned Integers ... 25
  - 2.3.1 Bitwise Logical Operations ... 25
  - 2.3.2 Testing, Setting, Clearing, and Toggling Bits ... 30
  - 2.3.3 Shifts and Rotates ... 33
  - 2.3.4 Comparisons ... 37
  - 2.3.5 Arithmetic ... 41
  - 2.3.6 Square Roots ... 52
- 2.4 What About Negative Integers? ... 54
  - 2.4.1 Sign-Magnitude ... 54
  - 2.4.2 One's Complement ... 55
  - 2.4.3 Two's Complement ... 55
- 2.5 Operations on Signed Integers ... 56
  - 2.5.1 Comparison ... 56
  - 2.5.2 Arithmetic ... 58
- 2.6 Binary-Coded Decimal ... 67
  - 2.6.1 Introduction ... 67
  - 2.6.2 Arithmetic with BCD ... 69

2.6.3 Conversion Routines 70
2.6.4 Other BCD Encodings 73
2.7 Chapter Summary 76
Exercises 77
References 79

3 Floating Point 81
3.1 Floating-Point Numbers 81
3.2 An Exceedingly Brief History of Floating-Point Numbers 84
3.3 Comparing Floating-Point Representations 85
3.4 IEEE 754 Floating-Point Representations 89
3.5 Rounding Floating-Point Numbers (IEEE 754) 97
3.6 Comparing Floating-Point Numbers (IEEE 754) 100
3.7 Basic Arithmetic (IEEE 754) 102
3.8 Handling Exceptions (IEEE 754) 105
3.9 Floating-Point Hardware (IEEE 754) 108
3.10 Binary Coded Decimal Floating-Point Numbers 110
3.11 Chapter Summary 113
Exercises 114
References 115

4 Pitfalls of Floating-Point Numbers (and How to Avoid Them) 117
4.1 What Pitfalls? 117
4.2 Some Experiments 119
4.3 Avoiding the Pitfalls 130
4.4 Chapter Summary 134
Exercises 135
References 135

Part II Other Representations

5 Big Integers and Rational Arithmetic 139
5.1 What is a Big Integer? 139
5.2 Representing Big Integers 140
5.3 Arithmetic with Big Integers 146
5.4 Alternative Multiplication and Division Routines 158
5.5 Implementations 167
5.6 Rational Arithmetic with Big Integers 171
5.7 When to Use Big Integers and Rational Arithmetic 177
5.8 Chapter Summary 180
Exercises 180
References 181

6 Fixed-Point Numbers 183
6.1 Representation (Q Notation) 183
6.2 Arithmetic with Fixed-Point Numbers 188
6.3 Trigonometric and Other Functions 194

6.4 An Emerging Use Case 204
6.5 When to Use Fixed-Point Numbers 211
6.6 Chapter Summary 212
Exercises 212
References 213

7 Decimal Floating Point 215
7.1 What is Decimal Floating-Point? 215
7.2 The IEEE 754-2008 Decimal Floating-Point Format 216
7.3 Decimal Floating-Point in Software 225
7.4 Thoughts on Decimal Floating-Point 232
7.5 Chapter Summary 233
Exercises 234
References 234

8 Interval Arithmetic 235
8.1 Defining Intervals 235
8.2 Basic Operations 237
8.3 Functions and Intervals 253
8.4 Implementations 258
8.5 Thoughts on Interval Arithmetic 262
8.6 Chapter Summary 263
Exercises 263
References 263

9 Arbitrary Precision Floating-Point 265
9.1 What is Arbitrary Precision Floating-Point? 265
9.2 Representing Arbitrary Precision Floating-Point Numbers 265
9.3 Basic Arithmetic with Arbitrary Precision Floating-Point Numbers 270
9.4 Comparison and Other Methods 273
9.5 Trigonometric and Transcendental Functions 274
9.6 Arbitrary Precision Floating-Point Libraries 278
9.7 Thoughts on Arbitrary Precision Floating-Point 290
9.8 Chapter Summary 291
Exercises 291
References 292

10 Other Number Systems 293
10.1 Introduction 293
10.2 Logarithmic Number System 293
10.3 Double-Base Number System 307
10.4 Residue Number System 324
10.5 Redundant Signed-Digit Number System 332
10.6 Chapter Summary 339
Exercises 340
References 341

Index 343

Part I
Standard Representations

Chapter 1
Number Systems

Abstract Computers use number bases other than the traditional base 10. In this chapter we take a look at number bases focusing on those most frequently used in association with computers. We look at how to construct numbers in these bases as well as how to move numbers between different bases.

1.1 Representing Numbers

The ancient Romans used letters to represent their numbers. These are the "Roman numerals" which are often taught to children,

I	1
II	2
III	3
IV	4 (1 before 5)
V	5
X	10
L	50
C	100
D	500
M	1000

By grouping these numbers we can build larger numbers (integers),

$$\text{MCMXCVIII} = 1998$$

The Romans built their numbers from earlier Egyptian numbers as seen in Fig. 1.1. Of course, we do not use either of these number systems for serious computation today for the simple reason that they are hard to work with.

For example, the Egyptians multiplied by successive doubling and adding. So, to multiply 17×18 to get 306 the Egyptians would make a table with two columns. The first column consists of the powers of two: $2^0, 2^1, 2^2, 2^3, 2^4 = 1, 2, 4, 8, 16$, etc. and the second column is the first number to be multiplied and its doublings.

R.T. Kneusel, *Numbers and Computers*, DOI 10.1007/978-3-319-50508-4_1

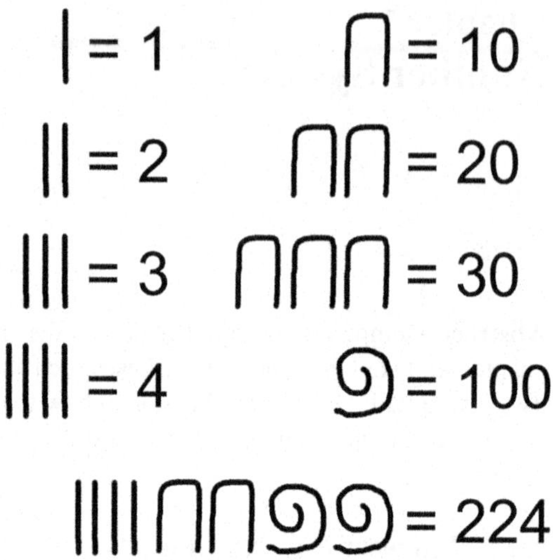

Fig. 1.1 Egyptian numbers. The ancient Egyptians wrote numbers by summing units, tens, and hundreds. There were also larger valued symbols which are not shown. Numbers were written by grouping symbols until the sum equaled the desired number. The order in which the symbols were written, largest first or smallest first, did not matter, but common symbols were grouped together. In this example, the smallest units are written first when writing from left to right to represent 224

In this case, `17` is the first number in the problem so the second column would be `17, 34, 68, 136, 272`, etc. where the next number in the sequence is two times the previous number. Therefore, the table would look like this (ignore the starred rows for the moment),

1	17	
2	34	*
4	68	
8	136	
16	272	*

Next, the Egyptians would mark the rows of the table that make the sum of the entries in the first column equal to the second number to multiply, in this case `18`. These are marked in the table already and we see that $2 + 16 = 18$. Lastly, the final answer is found by summing the entries in the second column for the marked rows, $34 + 272 = 306$.

This technique works for any pair of integers. If one were to ask the opposite question, what number times `17` gives `306`, one would construct the table as before and then mark rows until the sum of the numbers in the second column is `306`. The answer, then, is to sum the corresponding numbers in the first column to get `18`. Hence, division was accomplished by the same trick. In the end, since we have algebra at our disposal, we can see that what the Egyptians were really doing was applying the distributive property of multiplication over addition,

$$17(18) = 17(2 + 16) = 34 + 272 = 306$$

But, still, this is a lot of effort. Fortunately, there is another way to represent numbers: place notation (or, more formally, place-value notation). In place notation, each digit represents the number of times that power of the base is present in the number with the powers increasing by one for each digit position we move to the left. Negative powers of the base are used to represent fractions less than one. The digit values we typically use range from 0 to 9 with 10 as the base. Here we write 10 to mean $1 \times 10^1 + 0 \times 10^0$ thereby expressing the base of our numbers in the place notation using that base. Naturally, this fact is true regardless of the base, so 10_B is the base B for all $B > 1$. Computers do use base 10 numbers but not very often though we will see a few examples of this later in the book.

In general, if the base of a number system is B then numbers in that base are written with digits which run from 0 to $B-1$. These digits, in turn, count how many instances of the base raised to some integer power are present in the number. Lots of words, but an equation may clarify,

$$abcd_B = a \times B^3 + b \times B^2 + c \times B^1 + d \times B^0$$

for digits a, b, c, and d. Notice how the exponent of the base is counting down. This counting continues after zero to negative exponents which are just fractions,

$$B^{-n} = \frac{1}{B^n}$$

So, we can introduce a "decimal point" to represent fractions of the base. Formally, this is known as the *radix point* and is the character which separates the integer part of a number in a particular base from the fractional part. In this book, we will use the "." (period) character and indicate the base of the number with a subscript after the number,

$$abcd.efg_B = a \times B^3 + b \times B^2 + c \times B^1 + d \times B^0 + e \times B^{-1} + f \times B^{-2} + g \times B^{-3}$$

for additional digits e, f, g and base B. If no explicit base is given, assume the base is 10.

Before moving on to modern number systems let's take a quick look at two ancient number systems that used place notation. The first is the Babylonian number system which used base 60 (*sexagesimal*) but constructed its digits by grouping sets of symbols representing ones and tens. The second is the Mayan number system which used base 20 and constructed digits by combining dots for one and bars for five.

Figure 1.2 illustrates Babylonian numbers and performs some simple additions with them. One interesting thing to note is that the ancient Babylonians did not have a symbol for zero or the radix point. Instead, they used spaces and context from the text of the document to imply where the radix point should be inserted. Figure 1.2b shows examples of decimal numbers written in base 60 and as the Babylonians would have written them. Since it is difficult to work with the actual

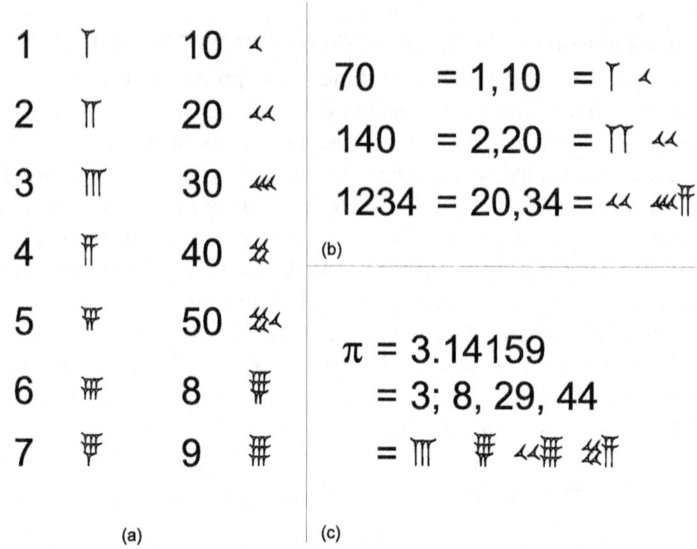

Fig. 1.2 (**a**) Sexagesimal numbers in ancient Babylon used groups of symbols for one and ten to be the digits of their base 60 numbers as opposed to our use of ten different symbols for our digits. (**b**) Decimal numbers written in sexagesimal. The *left column* is the decimal number, the *middle column* is the equivalent base 60 number using decimal digits and "," to separate the digits. The *right column* shows the Babylonian number which matches directly with the *middle column* representation. (**c**) The Babylonians were able to write numbers with fractional parts by using a space to separate it from the integer part. Here we show a three place representation of π

notation, scholars use a shorthand which combines our decimal notation with base 60. In this notation, a number is written with "," (comma) separating the digits and, if necessary, a ";" (semicolon) to serve as the radix point. For example,

$$1234_{10} = 20 \times 60^1 + 34 \times 60^0 = 20, 34$$

or

$$\begin{aligned} 3.14159 &= 3 \times 60^0 + 8 \times 60^{-1} + 29 \times 60^{-2} + 44 \times 60^{-3} \\ &= 3; 8, 29, 44 \end{aligned}$$

The fact that the ancient Babylonians had a place notation is all the more impressive given that they wrote and performed their calculations on clay tablets which were allowed to dry in the sun. This is a good thing as it preserved thousands of tablets for us to investigate today. Figure 1.3 is a particularly important mathematical tablet. This tablet shows a square with its diagonals marked. The side of the square is marked as `30` which is interpreted to be $\frac{1}{2}$. There are two numbers

Fig. 1.3 YBC 7289 showing the calculation of the diagonal of a square with a side of length $\frac{1}{2}$. Note that this calculation makes use of a particularly good value for $\sqrt{2}$. Image copyright Bill Casselman and is used with permission, see http://www.math.ubc.ca/~cass/Euclid/ybc/ybc.html. The original tablet is in the Yale Babylonian Collection

written by one of the diagonals. The first is `1, 24, 51, 10` and the one below it is `42, 25, 35`. The length of the diagonal of a square is $\sqrt{2}$ times the length of the side. If we multiply `30` by `1, 24, 51, 10` we do get `42, 25, 35`. This means that the Babylonians used `1, 24, 51, 10` as an approximation to $\sqrt{2}$. How good was this approximation? Squaring `1, 24, 51, 10` gives `1, 59, 59, 59, 38, 1, 40` which is, in decimal, about 1.99999830, a very good approximation indeed.

The Mayan people developed their own place notation in base 20 using combinations of dots (one) and bars (five) to form digits. The numbers were written vertically from bottom to top of the page. Figure 1.4 shows on the left an addition problem written with this notation. Unlike the Babylonians, the Mayans did use a symbol for zero, shaped like a shell, and this is seen in Fig. 1.4 on the right.

It is interesting to see how the Maya, like the Babylonians before them, did not create unique symbols for all possible base 20 digits but instead used combinations of likely base symbols. This might imply that place notation evolved from an early tally system, which is a reasonable things to imagine. Addition in the Mayan system worked by grouping digits, replacing five dots with a bar and carrying to the next higher digit if the total in the lower digit was twenty or more.

For more background on the fascinating history of numbers and number systems, including Mayan numbers, see [1] and [3]. For a detailed look at Babylonian numbers and mathematics, see [4] and [5]. Lastly, for a look at Egyptian numbers and mathematics see [2].

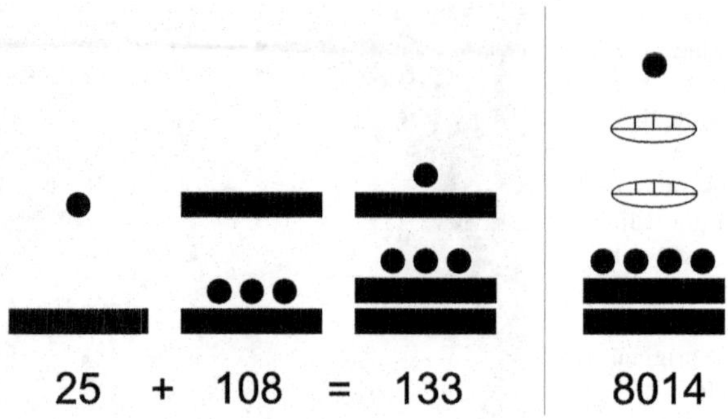

Fig. 1.4 Mayan numbers in base 20 written from bottom to top. *Bars* represent *five*, *dots* represent ones, with groups of *bars* and *dots* used to show the numbers 1 through 19. The *zero symbol*, shown on the *right*, is a shaped like a *shell* and holds empty places in the same way zero is used in base 10 so that the number on the *right* is $1 \times 20^3 + 0 \times 20^2 + 0 \times 20^1 + 14 \times 20^0 = 8014$

1.2 The Big Three (and One Old Guy)

In this book we are concerned, primarily, with three number bases: base 2 (binary), base 10 (decimal), and base 16 (hexadecimal). These are the "big three" number bases when it comes to computers. In the past, base 8 (octal) was also used frequently, so we will mention it here. This is the "one old guy". In this section we take a look at how to represent numbers in these bases and how to move a number between these bases. Operations such as arithmetic, will be covered in subsequent chapters from the point of view of how the computer works with numbers of a particular type.

Decimal Numbers Relatively little needs to be said about decimal numbers as we have been using them all our lives to the point where working with them is second nature. Of course, the reason for this frequent appearance of ten is that we have ten fingers and ten toes. Decimal numbers, however, are not a good choice for computers. Digital logic circuits have two states, "on" or "off". Since computers are built of these circuits it makes sense to represent the "on" state (logical `True`) as a 1 and the "off" state (logical `False`) as a 0. If your digits are 0 and 1 then the natural number base to pick is 2, or binary numbers.

Binary Numbers In the end, computers work with nothing more than binary numbers which is really quite interesting to think about: all the activities of the computer rely on numbers represented as ones and zeros. Following the notation above, we can see that a binary number such as 1011_2 is really,

$$1011_2 = 1 \times 2^3 + 0 \times 2^2 + 1 \times 2^1 + 1 \times 2^0 = 11$$

Thinking back to the example of ancient Egyptian multiplication above we can now see that the left-hand column was nothing more than the powers of two which are the place values of a binary number. Since each power of two appears only once in a given number (integer) it is always possible to find a sum of powers of two that equals any number. The 1 values in a binary integer are nothing more than tick marks indicating which power of two is present. We know that each power of two appears at most once in a given integer because it is possible to represent any integer in binary and the only digits allowed in a binary number are 0 and 1.

Hexadecimal Numbers If you work with computers long enough, you will eventually encounter hexadecimal numbers. These are base 16 numbers since "hex" refers to six and "dec" refers to ten. One question might immediately present itself: if we need ten separate symbols for the ten decimal digits, we need 16 symbols for hexadecimal digits. Zero through nine can be reused, but what should we use for 10 through 15? The flippant answer is, whatever symbols you want, you could even make up sixteen entirely new symbols, but typically we follow the convention the ancient Greeks used (see [1] Chap. 4) and take the first six letters of the Latin alphabet as the missing symbols,

A	10
B	11
C	12
D	13
E	14
F	15

In this case, counting runs as $1, 2, 3, 4, 5, 6, 7, 8, 9, A, B, C, D, E, F, 10$. It should be no surprise that when writing hexadecimal numbers one often sees lowercase letters, *a–f*, instead of *A–F*. Additionally, computer scientists, like most English speakers, tend to shorten words whenever possible. So, you will often hear or see hexadecimal numbers referred to as "hex" numbers. No reference to magic is intended unless it is the magic by which computers work at all.

If we break down a hex number,

$$FDED_{16} = 15 \times 16^3 + 13 \times 16^2 + 14 \times 16^1 + 13 \times 16^0 = 65005$$

we see that it is a compact notation compared to binary. This should not be surprising as in place of simply indicating which power of 2 is in a number, as binary does, we are here allowed up to 16 possible multipliers on each power of 16.

Octal Numbers While much less common than in the past, octal (base 8) numbers still show up from time to time. Since these are base 8 numbers we already know that we only need the digits $0 - 7$ to write numbers in this base. In Sect. 1.3 we will see why, if computers are really working in binary, people bother to use other number bases like octal or hexadecimal.

We should expect by now that octal numbers break down in a familiar way,

$$1234_8 = 1 \times 8^3 + 2 \times 8^2 + 3 \times 8^1 + 4 \times 8^0 = 668$$

and we will not be disappointed. In modern programming languages, octal numbers are typically reserved for character codes and will be rarely seen outside of that context.

1.3 Converting Between Number Bases

Representing numbers is a good thing but sometimes we need to convert a number between bases. In this section we look at converting between the bases we introduced above. Table 1.1 will be handy when converting numbers between bases. It shows the first 16 integers in binary, octal, hexadecimal and decimal.

Hexadecimal and Binary Hexadecimal numbers are base 16 while binary numbers are base 2. We see that $2^4 = 16$ so the conversion between hexadecimal and binary, either direction, will involve groups of four binary digits. To convert a binary number to hexadecimal, simply separate the digits into groups of four, from right to left, and write each group as a hex digit referring to Table 1.1 if necessary,

$$\begin{aligned} \texttt{101110010011110}_2 &= \texttt{101 1100 1001 1110} \\ &= \texttt{5 \quad C \quad 9 \quad E} \\ &= \texttt{5C9E}_{16} \end{aligned}$$

Table 1.1 The first 16 integers in decimal, binary, octal, and hexadecimal

Decimal	Binary	Octal	Hexadecimal
0	0000	0	0
1	0001	1	1
2	0010	2	2
3	0011	3	3
4	0100	4	4
5	0101	5	5
6	0110	6	6
7	0111	7	7
8	1000	10	8
9	1001	11	9
10	1010	12	A
11	1011	13	B
12	1100	14	C
13	1101	15	D
14	1110	16	E
15	1111	17	F

Turning a hexadecimal number into binary is simply the reverse process,

$$
\begin{aligned}
\mathtt{FDDA}_{16} &= \mathtt{F} \quad \mathtt{D} \quad \mathtt{D} \quad \mathtt{A} \\
&= \mathtt{1111\ 1101\ 1101\ 1010} \\
&= \mathtt{1111110111011010}_2
\end{aligned}
$$

This conversion process is simple enough, but why does it work? The key is in the fact already noted, that $2^4 = 16$. Recall that in a place notation system each digit position to the left is one power of the base larger just as the hundreds place in decimal is a factor of ten, the base, larger than the tens place. If we look at base 2 and instead of moving one digit to the left, which is two times larger than the original digit value, we move four digits we are now $2\times2\times2\times2$ larger or 16. But, in base 16, each digit position is 16 times larger than the next place to the right. Therefore, we have an equivalence between moving one digit to the left in a hexadecimal number and four digits to the left in a binary number. This is why we can group the binary digits in sets of four to get the equivalent hexadecimal digit.

Octal and Binary If $2^4 = 16$ implies we group binary digits in sets of four, then $2^3 = 8$ implies we should group binary digits in sets of three to convert between binary and octal, and this is indeed the case,

$$
\begin{aligned}
\mathtt{101110010011110}_2 &= \mathtt{101\ 110\ 010\ 011\ 110} \\
&= \mathtt{5} \quad \mathtt{6} \quad \mathtt{2} \quad \mathtt{3} \quad \mathtt{6} \\
&= \mathtt{56236}_8
\end{aligned}
$$

Likewise, we can reverse the process to get,

$$
\begin{aligned}
\mathtt{52431}_8 &= \mathtt{5} \quad \mathtt{2} \quad \mathtt{4} \quad \mathtt{3} \quad \mathtt{1} \\
&= \mathtt{101\ 010\ 110\ 011\ 001} \\
&= \mathtt{101010110011001}_2
\end{aligned}
$$

Lastly, should the need arise, conversion between hexadecimal and octal, or vice versa, is easily accomplished by first moving to binary and then to the other base from binary.

Decimal and Binary The simple conversion technique above falls apart when moving between decimal and binary. We already know the reason, $2^x = 10$ does not have an integer solution so we cannot simply group digits. Instead, we handle the conversions in a more cumbersome way.

Decimal to Binary Perhaps the simplest way to convert a decimal number to binary by hand is to repeatedly divide the number by two and keep track of the quotient and remainder. The quotient gives the next number to divide by two and the remainder, always either 1 or 0, is one binary digit of the result. If the numbers are not too big, dividing by two is easy enough to do without paper and pencil.

For example, to convert `123` to binary,

$$\begin{aligned}
123 \div 2 &= 61 \text{ r } 1\\
61 \div 2 &= 30 \text{ r } 1\\
30 \div 2 &= 15 \text{ r } 0\\
15 \div 2 &= 7 \text{ r } 1\\
7 \div 2 &= 3 \text{ r } 1\\
3 \div 2 &= 1 \text{ r } 1\\
1 \div 2 &= 0 \text{ r } 1
\end{aligned}$$

Now, to get the binary representation, read from bottom to top: $123 = 1111011_2$.

The conversion works but it is good to know *why* it works so let's take a closer look. If we write a seven digit binary number, $b_6b_5b_4b_3b_2b_1b_0$, with b_k the k-th digit, in expanded form we get,

$$b_0 \times 2^0 + b_1 \times 2^1 + b_2 \times 2^2 + b_3 \times 2^3 + b_4 \times 2^4 + b_5 \times 2^5 + b_6 \times 2^6$$

which, since $2^0 = 1$ and every other term is a multiple of two, can be written as,

$$b_0 + 2(b_1 \times 2^0 + b_2 \times 2^1 + b_3 \times 2^2 + b_4 \times 2^3 + b_5 \times 2^4 + b_6 \times 2^5)$$

If we divide the above by 2 we get the quotient,

$$b_1 \times 2^0 + b_2 \times 2^1 + b_3 \times 2^2 + b_4 \times 2^3 + b_5 \times 2^4 + b_6 \times 2^5$$

and a remainder of b_0. Therefore, it is clear that performing an integer division of a number by two leads to a new quotient and a remainder of either zero or one since b_0 is a binary digit. Further, this remainder is in fact the right-most digit in the binary representation of the number. If we look at the quotient above it has the same form as the initial binary representation of the number but with every binary digit now shifted to the next lowest power of two. Therefore, if we take this quotient and divide it by two we will repeat the calculation above and get b_1 as the remainder. This is then the next digit in the binary representation of the number. Lastly, if we repeat this over and over until the quotient is simply zero we will generate all the digits in the binary representation of the number in reverse order. Finally, writing the stream of binary digits backwards will put them in the proper sequence and give the desired binary representation.

Another method for converting a decimal number to binary involves making a table of the powers of two and asking repeated questions about sums from that table. This method works because the conversion process is really tallying the powers of two that add up to the decimal number.

For example, to convert 79 to binary make a table like this,

64	32	16	8	4	2	1

with the powers of two written highest to lowest from left to right. Next, ask the question, "Is $64 > 79$?" If the answer is yes, put a 0 under the 64. If the answer is no, put a 1 under the 64 because there is a 64 in 79,

64	32	16	8	4	2	1
1						

Now add a new line to the table and ask the question "Is $64 + 32 > 79$?" Since the answer is yes, we know that there is no 32 in the binary representation of 79 so we put a 0 below the 32,

64	32	16	8	4	2	1
1						
	0					

We continue the process, now asking "Is $64 + 16 > 79$?" In this case, the answer is again yes so we put a 0,

64	32	16	8	4	2	1
1						
	0					
		0				

We continue for each remaining power of two asking whether that power of two, plus all the previous powers of two that have a 1 under them, is greater or less than 79,

64	32	16	8	4	2	1
1						
	0					
		0				
			1			
				1		
					1	
						1

To complete the conversion we collapse the table vertically and write the binary answer, $79 = 1001111_2$.

Binary to Decimal If a binary number is small the fastest way to convert it to a decimal is to simply remember Table 1.1. Barring that, there is a straightforward iterative way to convert binary to decimal. If we have a k digit binary number,

$b_4b_3b_2b_1b_0$, where in this case $k = 5$, we can use the following recurrence relation to find the decimal equivalent,

$$d_0 = 0$$
$$d_i = b_{k-i} + 2d_{i-1}, i = 1 \ldots k$$

For example, let's convert 11001_2 to decimal using the recurrence relation. In this case we get,

i	d_i
0	0
1	$1 + 2(0) = 1$
2	$1 + 2(1) = 3$
3	$0 + 2(3) = 6$
4	$0 + 2(6) = 12$
5	$1 + 2(12) = 25$

This technique is easily implemented in a computer program. Why does it work? Consider the five digit binary number we started this section with, $b_4b_3b_2b_1b_0$. If we write the algebraic expression that represents the recurrence relation after five iterations we get,

$$d_5 = b_0 + 2(b_1 + 2(b_2 + 2(b_3 + 2(b_4 + 2(0)))))$$

Which, if we work it out step by step, becomes,

$$\begin{aligned} & b_0 + 2(b_1 + 2(b_2 + 2(b_3 + 2(b_4 + 2(0))))) \\ & \quad = b_0 + 2(b_1 + 2(b_2 + 2(b_3 + 2b_4))) \\ & \quad = b_0 + 2(b_1 + 2(b_2 + 2b_3 + 2^2 b_4)) \\ & \quad = b_0 + 2(b_1 + 2b_2 + 2^2 b_3 + 2^3 b_4) \\ & \quad = b_0 + 2b_1 + 2^2 b_2 + 2^3 b_3 + 2^4 b_4 \\ & \quad = b_0 \times 2^0 + b_1 \times 2^1 + b_2 \times 2^2 + b_3 \times 2^3 + b_4 \times 2^4 \end{aligned}$$

which is the expanded form of a binary number with digits $b_4b_3b_2b_1b_0$.

Others to Decimal The recurrence relation for converting binary numbers to decimal works for any number base. For example,

$$d_0 = 0$$
$$d_i = b_{k-i} + Bd_{i-1}, i = 1 \ldots k$$

for a base B number with digits b_i.

Armed with this knowledge it becomes straightforward to convert hexadecimal or octal numbers to decimal. Simply use $B = 16$ for hexadecimal and $B = 8$ for octal. In the case of hexadecimal, which is a base greater than 10, the additional digits, $A - F$, take on their actual decimal values $10 - 15$.

As an example, let's convert $56AD_{16}$ to decimal. Since this is a four digit number $k = 4$ and the recurrence relation becomes,

$$d_0 = 0$$
$$d_i = b_{k-i} + 16d_{i-1}, i = 1 \ldots k$$

and calculation proceeds as,

i	d_i
0	0
1	$5 + 16(0) = 5$
2	$6 + 16(5) = 86$
3	$10 + 16(86) = 1386$ *(since* $A_{16} = 10$*)*
4	$13 + 16(1386) = 22189$ *(since* $D_{16} = 13$*)*

so $56AD_{16} = 22189$.

Finally, a similar calculation works for octal numbers. Convert 7502_8 to decimal. Again, $k = 4$ and the recurrence relation is now,

$$d_0 = 0$$
$$d_i = b_{k-i} + 8d_{i-1}, i = 1 \ldots k$$

giving,

i	d_i
0	0
1	$7 + 8(0) = 7$
2	$5 + 8(7) = 61$
3	$0 + 8(61) = 488$
4	$2 + 8(488) = 3906$

with a final result of $7502_8 = 3906$.

Frequent use of hexadecimal and octal numbers will quickly teach you to simply remember the first few powers of each base. Embedded systems programmers learn this early on for at least the first four digits. For hexadecimal, the powers of 16 are,

16^0	1
16^1	16
16^2	256
16^3	4096

and similarly for octal,

8^0	1
8^1	8
8^2	64
8^3	512

with these, small hexadecimal and octal integers can be converted to decimal quickly,

$$\begin{aligned} 1234_{16} &= 1 \times 4096 + 2 \times 256 + 3 \times 16 + 4 \times 1 = 4660 \\ 1234_8 &= 1 \times 512 + 2 \times 64 + 3 \times 8 + 4 \times 1 \qquad = 668 \end{aligned}$$

With a little practice such conversion can be performed mentally and quickly.

1.4 Chapter Summary

In this chapter we briefly reviewed numbers, how they were represented historically, how place notation works, and how to manipulate numbers in the bases most frequently used by computers. It is helpful to become comfortable working with the conversion between number bases and it is especially helpful to memorize the binary representation of the first 16 integers. The exercises will help build this confidence and the skill will be used frequently in the remainder of this book.

Exercises

1.1 Using Egyptian notation and method calculate 22×48 and $713 \div 31$.

1.2 Using Babylonian notation calculate $405 + 768$. (*Hint:* carry on sixty, not ten.)

1.3 Using Mayan notation calculate $419 + 2105$. (*Hint:* carry on twenty, not ten.)

1.4 Convert the following hexadecimal numbers to binary: $A9C1_{16}$, $20ED_{16}$, $FD60_{16}$.

1.5 Convert the following octal numbers to binary: 773_8, 203_8, 656_8.

1.6 Using the table method, convert 6502 to binary.

1.7 Using the division method, convert 8008 to binary.

1.8 Write $e = 2.71828182845904\ldots$ to at least four decimal places using Babylonian notation. (*Hint:* You will need to use three negative powers of sixty.) **

1.9 Write a program to convert a number in any base < 37 to its decimal equivalent using the recurrence relation. Use a string to hold the number itself and use `0` – `9` and capital letters `A` – `Z` for the digits. *

References

1. Boyer, C.B.: A History of Mathematics. Princeton University Press, Princeton, New Jersey (1985)
2. Gillings, R.: Mathematics in the Time of the Pharaohs. Dover Publications, New York (1982)
3. Menninger, K.: Number Words and Number Symbols. Dover Publications, New York (1992)
4. Neugebauer, O.: Mathematical Cuneiform Texts. American Oriental Series, vol. 29. New Haven: American Oriental Society (1945)
5. Neugebauer, O.: The Exact Sciences in Antiquity. Princeton University Press, Princeton, New Jersey (1952)

Chapter 2
Integers

Abstract Integers are perhaps the most important class of numbers. This is certainly true in the case of computers. In this chapter we dive into the integers and how computers represent and operate on them. Without these operations, digital computers would not function. We begin with some preliminary notation and terminology. Next we take a detailed look at the unsigned integers. We follow this with an examination of negative integers and their operations. Lastly, we finish with a look at binary coded decimals.

2.1 Bits, Nibbles, Bytes, and Words

Our tour of integers will be easier if we get some terminology out of the way first. This will make it just that much easier to talk about the way numbers are actually represented inside a computer. We'll start small and work our way up.

Bits As we have seen, computers work with binary numbers. A single binary digit is known affectionately as a *bit*. This bit is either zero or one, off or on, false or true. The word *bit* is short for *binary digit* and appears in Claude Shannon's classic 1948 paper *A Mathematical Theory of Communication* [2]. Shannon attributed *bit* to John Tukey who used it in a Bell Labs memo dated January 9, 1947. Note, throughout this book we will play fast and loose with the bit terminology and freely alternate between "on" and "true" for when a bit is set to one and "off" or "false" when a bit is set to zero. The choice of word will depend on the context.

Since a bit is so simple, just a zero or a one, it is the smallest unit of information in the digital world. Still, this small unit can be very important. A status bit set to one may indicate that the rocket is ready to launch while a status bit set to zero indicates a fault. Indeed, some microcontroller devices, like the 8051, make excellent use of their limited resources and support single bit data. We will make extensive use of bits in the remainder of this chapter. Zero or one, how hard can it be? Still, as useful as a bit is, it is just, well, a *bit*. You need to group them together to get more expressive numbers.

Nibbles A *nibble* (sometimes *nybble*) is a four bit number. As a four bit number a nibble can represent the first `16` integers with $\texttt{0000}_2 = \texttt{0}$ and $\texttt{1111}_2 = \texttt{15}$. And, as we now know, the numbers `0` through `15` are exactly the digits of a hexadecimal

R.T. Kneusel, *Numbers and Computers*, DOI 10.1007/978-3-319-50508-4_2

number. So, a nibble is a single hexadecimal digit. This makes sense since with four bits we can represent $2^4 = 16$ possible values. A nibble is also the amount of information necessary to represent a single binary-coded decimal digit but we are getting ahead of ourselves and will talk about binary-coded decimal later in this chapter. The origin of the word *nibble* is related to the origin of the word *byte* so we will expand on it below.

Modern computers do not use nibbles as a unit of data for processing. However, the very first microprocessors, like the TMS 1000 by Texas Instruments and the 4004 by Intel, both introduced in 1971, were 4-bit devices and operated on 4-bit data. This makes the nibble a bit like octal numbers: appearing in the shadows but not a real player in modern computing.

Bytes For a modern computer, the smallest unit of data that it will work with is the *byte* which is an eight bit binary number. This means that the integers from $0 \ldots 255$ can be represented in a single byte. We'll ignore the concept of negative numbers for the time being.

The term *byte* was first used by Werner Buchholz in July 1956 during the design of the IBM Stretch computer and was an intentional misspelling of "bite" to avoid confusion with "bit" [3]. Since *byte* came from "bite" it is natural to be tempted to call half a byte, which is a four bit number, a "nibble" since a nibble is a small bite. Hence the origin of *nibble* above.

Buchholz intended a byte to represent a character. We still use bytes to represent characters in ASCII, but have since moved on to larger numbers for characters in Unicode. In this book, however, to keep things simple, we will make an implicit assumption that characters are bytes. So, a text that contains 1000 characters will take 1000 bytes of memory to store. The universal use of bytes when working with computers leads to the frequent appearance of the numbers $2^8 = 256$ and $2^8 - 1 = 255$. For example, colors on computers are often represented as triplets of numbers to indicate the amount of red, green, and blue that make up the color. Typically, these triplets are given as a set of three bytes to indicate the amount so that 000000_{16} is black and $00FF00_{16}$ is bright green. This is just one example, bytes are everywhere. The relationship between bits, nibbles and bytes is visualized in Fig. 2.1. Here we see that a byte can be thought of as two nibbles, each a hexadecimal digit, or as eight bits, each a binary digit.

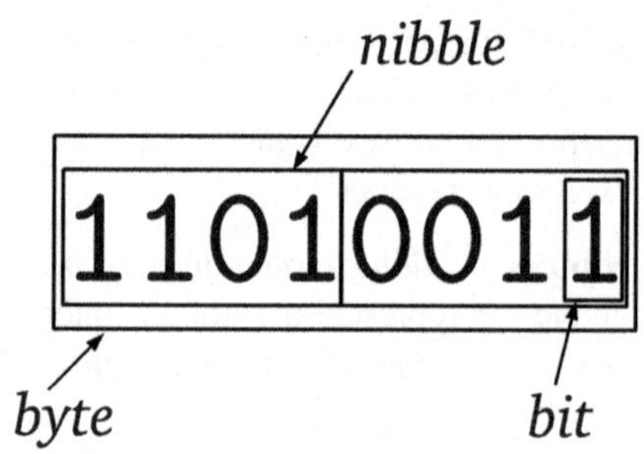

Fig. 2.1 A byte which consists of two nibbles and eight bits. The number is 211 which is made up of two nibbles, D_{16} and 3_{16} giving $D3_{16}$, or eight bits 11010011_2

Words A byte is two nibbles, so, is a *word* two bytes? Well, sometimes, yes. And sometimes it is four bytes, or eight, or some other value. Unfortunately, the term *word* is not generic like bit and byte, instead it is tied to a particular computer architecture and describes the basic data size of the architecture.

Historically, the word size of a computer could be anywhere from four bits for early microprocessors (Intel 4004) to 60 bits for early mainframes (CDC 6600) but modern computers have settled on powers of two for their word sizes, typically 32 or 64 bits with some second generation microprocessors using 16 bit words (Intel 8086, WDC 65816). For our purposes, we can assume a 32 bit word size when describing data formats. On occasion we will need to be more careful and explicitly declare the word size we are working with.

With these preliminaries now under our belt, it is time to start working with actual data representations. We start naturally with unsigned integers.

2.2 Unsigned Integers

Unsigned integers are our first foray into the depths of the computer as far as numbers are concerned. These are the basic numbers, the positive integers, that find frequent use in representing characters, as counters, or as constants, really anything that can be mapped to the set $\{0, 1, 2, \ldots\}$. Of course, computers have finite memory so there is a limit to the size of an unsigned integer and we will get to that in the next section.

2.2.1 Representation

Integers are stored in memory in binary using one or more bytes depending on the range of the integer. The example in Fig. 2.1 is a one byte unsigned integer. If we are working in a typed language like C we need to declare a variable that will store one byte. Typically, this would mean using an *unsigned char* data type,

```
unsigned char myByte;
```

which can store a positive integer from $\texttt{00000000}_2 = \texttt{0}$ to $\texttt{11111111}_2 = \texttt{255}$. This is therefore the *range* of the `unsigned char` data type. Note how C betrays its age by referring to a byte number as a character. Unsigned integers larger than `255` require more than one byte, but we will get to that in the next section.

Table 2.1 shows standard C types for unsigned integers and the allowed range for that type. These are fixed in the sense that numbers must fit into this many bits at all times. If not, an underflow or overflow will occur. Languages like Python abstract integers for the programmer and only support the concept of integer as opposed to a floating-point number. The unsigned integer operations discussed later in the chapter work nicely with Python but the concept of range is a little nebulous in that case since Python will move between internal representations as necessary.

Table 2.1 Unsigned integer declarations in C

Declaration	Minimum	Maximum	Number of bits
`unsigned char`	0	255	8
`unsigned short`	0	65,535	16
`unsigned int`	0	4,294,967,295	32
`unsigned long`	0	4,294,967,295	32
`unsigned long long`	0	18,446,744,073,709,551,615	64

The declaration, minimum value, maximum value and number of data bits are given

If a number like $11111111_2 = 255$ is the largest unsigned number that fits in a single byte how many bytes will be needed to store `256`? If we move to the next number we see that we will need *nine* bits to store `256` which implies that we will need two bytes. However, there is a subtlety here that needs to be addressed. This is, if the number is to be stored in the computer's memory, say starting at address `1200`, how should the individual bits be written to memory location `1200` and the one following (`1201`, since each memory location is a byte)? This question has more than one answer.

2.2.2 *Storage in Memory: Endianness*

Addresses In order to talk about how we store unsigned integers in computer memory we have to first talk a bit about how memory is addressed. In this book, we have a working assumption of a 32-bit Intel-based computer running the Linux operating system. Additionally, we assume `gcc` to be our C compiler. In this case, we have *byte-addressable* memory meaning that even though the word size is four bytes, we can refer to memory addresses using bytes. For example, this small C program,

```
#include <stdio.h>

int main() {
    unsigned char *p;
    unsigned short n = 256;

    p = (unsigned char *)&n;
    printf("address of first byte  = %p\n", &p[0]);
    printf("address of second byte = %p\n", &p[1]);

    return 0;
}
```

when compiled and run produces output similar to,

```
address of first byte  = 0xbfdafe9e
address of second byte = 0xbfdafe9f
```

where the specific addresses, given in hexadecimal and starting with the `0x` prefix used by C, will vary from system to system and run to run. The key point is that the difference between the addresses is *one* byte. This is what is meant by a byte-addressable memory. If you don't know the particulars of the C language, don't worry. The program is defining a two byte number in line 5 (n) and a pointer to a single byte number in line 4 (p). The pointer stores a memory address which we set to the beginning of the memory used by n in line 7. We then ask the computer to print the numeric address of the memory location (line 8) and of the next byte (line 9) using the `&` operator and indexing for the first (`p[0]`) and second (`p[1]`) bytes. With this in mind, let's look at actually storing unsigned integers.

Bit Order To store the number $11011101_2 = 221$ in memory we use the eight bits of the byte at a particular memory address. The question then becomes, which bits of `221` map to which bits of the memory? If we number the bits from `0` for the lowest-order bit, which is the right-most bit when writing the number in binary, to `7` for the highest-order bit, which is the left-most bit when writing in binary, we get a one-to-one mapping,

7	6	5	4	3	2	1	0
1	1	0	1	1	1	0	1

so that the highest-order bit of the number is put into the highest-order bit of the memory address. This seems sensible enough but one could imagine doing the reverse as well,

7	6	5	4	3	2	1	0
1	0	1	1	1	0	1	1

So, which is correct? The answer depends on how bits are read from the memory location. In modern computer systems, the bits are read low to high meaning the low-order bit is read first (bit `0`) followed by the next higher bit (bit `1`) so that the computer will use the first ordering of bits that maps `76543210` to `11011101`.

Byte Order Within a byte, bits are stored low to high. What about the bytes of a multibyte integer? We know that the number $100000000_2 = 256$ requires two bytes of memory, one storing the low set of eight bits 00000000_2 and another storing the high set of eight bits 00000001_2 where leading zeros are used to indicate that it is the first bit position set and the rest of the bits in the byte are clear (set to zero). But, what order in memory should we use,

Memory address:	1200	1201
Low first:	00000000	00000001
High first:	00000001	00000000

low first or high first? The answer is either. If we put the low order bytes of a multibyte number in memory first we are using *little-endian* storage. If we put the high order bytes first we are using *big-endian* or *network order*. The choice is the *endianness* of the number and both are in use typically as a function of the microprocessor, which has a preferred ordering. The terms little-endian and big-endian are derived from the book *Gulliver's Travels* by satirist Jonathan Swift [4]. Published in 1726 the book tells the story of Gulliver and his travels throughout a fictional world. When Gulliver comes to the land of Lilliput he encounters two religious sects who are at odds over whether to crack open their soft-boiled eggs little end first or big end first. Intel based computers use little-endian when storing multibyte integers. Motorola based computers use big-endian. Additionally, big-endian is called network order because the Internet Protocol standard [5] requires that numeric values in packet headers be stored in big-endian order.

Let's look at some examples of little and big endian numbers. To save space, we will use a 32-bit number but represent the numbers showing only the hexadecimal values of the four bytes such a number would occupy. For example, if the 32-bit number is,

$$\texttt{10101110111100000101011010010110}_2 = \texttt{AE}_{16}\ \texttt{F0}_{16}\ \texttt{56}_{16}\ \texttt{96}_{16}$$

we will write it as `AEF05696` dropping the `16` base indicator for the time being. Putting this number into memory starting with byte address zero and using big-ending ordering gives,

Address:	0	1	2	3
Value:	AE	F0	56	96

where the first byte one would read, pulling bytes from address zero first, is the high-order byte. If we want to use little-endian we would instead put the lowest order byte first to fill in memory with,

Address:	0	1	2	3
Value:	96	56	F0	AE

which will give us the low-order byte first when reading from address zero. Note, within a particular byte we always still use the bit ordering that maps the high-order bit (bit 7) to the high-order bit of the number. Reversing the bits within a byte when using little-endian is an error.

When we are working with data on a single computer within a single program, we typically do not need to pay attention to endianness. However, when transmitting data over the network or using data files written on other machines it is a good idea to consider what the endianness of the data is. For example, when reading sensor values from a CCD camera, which are typically 16-bit numbers requiring two bytes to store, it is essential to know the endianness otherwise the image generated from the raw data will look strange because the high-order bits which contain the majority of the information about the image will be in the wrong place. From the examples above, we see that correcting for endianness differences is straightforward, simply flip the order of the bytes in memory to convert from little-endian to big-endian and vice versa.

2.3 Operations on Unsigned Integers

Binary and unary operations on unsigned integers is the heart of what computers do. In this section we go over all the common operations, bit by bit, and mention some of the key things to look out for when working with unsigned integers of fixed bit width. Examples will use byte values to keep it simple but in some cases multibyte numbers will be shown. Specific operations will be shown in C and Python. Though the Python syntax is often the same as C, the results may differ because the Python interpreter will change the internal representation as necessary.

2.3.1 Bitwise Logical Operations

Bitwise operators are binary operators, meaning they operate on two numbers (called operands), to produce a new binary number. In the previous sentence the first instance of "binary" refers to the number of operands or arguments to the operation while the second instance of "binary" refers to the base of the operands themselves. This is an unfortunate but common abuse of notation.

Digital logic circuits, which are fascinating to study but well beyond the purview of this book, are built from logic gates which implement in hardware the logical operations we are discussing here. The basic set of logic operators are given names which relate to Boolean logic: `AND`, `OR`, `XOR`, and the unary `NOT`. Negated versions of `AND` and `OR`, called `NAND` and `NOR`, are also in use but these are easily constructed by taking the output of `AND` and `OR` and passing it through a `NOT` so we will ignore them here.

Logical operations are most easily understood by looking at their *truth tables* which illustrate the function of the operator by enumerating the possible set of inputs and giving the corresponding output. They are called truth tables because of the correspondence between the `1` and `0` of a bit and the logical concept of `true` and `false` which goes back to Boolean algebra developed by George Boole in the nineteenth century [6].

A truth table shows, for each row, the explicit inputs to the operator followed by the output for those inputs. For example, the truth table for the `AND` operator is,

AND		
0	0	0
0	1	0
1	0	0
1	1	1

which says that if the two inputs to the `AND`, recall the inputs are single bits, are `0` and `0` that the output bit will also be `0`. Likewise, if the inputs are `1` and `1` the output will also be `1`. In fact, for the `AND` operation the only way to get an output of `1` is for both input bits to be set. We can think of this as meaning "only both".

The two other most commonly used binary logic operators are `OR` and `XOR`. The latter is a called exclusive-`OR` and is sometimes indicated with `EOR` instead of `XOR`. Their truth tables are,

OR		
0	0	0
0	1	1
1	0	1
1	1	1

XOR		
0	0	0
0	1	1
1	0	1
1	1	0

where we see that `OR` means "at least one" and `XOR` means "one or the other but not both". The last of the common bitwise logical operators is `NOT` and it has a very simple truth table,

NOT	
0	1
1	0

where the operation is simply turn `1` to `0` and `0` to `1`.

If we look again at the truth tables for `AND`, `OR` and `XOR` we see that there are four rows in each table which match the four possible sets of inputs. Looking at these rows we can interpret the four bits in the output as a single four bit number. A four bit number has sixteen possible values from 0000_2 to 1111_2 which implies that there are sixteen possible truth tables for a binary bitwise operator. Three of them are `AND`, `OR` and `XOR` and another three are the tables made by negating the output of these operators. That leaves ten other possible truth tables. What do these correspond to and are they useful in computer programming? The full set of possible binary truth tables is given in Table 2.2 along with an interpretation of each operation. While all of these operators have a name and use in logic many are clearly of limited utility in terms of computer programming. For example, always outputting `0` or `1` regardless of the input (`FALSE` and `TRUE` in the table) is of no practical use in helping the programmer. The negated versions of the three main operators, `NAND`, `NOR` and `XNOR`, are naturally built by adding a "not" word to the positive versions and are therefore useful.

Now that we know the basic operators and why this set is most useful to computer programmers let's look at a few examples. First the `AND` operator,

	10111101	=	189
AND	11011110	=	222
	10011100	=	156

where the binary representation of the number is in the first column and the decimal is in the second. We see that for each bit in the two operands, `189` and `222`, the output bit is on only if the corresponding bits in the operand were on. If the operands were simply zero or one the output would be one only if the inputs were both one. Since this is exactly what we mean when we use "and" in a sentence we can use this operator to decide if two things are both true. A natural place to use this ability is in

Table 2.2 The sixteen possible binary bitwise logical operators

p:	0	0	1	1	
q:	0	1	0	1	
FALSE	0	0	0	0	always false
AND	0	0	0	1	*p* and *q*
	0	0	1	0	if *p* then not *q* else false
	0	0	1	1	*p*
	0	1	0	0	if not *p* and *q*
	0	1	0	1	*q*
XOR	0	1	1	0	*p* or *q* but not (*p* and *q*)
OR	0	1	1	1	*p*, *q* or (*p* and *q*)
NOR	1	0	0	0	not OR
XNOR	1	0	0	1	not XOR
	1	0	1	0	not *q*
	1	0	1	1	if *q* then *p* else true
	1	1	0	0	not *p*
	1	1	0	1	if *p* then *q* else true
NAND	1	1	1	0	not AND
TRUE	1	1	1	1	always true

If a common name exists for the operation it is given in the first column. The top two rows labeled *p* and *q* refer to the operands or inputs to the operator. A `0` is considered false while `1` is true. A description of the operator is given in the last column. In the description the appearance of *p* or *q* in an if-statement implies that the operand is true

an if-statement and many programming languages, C and Python included, use the concept that zero is false and one is true. Actually, C and Python extend this concept and declare anything that is not zero to be true. It should be noted that Python also supports Boolean variables and treats the keywords `True` and `False` as true and false respectively.

We look next at the OR and XOR operators. For OR we have,

```
      10111101  =  189
OR    11011110  =  222
----------------------
      11111111  =  255
```

which means at every bit position at least one of the operands has a bit set. The XOR operator will give us,

```
      10111101  =  189
XOR   11011110  =  222
----------------------
      01100011  =   99
```

One useful property of the XOR operator is that it undoes itself if applied a second time. Above, we calculated `189 XOR 222` to get `99`. If we now perform `99 XOR`

222 we get, giving us back what we started with. As an aside, this is a useful

	01100011	=	99
XOR	11011110	=	222
	10111101	=	189

simple way to encrypt data. If Alice wishes to encrypt a stream of n bytes (M), which may represent any data at all, so that she can send it (relatively) securely to Bob, all she need do is generate a stream of n random bytes (S) and apply XOR, byte by byte, to the original stream M to produce a new stream, M'. Alice can then transmit M' to Bob any way she wishes knowing that if Carol intercepts the message she cannot easily decode it without the same random stream S. Of course, Bob must somehow have S as well in order to recover the original message M. If S is truly random, used only once and then discarded, this is known as a *one-time pad* encryption. The keys are that S is truly random and that it is only used once. Of course, this does not cover the possibility that Carol may get her hands on S as well, but if she cannot, there is (relatively) little chance she will be able to recover M from just M'.

Another handy use for XOR is that it allows us to swap two unsigned integers without using a third variable. So, in C, instead of,

```
unsigned char a,b,t;
...
t = a;
a = b;
b = t;
```

we can use,

```
unsigned char a,b;
...
a ^= b;   // a = a ^ b;
b ^= a;
a ^= b;
```

where ^ is the C operator for XOR and `a ^= b` is shorthand for `a = a ^ b`. Why does this work? If we write out an example in binary, we will see,

a = 01011101	
b = 11011011	
a ^= b	01011101 ^ 11011011 → a=10000110
b ^= a	10000110 ^ 11011011 → b=01011101
a ^= b	10000110 ^ 01011101 → a=11011011

As mentioned above, if we do a XOR $b \rightarrow c$ and then c XOR b we will get a back. If we instead do c XOR a we will get b back. So, the trick makes use of this by first doing a XOR b and then using that result, stored temporarily in a, to get a back but this time storing it in b and likewise for getting b back and storing it in a. This, then, swaps the two values in memory. This trick works for unsigned integers of any size. Another common use of `XOR` is in parity calculations. The parity of an integer is the number of `1` bits in its binary representation. If the number is odd, the parity is `1`, otherwise it is `0`. This can be used as a simple checksum when transmitting a series of bytes. A checksum is a indicator of the integrity of the data. If the checksum calculated on the receiving end does not match then there has been an error in transmission. If the last byte of data is the running `XOR` of all previous bytes the parity of this byte can be used to determine if a bit was changed during transmission. In order to see how to use `XOR` for parity calculations we need to wait a little bit until we talk about shifts below.

The effect of `NOT` is straightforward to illustrate,

NOT	10111101	=	189
	01000010	=	66

it simply flips the bits from `0` to `1` and `1` to `0`. We will see this operation again when we investigate signed integers.

Figure 2.2 gives a C program that implements the `AND`, `OR`, `XOR` and `NOT` logical operators. This illustrates the syntax. Note that the C example uses the `unsigned char` data type. This is an 8-bit unsigned integer like we saw in the examples above. The operators work with any size integer type so we could just as well have used `unsigned short` or `unsigned int`, etc. Figure 2.3 gives the corresponding Python code.

The standard C language does not support direct output of numbers in binary so the example in Fig. 2.2 instead outputs in decimal and hexadecimal. Recall that in Chap. 1 we learned how to easily change hexadecimal numbers into binary by replacing each hexadecimal digit with four bits representing that digit. The Python

```c
#include <stdio.h>

void pp(unsigned char z) {
    printf("%3d (%02x)\n", z, z);
}

int main() {
    unsigned char z;

    z = 189 & 222;   pp(z); // '&' is AND
    z = 189 | 222;   pp(z); // '|' is OR
    z = 189 ^ 222;   pp(z); // '^' is XOR
    z = z ^ 222;     pp(z); // returns 189
    z = ~z;          pp(z); // '~' is NOT

    return 0;
}
```

Fig. 2.2 Bitwise logical operators in C

Fig. 2.3 Bitwise logical operators in Python

```
def pp(z):
    print "{0:08b}".format(z & 0xff)

def main():
    z = 189 & 222;    pp(z) # '&' is AND
    z = 189 | 222;    pp(z) # '|' is OR
    z = 189 ^ 222;    pp(z) # '^' is XOR
    z = z ^ 222;      pp(z) # returns 189
    z = ~z;           pp(z) # '~' is NOT

main()
```

code makes use of standard string formatting language features to output in binary directly. As Python is itself written in C and the language was designed to show that heritage, the syntax for the bitwise logical operators is the same. There is one subtlety with the Python code you may have noticed. The function `pp()` uses "`z & 0xff`" in the `format()` method call where you might have expected only "`z`". If we do not AND the output value with $\texttt{FF}_{16}$ we will output a full range integer and as we will see later in this chapter, that would be a negative number to Python. The AND keeps only the lowest eight bits and gives us the output we would expect matching the C code in Fig. 2.2.

2.3.2 *Testing, Setting, Clearing, and Toggling Bits*

A common operation on unsigned integers is the setting, clearing and testing of particular bits. Setting a bit means to turn that bit on (`1`) while clearing is to turn the bit off (`0`). Testing a bit simply returns its current state, `0` or `1`, without changing its value. In the embedded computing world this is often done to set an output line high or low or to read the state of an input line. For example, a microcontroller uses specific pins on the device as digital inputs or outputs. Typically, this requires setting bits in a control register or reading bits in a register to know whether a voltage is present or not on pins of the device.

The bitwise logical operators introduced above are the key to working with bits. As Fig. 2.3 already alluded to, the AND operator can be used to mask bits. Masking sets certain bits to zero and some examples will make this clear. First, consider a situation similar to the Python code of Fig. 2.3 that keeps the lower nibble of a byte while it clears the upper nibble. To do this, we AND the byte with $\texttt{0F}_{16}$,

	10111101	=	BD
AND	00001111	=	0F
	00001101	=	0D

where we now display the binary and hexadecimal equivalent. The output byte will have all zeros in the upper nibble because AND is only true when both operands are one. Since the second operand has all zeros in the upper nibble there is no situation where the output can be one. Likewise, the lower nibble is unaffected by the AND operation since every position where the first operand has a one the second operand will also have a one leading to an output of one but every position where the first operand has a zero the second operand will still have a one leading to an output of zero, which is just what the first operand has in that position. A quick look back at the truth table for AND (above) will clarify why this masking works.

This property of AND can be used to test if a bit is set. Say we wish to see if bit 3 of $\texttt{BD}_{16}$ is set. All we need to do is create a mask that has a one in bit position 3 and zeros in all other bit positions (recall that bit positions count from zero, right to left),

	10111101	=	BD
AND	00001000	=	08
	00001000	=	08

if the result of the AND is not zero we know that the bit in position 3 of $\texttt{BD}_{16}$ is indeed set. If the bit was clear, the output of the entire operation would be exactly zero which would indicate that the bit was clear. Since C and Python treat nonzero as true the following C code would output "bit 3 on",

```
if (0xBD & 0x08) {
    printf("bit 3 on\n");
}
```

as would this Python code,

```
if (0xBD & 0x08):
    print "bit 3 on"
```

We use AND to test bits and mask bits but we use OR to actually set bits. For example, to actually set bit 3 and leave all other bits as they were we would do something like this, which works because OR returns true when either operand

	10110101	=	B5
OR	00001000	=	08
	10111101	=	BD

or both have a one in a particular bit position. Just as AND with an operand of all ones will return the other operand, so an OR with all zeros will do likewise. In order to set bits, then, we need OR, a bit mask, and assignment as this C code shows,

```
#include <stdio.h>

int main() {
    unsigned char z = 0xB5;

    z = z | 0x08;   // set bit 3
    printf("%x\n", z);
```

```
    z |= 0x40;      // set bit 6
    printf("%x\n", z);

    return 0;
}
```

which outputs $\texttt{BD}_{16}$ and $\texttt{FD}_{16}$ after setting bits 3 and 6. Since we updated `z` each time the set operations are cumulative.

We've seen how to test and set bits now let's look at clearing bits without affecting other bits. We know that if we AND with a mask of all ones we will get back our first operand. So, if we AND with a mask of all ones that has a zero in the bit position we want to clear we will turn off that bit and leave all other bits as they were. For example, to turn off bit 3 we do something like this, where the mask

	10111101	=	BD
AND	11110111	=	F7
	10110101	=	B5

value for AND is $\texttt{F7}_{16}$. But, this begs the question of how we get the magic $\texttt{F7}_{16}$ in the first place? This is where NOT comes into play. We know that to set bit 3 we need a mask with only bit 3 on and all others set to zero. This mask value is easy to calculate because the bit positions are simply powers of two so that the third position is $2^3 = 8$ implying that the mask is just $8 = \texttt{00001000}_2$. If we apply NOT to this mask we will invert all the bits giving us the mask we need: $\texttt{11110111}_2$. This is readily accomplished in C,

```
unsigned char z = 0xBD;
z = z & (~0x08);
printf("%x\n", z);
```

which outputs $\texttt{B5}_{16} = \texttt{10110101}_2$ with bit 3 clear. The same syntax will also work in Python.

Our last example toggles bits. When a bit is toggled its value is changed from off to on or on to off. We can use NOT to quickly toggle all the bits, since this is what NOT does, but how to toggle only one bit and leave the rest unchanged? The answer lies in using XOR. The XOR operator returns true only when the operands are different. If both are zero or one, the output is zero. This is what we need. If we XOR our value with a mask that has a one in the bit position we desire to toggle we will get something like this for bit 1,

	10111101	=	BD
XOR	00000010	=	02
	10111111	=	BF

where we have turned bit 1 on when previously it was off. If bit 1 was already on we would have 1 XOR 1 = 0 which would turn it off in the output. Clearly, we can toggle multiple bits by setting the positions we want to toggle in the mask and then apply the XOR. In C we have,

```
unsigned char z = 0xBD;
z = z ^ 0x02;
printf("%x\n", z);
```

which will output $\texttt{BF}_{16} = \texttt{10111111}_2$ as expected.

2.3.3 *Shifts and Rotates*

So far we have looked at operations that manipulate bits more or less independently of other bits. Now we take a look at sliding bits from one position to another within the same value. These manipulations are accomplished through the shift and rotate operators. A *shift* is as straightforward as it sounds, just move bits from lower positions in the value to higher, if shifting to the left, or from higher positions to lower if shifting to the right. When shifting, bits simply "fall off" the left or right if they hit the end of the integer. This implies something, namely, that we impose a specific number of bits on the integer. For our examples we will stick with 8-bit unsigned integers though all of these operations work equally well on integers of any size. Let's look at what happens when we shift a value to the left one bit position using binary notation,

$$\texttt{10101111} \leftarrow \texttt{1} = \texttt{01011110}$$

where we use the $\leftarrow$ symbol to mean shift to the left and the `1` is the number of bit positions. The leading `1` drops off the left end and a zero moves in from the right end to fill in the empty space. All other bits move up one bit position. Now, what happens when only a single bit is set and we shift one position to the left,

$$\texttt{00000010} \leftarrow \texttt{1} = \texttt{00000100}$$

we see that we started with a value of $2^1 = 2$ and we ended with a value of $2^2 = 4$. Therefore, a single position shift to the left will move each bit to the next highest bit position which is the same as multiplying it by two. Since this will happen to all bits, the net effect is to multiply the number by two. Of course, if bits that were set fall off the left end we will lose precision but the remaining bits will be multiplied by two. For example,

$$\texttt{00101110} \leftarrow \texttt{1} = \texttt{01011100}$$

which takes $\texttt{00101110}_2 = \texttt{46}$ to $\texttt{01011100}_2 = \texttt{92}$ which is 46×2. Shifting to the left by more than one bit position is equivalent to repeated shifts by one position so shifting by two positions will multiply the number by $2 \times 2 = 4$,

$$\texttt{00101110} \leftarrow \texttt{2} = \texttt{10111000}$$

giving $\texttt{10111000}_2 = \texttt{184}$ as expected.

It is natural to think that if a left shift multiplies by two a right shift would divide by two and this is indeed the case,

$$\texttt{00101110} \rightarrow \texttt{1} = \texttt{00010111}$$

gives $00010111_2 = 23$ which is $46 \div 2$. Just as bits falling off the left end of the number will result in a loss of precision so will bits falling off the right end with one significant difference. If a bit falls off the right end of the number it is lost from the ones position. If the ones position is set, that means the number is odd and not evenly divisible by two. If the bit is lost the result is still the number divided by two but the division is *integer division* which ignores any remainder. Information is lost since a right shift followed by a left shift results in a number that is one less than what we started with if the original number was odd. We can see this by shifting two positions to the right,

$$\texttt{00101110} \rightarrow 2 = \texttt{00001011}$$

to get $00001011_2 = 11$ which is $46 \div 4$ ignoring the remainder of 2.

Both C and Python support shifts using the << and >> operators for left and right shifts respectively. The « operator is frequently used with an argument of `1` in order to quickly build bit masks,

$$\texttt{1 << 3} = \texttt{00001000}_2$$
$$\texttt{1 << 5} = \texttt{00100000}_2$$

which is a pretty handy way to move bits into position.

Before we leave shifts, let's return to the parity calculation mentioned above. Recall that the parity of a number is determined by the number of `1` bits in its binary representation. If odd, the parity is `1`, otherwise it is `0`. The key observation here is that `XOR` preserves the parity of its arguments. For example, if, in binary, `a = 1101` and `b = 0111` then the parity of the two together is zero since there are a total of six on bits. If we apply `XOR` we get `a XOR b = 1010` which also has an even number of on bits and therefore has the same parity. This suggests the trick. If we `XOR` a number with itself but first shift the number half its bit width the resulting bits of the lower half of the output of `XOR` will have the same parity as the original number. We can see this if we look at `a = 01011101` and `XOR` it with itself after shifting down by four bits, which is half the width of the number,

```
     01011101
XOR  00000101
-------------
     xxxx1000
```

where we are ignoring the upper four bits. The original number has 5 on bits therefore the parity is `1`. If we look at the lower four bits after the `XOR` we see it also has a parity of `1`. If we repeat the process but using the new value and this time shifting by half its effective width, which is 4 bits, we will end up with a number that has the same parity as we started with and the same parity as the original number. If we repeat this all the way the final result will be a one bit number that is the parity of the original number. Therefore, for `01011101`,

original	`01011101`
right shift 4	`00000101`
XOR	`01011000`
right shift 2	`00010110`
XOR	`01001110`
right shift 1	`00100111`
XOR	`01101001`
parity bit	`xxxxxxx1`

For an 8-bit number we can define a parity function in C as,

```
unsigned char parity(unsigned char x) {
    x = (x >> 4) ^ x;
    x = (x >> 2) ^ x;
    x = (x >> 1) ^ x;

    return x & 1;
}
```

where the `return x & 1;` line returns only the lowest order bit since all other bits are meaningless at this point. The extension of this function to wider integers is straightforward.

Shifting moves bits left or right and drops them at the ends. This is not the only option. Instead of dropping the bits we might want to move the bits that would have fallen off to the opposite end. This is a *rotation* which like a shift can be to the left or right. Unfortunately, while many microprocessors contain rotate instructions as primitive machine language operations neither C nor Python support rotations directly. However, they can be simulated easily enough in code and an intelligent compiler will even turn the simulation into a single machine instruction (eg, `gcc`).

Again, to keep things simple, we will work with 8-bit numbers. For an 8-bit number, say $\texttt{AB}_{16} = \texttt{10101011}_2$, a rotation to the right of one bit looks like this,

$$\texttt{10101011} \Rightarrow 1 = \texttt{11010101}$$

where we introduce the $\Rightarrow$ symbol to mean rotation to the right instead of simple shifting ($\rightarrow$). Rotation to the left one bit position gives,

$$\texttt{10101011} \Leftarrow 1 = \texttt{01010111}$$

where we see that the bit that would have fallen off at the left end has now moved around to the right side.

To simulate rotation operators in C and Python we use a combination of $<<$ and $>>$ with an `OR` operator. We need to pay attention to the number of bits in the data type in this case. For example, in C, we can define rotations to the right by any number of bits using,

```
unsigned char rotr(unsigned char x, int n) {
    return (x >> n) | (x << 8 - n);
}
```

with a similar definition for rotations to the left,

```
unsigned char rotl(unsigned char x, int n) {
    return (x << n) | (x >> 8 - n);
}
```

where we have changed $<<$ to $>>$ and vice versa. One thing to note is the `8` in the second line of these functions. This number represents the number of bits used to store the data value. Since we have declared the argument `x` to be of type `unsigned char` we know it uses eight bits. If we wanted to modify the functions for 32-bit integers we would replace the `8` with `32` or, in general, use `sizeof(x) * 8` to convert the byte size to bits.

The Python versions of the C rotation functions are similar.

```
def rotr(x,s):
    return ((x >> s) | (x << 8 - s)) & 0xff
```

and,

```
def rotl(x,s):
    return ((x << s) | (x >> 8 - s)) & 0xff
```

where the only change is that the return value is AND'ed with $\texttt{FF}_{16}$ which as we have seen will keep the lowest eight bits and set all the others to zero. This is again because Python uses 32-bit integers internally and we are interested in keeping the output in the 8-bit range. If we wanted these functions to work with 16-bit integers, we would replace the `8` with `16`, as in the C versions, but we would also need to make the mask keep the lowest 16 bits by replacing `0xff` with `0xffff`.

The rotation functions are helpful, but why do they work? Let's take a look by breaking the operations up individually and seeing how they combine to produce the final result. We start with the original input, $\texttt{AB}_{16} = \texttt{10101011}_2$, and show, row by row, the first shift to the left, then the second shift to the right, and finally the OR to combine them. The values between the vertical lines are those that fit within the 8-bit range of the number, the other bits are those that are lost in the shift operations,

	10101011		AB_{16}
	01010101	1	$\mathrm{AB}_{16} >> 1$
1010101	10000000		$\mathrm{AB}_{16} << 8-1$
	11010101		OR

where we see that the bit lost when shifting to the right one position has been added back at the front of the number by the shift to the left and the OR operation. Since the shifts always introduce a zero bit the OR will always set the output bit to the proper value because `1 OR 0 = 1` and `0 OR 0 = 0`. This works for any number of bits to rotate,

	10101011		AB_{16}
	00101010	11	$\mathrm{AB}_{16} >> 2$
1010101	11000000		$\mathrm{AB}_{16} << 8-2$
	11101010		OR

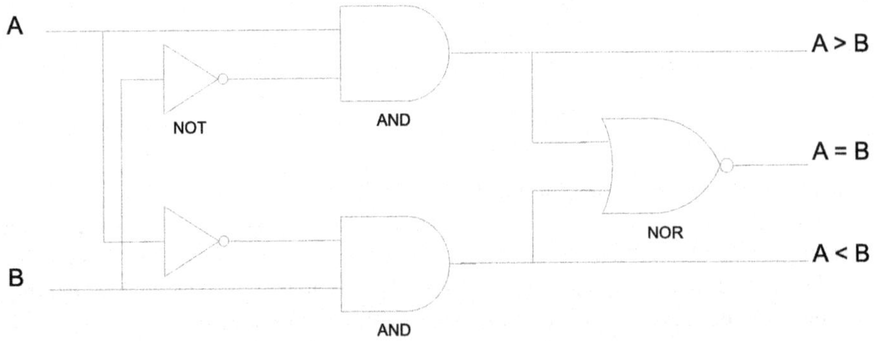

Fig. 2.4 A 1-bit digital comparator. The three output values represent the three possible relationships between the input values `A` and `B`. Recall that input values are `0` or `1`. The output is `1` for the relationship that is true and `0` for those that are not meet. Cascades of this basic form can be used to create multi-bit comparators

2.3.4 Comparisons

Magnitude comparison operators take two unsigned integers and return a truth value about whether or not the relationship implied by the operator holds for the operands. Here we look at the basic three, equality ($A = B$), greater than ($A > B$), and less than ($A < B$). There is a second set which can be created easily from the first: not equal ($A \neq B$), greater than or equal ($A \geq B$), and less than or equal ($A \leq B$) by using `NOT` and `AND` operators.

At its core, a computer uses digital logic circuits to compare bits. Figure 2.4 illustrates the layout of a 1-bit comparator. Comparators for larger numbers of bits are repeated instances of this basic pattern. As this not a book on digital logic we will go no further down this avenue but will instead talk about comparing unsigned integers from a more abstract point of view.

Most microprocessors have primitive instructions for comparison of integers as this operation is so fundamental to computing. In addition to direct comparison, many instructions affect processor flags based on privileged numbers like zero. For example, to keep things simple, the 8-bit 6502 microprocessor, which has a single accumulator, `A`, for arithmetic, performs comparisons with the `CMP` instruction but also sets processor status flags whenever a value is loaded from memory using the `LDA` instruction. There are other registers and instructions, of course, but we focus on the accumulator to keep the example simple. The 6502 uses branch instructions like `BEQ` and `BNE` to branch on equal or not equal respectively. This also applies to the implied comparison with the special value `0` which happens when the `LDA` instruction loads the accumulator.

Armed with this brief review of an old 8-bit microprocessor we can see that the following set of instructions would indeed perform an 8-bit comparison between 22_{16} already in the accumulator via the first `LDA` instruction and $1A_{16}$ stored in memory location 2035_{16} and branch if they are not equal. Additionally, we will also perform an implicit comparison of the value of memory location $20FE_{16}$ with zero and branch if it is,

```
LDA #$22        ; A = $22
CMP $2035       ; compare to location $2035
BNE noteql      ; branch to "noteql" if not equal

LDA #$20FE      ; A = contents of $20FE
BEQ iszero      ; branch to "iszero" if zero
```

where we use the classic notation of $\$22 = 22_{16}$.

Why bring up this example? In part to show that comparison is very fundamental to computers and is done as efficiently as possible in hardware and to set the stage for our less efficient alternatives to digital logic. The comparisons we are implementing in code are pure hardware even in the simplest of microprocessors.

Since for any two integers A and B exactly one of the following is true: A $=$ B, A $<$ B, or A $>$ B, it follows that if we know how to test for any two the last condition is simply when neither of the two we can test for applies. In our case, we look at the situation where we know how to test for equality (A $=$ B) and greater than (A $>$ B). We do this with two predicate functions that simply return 1 if the relationship holds for the arguments and 0 otherwise. Let's call these predicates isZero(A) and isGreater(A,B) and see how we might implement them directly in C for 8-bit values using the unsigned char data type.

You may be wondering why we chose to use isZero(A) instead of the perhaps more obvious isEqual(A,B). If so, good, you are paying attention. Given our experience with the XOR operator we now know that,

$$\texttt{a XOR a} \rightarrow 0$$

so we can immediately see that,

$$\texttt{isEqual(A,B) = isZero(A XOR B)}$$

but how do we implement isZero(A)? One approach in code would be to shift the bits and test the lowest order one. If we find one that is not zero then the number is not zero. The test is via an OR which is only zero when both operands are zero. We use AND to do the bit comparison and then shift the result down so that the compared bit is in the lowest position. Then, the OR of all these tests will be 1 if any bits are set and 0 if not. This is the exact opposite of what we want so we add a NOT to reverse the sense of the logic and a final AND with 1 to mask out all other bits and return the state of the lowest bit only. This, then, is the full predicate function isZero(x),

```
unsigned char isZero(unsigned char x) {
    unsigned char ans;

    return ~(((x & (1<<7)) >> 7) |   // test bit 7
             ((x & (1<<6)) >> 6) |
             ((x & (1<<5)) >> 5) |
             ((x & (1<<4)) >> 4) |
             ((x & (1<<3)) >> 3) |
             ((x & (1<<2)) >> 2) |
             ((x & (1<<1)) >> 1) |
              (x & 1)) & 1;           // test bit 1
}
```

Notice that there are no actual comparison operators in this function, only logical bitwise operators. With this we can quickly implement `isEqual(x,y)`,

```
unsigned char isEqual(unsigned char x,
                      unsigned char y) {
    return isZero(x ^ y);
}
```

We now need `isGreater(A,B)` which is implemented with bit operators, shifts, and a call to `isZero(x)`. This is why we started with `isZero(x)` instead of `isEqual(x,y)`. The C code for our function is given first followed by the explanation of why it works,

```
 1 unsigned char isGreater(unsigned char a,
 2                         unsigned char b) {
 3     unsigned char x,y;
 4
 5     x = ~a & b;
 6     y = a & ~b;
 7
 8     x = x | (x >> 1);
 9     x = x | (x >> 2);
10     x = x | (x >> 4);
11
12     return ~isZero(~x & y) & 1;
12 }
```

In order to tell if `a` is greater than `b` we need to know the first place where their respective bits do not agree. Once we know this bit position we know that `a` is greater than `b` if at that bit position `a` has a `1` while `b` has a `0`. So, we need to find the locations of where the bits differ. To make the example concrete, we let `a` be $\texttt{00011101}_2 = \texttt{29}$ and `b` be $\texttt{00011010}_2 = \texttt{26}$. If we look at line 5 we see,

```
x = ~a & b;
```

which sets `x` to the `AND` of the `NOT` of `a` and `b`. This leaves `x` with a `1` in all the positions where the bit in `a` is less than the same bit in `b`. For our example this sets `x` to $\texttt{00000010}_2$ which tells us that the only bit position in `a` that is less than the corresponding bit position in `b` is bit 1. Likewise, line 6 asks where are the bit positions where `a` is greater than `b`? In this case, we set `y` to $\texttt{00000101}_2$ to indicate that in bit 0 and bit 2 the value of `a` is greater than `b`. In order to see if `a` is greater than `b` we need to find the highest bit position where the two differ and see if that bit is set. We can do this if we take the value in `x`, which tells us where `a` bits are less than `b` bits, and build a mask which is `1` for all bit positions at or below the highest bit position where `a` is first less than `b`. We do this with lines 8 through 10. This operation which `OR`s the value with shifted versions of itself duplicates the highest `1` bit among all the lower bits. In this case,

```
x                      →   00000010
x = x | (x >> 1)       →   00000011
x = x | (x >> 2)       →   00000011
x = x | (x >> 4)       →   00000011
```

where the last two steps add nothing new since `x >> 2` and `x >> 4` both result in zero which will set no new bits. We now have a mask in `x` that tells us all the bit positions below the first place where the bit in `a` is less than the bit in `b`. If we NOT this mask, $\neg$ `00000011` $\rightarrow$ `11111100`, we can use the new mask to tell us all the bit positions where `a` was not less than `b`. Lastly, with this mask and the value in `y` which tells us where the bits in `a` were greater than the bits in `b`, we can perform one final AND, `~x & y`, which will result in zero if `a` $\leq$ `b` since no bits will be set in `y` in the region where the bits of `a` were greater than those of `b`, or a nonzero value since at least one bit will be set in `y` in that region. Line 12, then, asks if this result is zero by passing the output to `isZero`. It then applies NOT to change the output of `isZero` since the result is zero when `a` $\leq$ `b` and not zero when `a` $>$ `b`. The final AND with `1` gives us only the final bit since the NOT will change all the bits of the result of `isZero`.

We are nearly finished with our comparison operators. We have equality (`isEqual`) and greater than (`isGreater`). With these we see that `isLess` would be,

```
unsigned char isLess(unsigned char x,
                     unsigned char y) {
    return (!isEqual(x,y)) && (!isGreater(x,y));
}
```

which is reasonable since for any two unsigned integers `A` and `B`, if `A` $\not\geq$ `B` then `A` $<$ `B` must be true. Testing for not equal is even simpler,

```
unsigned char isNotEqual(unsigned char x,
                         unsigned char y) {
    return !isEqual(x,y);
}
```

since the only way for `A` to *not* not equal `B` is if the two are indeed equal. Less than or equal and greater than or equal follow directly from the functions already defined,

```
unsigned char isLessOrEqual(unsigned char x,
                            unsigned char y) {
    return isEqual(x,y) || isLess(x,y);
}
```

and

```
unsigned char isGreaterOrEqual(unsigned char x,
                               unsigned char y) {
    return isEqual(x,y) || isGreater(x,y);
}
```

which completes our implementation of comparison operators using only bitwise operations.

2.3.5 Arithmetic

Just as comparison is a fundamental microprocessor operation, so is arithmetic. In this section we look at arithmetic with unsigned binary integers, first from the point of view of doing it "by hand" and then from the point of view of a simple 8-bit microprocessor. These two approaches should illustrate the mechanism behind the operations. We will not, however, attempt to implement these operations in C as we did above for the comparison operators though we will use C to demonstrate overflow and underflow conditions.

The addition facts in binary are,

$$
\begin{aligned}
0 &+ 0 = 0 \\
0 &+ 1 = 1 \\
1 &+ 0 = 1 \\
1 &+ 1 = 0 \text{ carry } 1
\end{aligned}
$$

from which we see that one of them produces a carry since it results in a two digit number. Just as in decimal, the carry is applied to the next digit over to the left. So, to add two unsigned 8-bit binary numbers we move right to left, bit by bit, adding and moving any carry to the next digit to the left,

```
   11111      ← carry
   01101110   ← first operand
+  00110101   ← second operand
-----------------------------
   10100011   ← answer
```

where the carry from the second to leftmost bit does not cause difficulty since the highest bits of each number are zero. However, what would happen if there was a carry from the leftmost bit? In mathematics, nothing special would happen, there would simply be a new leftmost bit, but in computers this is not the case. Recall that we are working with 8-bit unsigned numbers which means that all numbers fit in eight bits in memory. If we use eight bits for numbers we have no place in which to put any final carry bit. This results in an *overflow* condition. The computer simply discards this new highest value bit and retains the lowest eight bits which fit in memory,

$$
\begin{array}{lrr}
 & 11111 & \\
 & 11101110 & EE_{16} \\
+ & 00110101 & 35_{16} \\
\hline
 & 1\ 00100011 & 123_{16}
\end{array}
$$

which is stored as 23_{16} discarding the carry on the leftmost bit. This is precisely what we see with the following C code,

```
#include <stdio.h>

int main() {
    unsigned char x, y, z;

    x = 0xEE;
    y = 0x35;
    z = x + y;

    printf("%x\n", z);
}
```

Let's take a look at how a simple 8-bit microprocessor would implement an addition. In this case, we look at an unsigned 16-bit addition which requires two addition operations. Working again with the 6502 processor mentioned above we see that an 8-bit addition between a value in memory locations 23_{16} and 25_{16} will involve a load into the accumulator (`LDA`), a clearing of the carry flag which catches any overflow bit (`CLC`) and an addition with memory (`ADC`). Specifically, we assume memory location 23_{16} contains EE_{16}, memory location 25_{16} contains 35_{16}. We then load the accumulator, clear the carry and add,

`LDA $23`	$A \leftarrow \mathrm{EE}$
`CLC`	$C \leftarrow 0$
`ADC $25`	$A \leftarrow A + 35 + C$
	$A \leftarrow 23, C \leftarrow 1$

with the accumulator set to 23_{16}, the lowest eight bits of the sum of EE_{16} and 35_{16}, and the carry flag set to `1` to indicate an overflow happened. This setting of the carry flag is the key to implementing multibyte addition. In C, we would simply declare the variables to be of data type `unsigned short` which is 16-bits and add as before. For example, in C we have,

```
#include <stdio.h>

int main() {
    unsigned short x, y, z;

    x = 0xEE;
    y = 0x35;
    z = x + y;

    printf("%x\n", z);
}
```

which gives us a 16-bit answer of 123_{16}. In memory, using little-endian representation for multibyte numbers, we have,

memory location \$24	:	`23`
\$25	:	`01`

since we store the lowest byte first.

In the simpler world of the 8-bit microprocessor we store the lowest part of the sum, the low byte, and add the high bytes without clearing the carry. Assuming memory is set to

memory location \$23	:	EE
\$24	:	00
\$25	:	35
\$26	:	00

we clear the carry flag, add the two low bytes, store the partial sum, add the high bytes with any carry, and store the final part of the sum like this,

`LDA $23`	$A \leftarrow \mathrm{EE}$
`CLC`	$C \leftarrow 0$
`ADC $25`	$A \leftarrow A + 35 + C,\ \ A{=}23$
`STA $27`	$\$27 \leftarrow 23, \mathrm{C} \leftarrow 1$
`LDA $24`	$A \leftarrow 0$
`ADC $26`	$A \leftarrow A + 0 + C,\ \ A{=}1$
`STA $28`	$\$28 \leftarrow 1$

where we have introduced a new instruction, `STA`, which stores the accumulator in memory. When this sequence of instructions is complete we have the final answer in memory locations \$27 and \$28 as 23_{16} and 1_{16} respectively, as we expect for a little-endian number.

The addition above is equivalent to this single 16-bit addition

```
           1 11111
    00000000 11101110     EE16
+   00000000 00110101     3516
-------------------------------
    00000001 00100011    12316
```

where we have separated the upper eight bits of the high byte from the lower eight bits of the low byte.

Addition of unsigned binary numbers is by now straightforward. We add, left to right, bit by bit with carry bit when necessary. If the result is too large, we overflow and retain the lowest *n* bits where *n* is the width of the number in bits. We now move on to subtraction of unsigned binary numbers.

The subtraction facts in binary are,

$$
\begin{aligned}
0 &- 0 = 0 \\
0 &- 1 = 1, \textit{underflow} \\
1 &- 0 = 1 \\
1 &- 1 = 0
\end{aligned}
$$

where the *underflow* condition will require a borrow from the next higher bit position. Like overflow, *underflow* is the situation where we cannot properly represent the number using the number of bits we have to work with. In this case, the underflow happens when we attempt to subtract a larger number from a smaller and we have no way to represent the resulting negative number. We'll address this issue below.

To continue with the example we used for addition, we now evaluate $\texttt{EE}_{16} - \texttt{35}_{16}$ to get $\texttt{B9}_{16}$. In binary, using the subtraction facts, we have,

$$
\begin{array}{rlr}
 & \texttt{01} & \\
 & \texttt{0101} & \\
 & \texttt{11101110} & \texttt{EE}_{16} \\
- & \texttt{00110101} & \texttt{35}_{16} \\
\hline
 & \texttt{10111001} & \texttt{B9}_{16}
\end{array}
$$

with each borrow written above the bit and the next bit set to one less than it was. If a second borrow is necessary for a bit position, we write it above again. Let's look at the subtraction again, bit by bit, right to left, we are computing,

$$
\begin{array}{l|ccccl}
\textit{bit 0} & 10 & - & 1 & = & 1, \textit{borrow} \\
\textit{bit 1} & 0 & - & 0 & = & 0 \\
\textit{bit 2} & 1 & - & 1 & = & 0 \\
\textit{bit 3} & 1 & - & 0 & = & 1 \\
\textit{bit 4} & 10 & - & 1 & = & 1, \textit{borrow} \\
\textit{bit 5} & 10 & - & 1 & = & 1, \textit{borrow} \\
\textit{bit 6} & 0 & - & 0 & = & 0 \\
\textit{bit 7} & 1 & - & 0 & = & 1
\end{array}
$$

which, reading from bottom to top, gives $\texttt{10111001}_2 = \texttt{B9}_{16}$ as expected.

What happens if we need to borrow across more than one bit position? For example, in base 10 a problem like,

$$
\begin{array}{rr}
 & \texttt{7003} \\
- & \texttt{972} \\
\hline
 & \texttt{6031}
\end{array}
$$

involves borrowing across two digits in order to subtract the 7 of 972 which we can write as,

$$
\begin{array}{rr}
 & \texttt{69}^{1}\texttt{03} \\
- & \texttt{9 72} \\
\hline
 & \texttt{6031}
\end{array}
$$

where we change the 700 into $\texttt{69}^{1}\texttt{0}$ to subtract 9 7 giving the partial result 603. We subtracted one from the next two digits and added it in as a ten to the digit we were working with. The same thing happens in binary. Consider this subtraction problem,

$$
\begin{array}{rlr}
 & \texttt{01} & \\
 & \texttt{1010}\cancel{\texttt{10}}^{1}\texttt{01} & \texttt{A9}_{16} \\
- & \texttt{101001 10} & \texttt{A6}_{16} \\
\hline
 & \texttt{00000011} & \texttt{03}_{16}
\end{array}
$$

where we attempt to subtract `1` from `0` in the second bit position (bit 1, since we always count from zero and from right to left). We need to borrow from bit 2, but since this is also zero, we instead borrow from bit 3 and change `100` into $\texttt{01}^{1}\texttt{0}$ in order to do the subtraction.

As we are working with 8-bit unsigned integers one will eventually be tempted to ask what happens if we try to subtract `1` from `0` since we cannot represent `-1`. What happens is we get the result we would get if we had an extra bit at left-most position and borrowed from it like so,

$$\begin{array}{rr} & \texttt{0 1111111} \\ & \cancel{\texttt{1}}\ \cancel{\texttt{0000000}}^{1}\texttt{0} \\ - & \texttt{0000000 1} \\ \hline & \texttt{11111111} \end{array}$$

meaning that subtracting one from the smallest number we can represent, namely zero, results in the largest number we can represent which for an 8-bit unsigned integer is every bit set, $2^8 - 1 = 255$. Another way to think about it is that the numbers form a loop with `00000000` and `11111111` set next to each other. If we move one position down below zero we wrap around to the top and get `255`. Likewise, if we move one position up from `255` we will wrap around to the bottom and get zero. Moving down below zero is an underflow condition while moving up above `255` is an overflow.

Before we move on to multiplication and division of unsigned integers, let's look at one more subtraction example that would result in a negative number. We'll use our running example but swap the operands,

$$\begin{array}{rrr} & \texttt{1 00110101} & \texttt{35}_{16} \\ - & \texttt{11101110} & \texttt{EE}_{16} \\ \hline & \texttt{01000111} & \texttt{47}_{16} \end{array}$$

where we have indicated the implied `1` bit from which we can borrow. This implied `1` bit is in the bit position for $2^8 = 256$ which suggests another way to think about the answer we would expect if we go negative in a subtraction of unsigned numbers. For our example, $\texttt{35}_{16} - \texttt{EE}_{16} = \texttt{-185}$, but if we add in 256 which is essentially what we are doing in thinking there is an implied extra high bit, we get $\texttt{-185} + \texttt{256} = \texttt{71} = \texttt{47}_{16}$ which is the answer we found previously. We have been working with 8-bit wide unsigned integers. If we are using 16-bit integers the implied bit 16 (recall, we count bits from zero) is $2^{16} = 65536$ which means we would add 65536 to any negative value to get the result we would expect from unsigned subtraction.

The following C example demonstrates that what we have been discussing is indeed the case,

```
#include <stdio.h>

int main() {
    unsigned char x=0xEE, y=0x35, z;

    z = x - y;
```

```
        printf("%X\n", z);

        z = y - x;
        printf("%X\n", z);

        z = 0 - 1;
        printf("%X\n", z);
}
```

The output of this program is,

$$\begin{aligned} \texttt{B9}_{16} &= \texttt{10111001}_2 \\ \texttt{47}_{16} &= \texttt{01001110}_2 \\ \texttt{FF}_{16} &= \texttt{11111111}_2 \end{aligned}$$

which is exactly what we saw in the examples above.

Now that we know how to subtract in binary we can examine a useful trick involving AND. If one bit in a number is set this implies that the number is a power of two since every position in a binary number is, by definition, a power of two. If we know which bit we want to test for, which power of two, it is straightforward to use a mask and check that bit. But, what if we wanted to know if the number in question was any power of two? We can use AND here along with an observation about bits that are on in a number that is a power of two. For example, if the number we want to test is $\texttt{00100000}_2 = \texttt{32}$ we see that it is a power of two and only one bit is on. Now, subtract one from this number. In this case, we will get $\texttt{00011111}_2 = \texttt{31}$. What has happened is that the single bit that was on is now off and some of the bits that were off are now on. Finally, what happens if we AND these two values together? We get,

	00100000	=	32
AND	00011111	=	31
	00000000	=	00

which is exactly zero. From this we see that we will only get exactly zero when one of two conditions are met: either the number is itself zero or it is a power of two which had only one bit set. This is nicely captured in a simple C function,

```
unsigned char is_power_of_two(unsigned char n) {
    return (n == 0) ? 0
                    : (n & (n-1)) == 0;
}
```

which returns `1` if the argument is a power of two and `0` otherwise. The function checks if the argument is zero, if so, return `0`. If not, then check whether `n & (n-1)` is exactly `0`. If it is, the expression is true and the function returns `1` to indicate a power of two, otherwise it returns `0`. While written for `unsigned char` data type the function will work for any unsigned integer type.

We've looked in detail at addition and subtraction, now we turn our attention to multiplication and division. Modern microprocessors perform multiplication and division as operations in hardware. This is a very good thing but makes it difficult for us in a sense so we will, as before, look at more primitive approaches which might be used in small 8-bit microcontrollers that lack hardware instructions for multiplication and division. To keep things simple we will illustrate the algorithms in C even though this would never be done in practice.

Since multiplication is really repeated addition one approach to finding n × m, where n and m are unsigned integers, would be to add n to itself m times or vice versa. Naturally, it would make sense to run the loop as few times as possible so we would loop over the smaller of n or m adding the other number. In C we have,

```
unsigned short mult1(unsigned char n, unsigned char m) {
    unsigned char i;
    unsigned short ans = 0;

    if (n < m) {
        for(i=0; i < n; i++)
            ans += m;
    } else {
        for(i=0; i < m; i++)
            ans += n;
    }

    return ans;
}
```

which leaves the product of the 8-bit numbers in n and m in the now possibly 16-bit value p. Why is the product possibly 16-bits? Because the largest possible number we can get by multiplying two 8-bit numbers requires 16-bits to store it in memory since $255 \times 255 = 65025$ which is above $2^8 - 1 = 255$ meaning it needs more than 8-bits to store but is below $2^{16} - 1 = 65535$ which is the maximum for a 16-bit unsigned integer.

Is this really a good way to multiply numbers, however? Probably not. The loop needs to be repeated for the smaller of n or m which may be up to 255 times. Given we must do 16-bit addition inside the loop, recalling the example above, we see that the simple multiplication may turn into many thousands of individual machine instructions. Surely we can do better than this? To answer this question, let's look a little more closely at multiplication in binary as we might do it by hand,

$$\begin{array}{rrrcr}
 & 00010100 & 14_{16} & = & 20 \\
\times & 00001110 & 0E_{16} & = & 14 \\
\hline
 & 00000000 & & & \\
 & 00010100 & & & \\
 & 00010100 & & & \\
+ & 00010100 & & & \\
\hline
 & 00100011000 & 118_{16} & = & 280
\end{array}$$

in which we see that if the binary digit in the multiplier is 0 we simply copy down all zeros and if it is a 1 we copy the multiplicand lining it up beneath the multiplier bit as we would do in decimal. Then, again as in decimal multiplication, we add all the partial products to arrive at the final answer. For simplicity we did not write the leading zeros which would be present if showing all 16-bits of the result.

This method suggests a possible improvement over our existing multiplication function, `mult1`. Rather than repeatedly add the multiplier or multiplicand, we can copy the process just shown by shifting the multiplicand into position and adding it to the partial product if the multiplier bit is `1` otherwise ignore those that are `0`. This leads to a second multiplication function in C,

```
unsigned short mult2(unsigned char n, unsigned char m) {
    unsigned char i;
    unsigned short ans = 0;

    for(i=0; i < 8; i++) {
        if (m & 1) {
            ans += n << i;
        }
        m >>= 1;
    }

    return ans;
}
```

which, when compared to `mult1` and run ten million times proves to be about 1.6× faster. Let's look at what `mult2` is actually doing.

We are multiplying two 8-bit numbers so we need to look at each bit in the multiplier, `m`. This is the source of the `for` loop. The `if` statement AND's the multiplier with `1` to extract the lowest bit. If this bit is set, we want to add the multiplicand, `n`, to the partial product stored in `ans`. Note, though, that before we add the multiplicand, we need to shift it up to the proper bit position. Note also that this works because the result of the operation is a 16-bit value which will not lose any of the bits of `n` when we do the shift. Regardless of whether we add anything to the partial product we need to shift the multiplier down one bit so that in the next pass through the loop the `if` will be looking at the next highest bit of the original `m`. Lastly, we see that there are no side-effects to this function because C passes all arguments by value meaning `n` and `m` are copies local to the function.

The speed improvement between `mult1` and `mult2` becomes much more dramatic when we move from multiplying two 8-bit numbers to multiplying two 16-bit numbers. To do this, we take the source for `mult1` and `mult2` and replace all instances of `unsigned char` by `unsigned short` and all instances of `unsigned short` by `unsigned int`. Lastly, we change the loop limit in `mult2` from `8` to `16` since we are multiplying two 16-bit numbers. When this done we see that `mult2` is nearly 3500× faster than `mult1` for the same arguments (assuming both to be near the limit of `65535`).

What about division? We cover two operations with division since the algorithm returns the quotient and any remainder. The operations are integer division (/) which returns the quotient and modulo (%) which returns the remainder. For example, we need an algorithm that produces these answers,

```
123  /  4  =  30
123  %  4  =   3
```

since 123/4 = 30 with a remainder of 3.

We could implement division by repeated subtraction. If we count the number of times we can subtract the divisor from the dividend before we get a partial result that is less than the divisor we will have the quotient and the remainder. We might code this in C as,

```
unsigned char div1(unsigned char n,
                   unsigned char m,
                   unsigned char *r) {
    unsigned char q=0;

    *r = n;

    while (*r > m) {
        q++;
        *r -= m;
    }

    return q;
}
```

and test it with,

```
int main() {
    unsigned char n=123, m=7;
    unsigned char q,r;

    q = div1(n, m, &r);
    printf("quotient=%d, remainder=%d\n", q,r);
}
```

which prints `quotient=30, remainder=3` which is the answer we are expecting.

This function requires three arguments since we want to return the quotient as the function value and the remainder as a side-effect value. This is why we pass the remainder as a third argument using a pointer. Inside of `div1` we set the remainder (`r`) to our dividend and continually subtract the divisor (`m`) until we get a result less than the divisor. While doing this we keep count of the number of times we subtract in `q` which we return as the quotient.

Like our `mult1` example above, `div1` is an inefficient way to implement division. What happens if the dividend is large and the divisor is small? We must loop many times in that case before we are done. The problem is even worse if we use integers larger than 8-bits. What to do, then?

Just as we did for multiplication, let's look at binary division by hand. Unlike decimal long division, binary division is rather simple, either the divisor is less than or equal to the dividend in which case the quotient bit is 1, otherwise, the quotient bit is 0; there are no trial multiplications. Dividing $123 = 01111011_2$ by $4 = 100_2$ in this way gives,

```
      00011110
   100)01111011
       0
       -
       01
        0
       --
       011
         0
       ---
       0111
        100
       ----
        0111
         100
        ----
         0110
          100
         ----
          0101
           100
          ----
             11
              0
             --
             11
```

with $00011110_2 = 30$ and a remainder of $11_2 = 3$ as expected.

As we have seen several times already, modern microprocessors implement such a fundamental operation as division in hardware, but we can look at unsigned division in the way it might be implemented in a more primitive microprocessor or microcontroller. For simplicity, we again implement the algorithm in C. With all of this in mind, we get Fig. 2.5 which will require a bit of explanation.

The key to understanding what Fig. 2.5 is doing it to observe that binary division by hand is really a matter of testing whether or not we can subtract the divisor from the partial dividend. If so, we set a one in that bit of the quotient, otherwise we set a zero. The algorithm of Fig. 2.5 is setup for 8-bit division using 8-bit dividends and divisors, and by implication quotients. Therefore, we need to examine all eight bits of the dividend starting with the highest bit. To do this, and to save space as this algorithm is often implemented in hardware, we take advantage of the fact that C passes arguments by value and use `n` to be both dividend and quotient. When we have examined all eight bits the value in `n` will be the quotient. We can do this because as we look at each bit of the dividend we shift it out to the left while shifting in the new bit of the quotient from the right.

We store the remainder in `r` and pass it back out of the function by using a pointer. To start the division process we set `r` to zero and the quotient to the dividend in `n`. Since `n` already has the dividend there is no explicit operation to do this, we get it for free. If we call `div2(123, 4, &r)` to continue with our running division example, the state of affairs in binary after the execution of line 6 in Fig. 2.5 is,

```
 1 unsigned char div2(unsigned char n,
 2                    unsigned char m,
 3                    unsigned char *r) {
 4     unsigned char i;
 5
 6     *r = 0;
 7
 8     for(i=0; i<8; i++) {
 9         *r = (*r << 1) + ((n & 0x80) != 0);
10         n <<= 1;
11
12         if ((*r-m) >= 0) {
13             n |= 1;
14             *r -= m;
15         }
16     }
17
18     return n;
19 }
```

Fig. 2.5 Shift, test, and restore unsigned integer division

i	r	n	m
undefined	00000000	01111011	100

where the dividend is in n and the remainder is zero. We next hit the loop starting in line 8. This loop executes eight times, once for each bit of the dividend. Lines 9 and 10 perform a double left shift. This is the equivalent of treating r and n as a single 16-bit variable with r the high order bits. First we shift r one bit to the left (*r << 1) and then comes the rather cryptic expression,

```
((n & 0x80) != 0)
```

which tests whether the highest bit in n, our dividend and quotient is set. Recall our discussion of AND above. If it is set, we add it into r. This is because we are about to left shift n one bit and if the bit we are shifting out of n is set, we need to move it to r to complete the virtual 16-bit shift of r and n. We then shift n in line 10.

Line 12 checks whether or not we can subtract the divisor in m from the partial dividend which is being built, bit by bit, in r. If we can, we set the first bit of n, our quotient, in line 13 and then update the partial dividend by subtracting the divisor in line 14. Recall, we are examining the dividend one bit at a time by moving it into r. We are simultaneously storing the quotient in n one bit at a time by putting it in on the right side. Since we already shifted to the left the first bit in n is always zero, we only update it if the subtraction succeeds. After this first pass through the loop we have,

i	r	n	m
0	00000000	11110110	100

with the first bit of the quotient, a zero, in the first bit of `n`. If we continue through the loop we will get the following sequence of values,

i	r	n	m
1	00000001	11101100	100
2	00000011	11011000	100
3	00000011	10110001	100
4	00000011	01100011	100
5	00000010	11000111	100
6	00000001	10001111	100
7	00000011	00011110	100

where we end with a quotient of `30` in `n` which is the return value of the function and a remainder in `r` of `3`. Notice how `n` changes as we move through the loop. Each binary digit is shifted into `r` from the right as the new quotient bits are assigned from the left until all bits are examined. This algorithm, unlike the `div1` example, operates in constant time. There are a fixed number of operations needed regardless of the input values.

2.3.6 *Square Roots*

We briefly consider here a simple integer square root algorithm which makes use of an interesting mathematical fact. This algorithm works by counting the number of times an ever increasing odd number can be subtracted before reaching or going below zero. The algorithm itself is easy to implement in C,

```
unsigned char sqr(unsigned char n) {
    unsigned char c=0, p=1;

    while (n >= p) {
        n -= p;
        p += 2;
        c++;
    }

    return c;
}
```

where we again make use of the fact that C passes arguments by value which allows us to modify `n` in the function without changing it outside of the function. Our count, which will be the square root of `n`, is initialized to zero in `c`. We start our odd number in `p` at `1` and then move to `3`, `5`, `7`, and so on. The `while` loop is checking to see if our `n` value is still larger or the same as `p` and if so, we subtract `p` and count one more subtraction in `c`. When `n` is less than `p` we are done counting and return `c` as the square root. Naturally, this algorithm is only approximate by underestimating when `n` is not actually a perfect square.

If we call `sqr` with `64` as the argument, we get the following sequence of values in the `while` loop,

n	p	c
63	3	1
60	5	2
55	7	3
48	9	4
39	11	5
28	13	6
15	15	7
0	17	8

where the final value of `n` is zero since `64` is a perfect square and `c` is `8`, which is the square root of `64`. We see that the algorithm works, but why?

The trick is the observation that the sum of the sequence of odd numbers is always a perfect square. For example,

$$\begin{aligned} 1 &= 1 \\ 1+3 &= 4 \\ 1+3+5 &= 9 \\ 1+3+5+7 &= 16 \\ 1+3+5+7+9 &= 25 \end{aligned}$$

or more compactly,

$$\sum_{i=1}^{n} 2i - 1 = n^2$$

where n^2 is the argument to `sqr` and n is the square root.

Another way to see this, courtesy of Eric Spellman, is to consider the difference between n^2 and $(n+1)^2$. The latter is n^2+2n+1 which implies that the difference between a squared integer and the next integer squared is $2n+1$ which is, always, an odd number. For example, we can apply this same observation to computing the unsigned cube root of an integer. In this case, we have n^3 and $(n+1)^3 = n^3+3n^2+3n+1$ implying that the difference between cubes is $3n^2+3n+1$. With this in mind a simple modification of the square root code above will now return the cube root of the argument,

```
unsigned int cbroot(unsigned int n) {
    unsigned int c=0, p=1;

    while (n >= p) {
        n -= p;
        c++;
        p = 3*c*c+3*c+1;
    }

    return c;
}
```

where instead of adding 2 to p every iteration we compute $p = 3c^2 + 3c + 1$.

2.4 What About Negative Integers?

In the previous section we took a thorough look at unsigned integers and the sorts of operations computers typically perform on them. Without a doubt, unsigned integers are the mainstay of computers, but often it is necessary to represent quantities that are less than zero. What do we do about that? In this section we examine three options, two in detail, for tracking the sign of an integer and performing operations with signed integers. We will naturally build on what we have learned about unsigned integers and bear in mind that, as before, while we may show examples using 8-bit numbers for simplicity, everything immediately translates to numbers with more bits, be they 16, 32, or larger.

The most common techniques for handling signs are sign-magnitude, one's complement, and two's complement.

2.4.1 *Sign-Magnitude*

Perhaps the most natural way to represent the sign of an integer is to reserve one bit of its representation for the sign and this is precisely what early computers did. If we decide that we will keep the highest-order bit of the number for the sign we can use the remainder of the bits to represent the magnitude as an unsigned integer and this is the *sign-magnitude* form,

```
01111111  =   127
...
00000010  =     2
00000001  =     1
00000000  =     0
10000001  =    -1
10000010  =    -2
...
11111111  =  -127
```

This seems to be a perfectly reasonable way to store a signed integer but notice a few things,

- We lose range in terms of magnitude. An unsigned 8-bit number can take values from `0` to `255` while a sign-magnitude number is restricted to `-127` to `+127`. As we will see, keeping track of the sign always results in a loss of magnitude range.

- There are two ways to represent zero: `+0` = `00000000` and `-0` = `10000000`. This seems unnecessary and wasteful of a bit pattern.
- Arithmetic becomes more tedious since we need to bear the sign of the number in mind at all times. It would be nice to be able to do some arithmetic without requiring separate logic for the sign.

For the reasons given above, especially the additional hardware logic, the sign-magnitude representation for integers has been abandoned by modern computer systems. Let us now turn our attention to possible replacements.

2.4.2 One's Complement

Our first candidate for a suitable replacement to the sign-magnitude form is called *one's complement*. In this notation we represent negative numbers by taking the positive form and calculating the one's complement. The one's complement is simple to do, just negate (logical `NOT`) every bit in the positive form of the number. So, we have,

```
01111111  =   127
...
00000010  =     2
00000001  =     1
00000000  =     0
11111110  =    -1
11111101  =    -2
...
10000000  =  -127
```

which again seems good in that we can look at the highest order bit to see if the number is negative or not and as we will shortly see, we can use this notation for arithmetic without too much trouble, but we do still have two ways to represent zero since `00000000` → `11111111`.

2.4.3 Two's Complement

The one's complement of a positive number is the bit pattern we get when we change all the zero bits to one and all the one bits to zero. The *two's complement* of a positive number is the bit pattern we get when we take the one's complement and then add one to it. This is the notation that has been accepted as the way to represent negative integers and the advantages will be come clear when we look at operations on signed integers. As with one's complement integers, a positive two's complement integer is represented in just the same way as an unsigned integer. With two's complement we have,

```
01111111  =   127
...
00000010  =     2
00000001  =     1
00000000  =     0
11111111  =    -1
11111110  =    -2
...
10000000  =  -128
```

where we now have only one way to represent zero,

`00000000`	→	`11111111`	→	`00000000`
positive		*one's complement*		*two's complement*

since adding one to `11111111` maps back around to `00000000` with the overflow bit ignored. Additionally, we have increased our range by one since we can represent numbers from `-128` to `+127` instead of `-127` to `127` as with one's complement or sign-magnitude.

2.5 Operations on Signed Integers

We would like to be able to perform operations on signed integers. The bit level operations like AND, OR and NOT work the same way with signed integers as with unsigned integers. To these operators, the bits are just bits, the "fact" of a negative integer is just a convention forced on certain bit patterns. Since this is the case, we need only look at how to compare negative integers, how to perform arithmetic on negative integers, and, as a special operation, how to deal with the sign of a two's complement integer when changing the number of bits used to represent the number. Let us first start with comparing two signed integers.

2.5.1 Comparison

Comparison of two signed integers, A and B, implies determining which relational operator, $<$, $>$, or $=$, should be put between them. When we compared unsigned integers we looked at the bits from highest to lowest. We still do that for signed integers but we need to first consider the signs of the two numbers. If the signs differ we know very quickly the relationship between the numbers without considering all the bits. If the signs match, either both positive or both negative, we need to

```
1  signed char bset(signed char v,
2                   signed char n) {
3      return (v & (1 << n)) != 0;
4  }
5
6  signed char scomp(signed char a,
7                    signed char b) {
8      unsigned char i;
9
10     if ((bset(a,7) == 0) && (bset(b,7) == 1))
11         return 1;
12     if ((bset(a,7) == 1) && (bset(b,7) == 0))
13         return -1;
14
15     for(i=0; i<7; i++) {
16         if ((bset(a,6-i) == 0) && (bset(b,6-i) == 1))
17             return -1;
18         if ((bset(a,6-i) == 1) && (bset(b,6-i) == 0))
19             return 1;
20     }
21
22     return 0;
23 }
```

Fig. 2.6 Comparison of two signed integers `a` and `b`. The function returns `0` if `a` = `b`, `1` if `a` > `b` and `-1` if `a` < `b`

look at the magnitude bits to see where they might be different. This will tell us the relationship between the two numbers. Naturally, if the bits are all the same, position for position, then the two numbers are equal.

We can use a C function like the one in Fig. 2.6, with included helper function to determine whether or not a particular bit position is on, to compare two signed numbers in two's complement notation. Note that we are now working with variables of type `signed char` which are 8-bit signed integers.

The helper function (`bset`, lines 1–4) returns `1` if the nth bit of v is on, otherwise it returns a `0`. It uses the shift and AND mask trick we saw above to test a bit position value. The main function, `scomp`, looks first at the signs of the arguments (lines 10–13) and returns the relationship if they differ. If the sign bit of `a` is zero, `a` is positive. If the sign bit of `b` is one, then `b` is negative and `a` must be greater than `b` so return `1` to indicate `a` > `b`. If the signs are reversed, `a` is less than `b` so return `-1` to indicate `a` < `b`.

If the signs of `a` and `b` match, either positive or negative, we then look at the remaining bits from highest to lowest to see where there are any differences. This is the loop of lines 15 through 20 in Fig. 2.6. If we find a bit position where `a` has a zero and `b` has a one we know that `a` < `b` must be true so we return `-1`. Likewise, if we find that `a` is one and `b` is zero at that bit position we know that `a` > `b` so we return `1`. Finally, if we make it through all the bits and find no differences the two integers are equal so we return `0`. With a comparison function like `scomp` it is easy to create predicate functions checking for equal, less than and greater than. Recall that a predicate function is a function that returns true or false. For example,

```
unsigned char isEqual(signed char a,
                      signed char b) {
    return scomp(a,b) == 0;
}

unsigned char isLessThan(signed char a,
                         signed char b) {
    return scomp(a,b) == -1;
}

unsigned char isGreaterThan(signed char a,
                            signed char b) {
    return scomp(a,b) == 1;
}
```

define functions which return true (`1`) when $a = b$, $a < b$, and $a > b$, respectively.

2.5.2 *Arithmetic*

Let's take a look at the basic arithmetic operations ($+$,$-$,$\times$,$\div$, and %) as they apply to signed numbers. For addition and subtraction we will consider both one's and two's complement negative numbers to show why two's complement is often preferable. We focus on operations involving negative numbers as operations involving positive numbers follow the techniques described earlier in the chapter for unsigned integers.

Addition and Subtraction Addition in one's complement notation is nearly identical to unsigned addition with one extra operation should there be a final carry. To see this, let's take a look at adding two negative integers represented in one's complement,

```
     11000000   -63
+    11000010   -61
-------------------      (one's complement)
   1 10000010 -125
     10000011 -124
```

where the carry at the end, shown by the extra `1` on the left, is added back into the result to get the correct answer of `-124`. This adding in of any carry is called the *end-around carry* and is the extra twist necessary when adding one's complement numbers.

The same addition in two's complement notation produces a carry which we ignore,

```
    11000001    -63
+   11000011    -61     (two's complement)
  ---------------------
  1 10000100  -124
```

since we see that `10000100` is `-124` by making it positive,

```
  01111011
+ 00000001
-------------
  01111100  124
```

Addition of a positive and negative number works in the same way for both one's and two's complement numbers. For example, in one's complement we have,

```
+   01111100  124
    11000010  -61
  -----------------     (one's complement)
  1 00111110   62
    00111111   63
```

where we have again made use of the end-around carry to give us the correct answer. The two's complement version is similar,

```
+   01111100  124
    11000011  -61      (two's complement)
  -----------------
  1 00111111   63
```

where we again ignore the carry and keep only the lower eight bits. Recall, we are giving all examples as signed or unsigned 8-bit numbers. If we were working with 16-bit or 32-bit numbers we would keep that many bits in the answer.

Computers implement subtraction by negation and addition. This allows for only one set of hardware circuits to be used for both operations. With that in mind, subtraction becomes particularly simple. If we want to calculate $124 - 61 = 63$ we actually calculate $124 + (-61) = 63$ which is exactly the example calculated above. For calculation by hand it is helpful to think of subtraction as an actual operation but, as we see here, when done with the appropriate notation for negative numbers, subtraction is really an "illusion" and is nothing more than addition with a negative number.

While addition and subtraction are straightforward, especially with two's complement notation, we have to consider one question: what happens if the result of the operation does not fit in the number of bits we are working with? For our examples, this means that the result does not fit in eight bits taking the sign into account. Let's consider only two's complement numbers. We already saw in the examples above that we could ignore the carry to the 9th bit and saw that the lower eight bits were correct. Is this always the case?

A signed 8-bit two's complement number has a range from `-128` to `127`. If we attempt an operation that falls outside of this range we will not be able to properly represent the answer. We call this condition an *overflow*. How can we detect this? By following two simple rules,

1. *If the sum of two positive numbers is negative, overflow has happened.*
2. *If the sum of two negative numbers is positive, overflow has happened.*

We need not worry about the sum of a positive and negative number because both the positive and negative number are already in the allowed range and it is impossible, because of the difference in sign, for the sum to be outside the allowed range. This is why we ignored the last carry bit when adding −61 to 124. Let's look at cases that prove the rules. If we try to calculate $124+124 = 248$ we know we will have trouble because 248 is greater than 127 which is the largest 8-bit positive two's complement number. We get,

```
+ 01111100  124
  01111100  124      (two's complement)
  --------------
  11111000   -8
```

which is clearly a wrong answer. According to our rule for addition of two positive numbers we know overflow has happened because the sign bit, bit 7, is one, indicating a negative answer. Similarly, two large negative numbers added will prove our second rule,

```
+ 10000100  -124
  10000100  -124      (two's complement)
  ---------------
  00001000     8
```

where we have ignored the carry to the 9th bit. We see that the result is positive since bit 7 is zero. This proves our second rule and we know that overflow has happened.

Multiplication We now consider multiplication of signed integers. One approach to signed multiplication would be to make use of the rules for products to track the sign of the result. If we do this, we can make any negative number positive, do unsigned integer multiplication as described above in Sect. 2.3.5, and negate the result if necessary to make it negative. This approach will work for both one's and two's complement numbers. As we recall from school, when multiplying two numbers there are four possible scenarios related to the signs,

1. *positive* × *positive* = *positive*
2. *positive* × *negative* = *negative*
3. *negative* × *positive* = *negative*
4. *negative* × *negative* = *positive*

with this in mind it is simple to extend our `mult2` example from Sect. 2.3.5 to check the signs of the inputs and negate the negative numbers to make them positive before multiplying. Then, the result can be negated to make it negative if the result should be negative. In C this gives us Fig. 2.7
where the main loop in lines 18 through 21 has not changed but before we run it we check the signs of the inputs to see if we need to negate any negative numbers to

```
1  signed short signed_mult2(signed char n, signed char m) {
2      unsigned char i, s=0;
3      signed short ans=0;
4  
5      if ((n > 0) && (m < 0)) {
6          s = 1;
7          m = -m;
8      }
9      if ((n < 0) && (m > 0)) {
10         s = 1;
11         n = -n;
12     }
13     if ((n < 0) && (m < 0)) {
14         n = -n;
15         m = -m;
16     }
17 
18     for(i=0; i < 8; i++) {
19         if (m & 1) ans += n << i;
20         m >>= 1;
21     }
22 
23     if (s) ans = -ans;
24     return ans;
25 }
```

Fig. 2.7 Unsigned integer multiplication modified for signed numbers

make them positive. The variable `s` holds the flag to tell us that the answer needs to be negative. We initially set it to `0` with the assumption that the inputs, `n` and `m`, will both be positive. In lines 5 through 16 we check this assumption. If `n` is positive and `m` is negative, we set `s` in line 6 and make `m` positive. Likewise, in line 9 we check to see if `n` is negative and `m` is positive and make `n` positive if this is the case. We also set the flag in `s` since we know the answer needs to be negative. Lastly, if both `n` and `m` are negative we make them both positive and leave the flag in `s` unset. Then we multiply as before. In line 23 we check to see if the negative flag is set, if so, we make the answer negative before returning it.

In this example we have taken advantage of the fact that the C compiler will properly negate a value as in line 7 regardless of the underlying notation used for negative numbers. We know, however, in practice that this will typically be two's complement. Can we multiply numbers directly in two's complement? Yes, in fact, there are several existing algorithms which to exactly that. Let's consider one of the more popular of them, the Booth algorithm [1] which was developed by Andrew Booth in 1950. A C implementation of this algorithm for multiplying two signed 8-bit integers is given in Fig. 2.8. Let's take a look at what it is doing.

Booth's essential insight was that when we multiply two binary numbers a string of ones can be replaced by a positive and negative sum in the same way that $16 \times 6 =$

```
1  signed short multb8(signed char m, signed char r) {
2      signed int A, S, P;
3      unsigned char i;
4
5      A = m << 9;
6      S = (-m) << 9;
7      P = (r & 0xff) << 1;
8
9      for(i=0; i < 8; i++) {
10         switch (P & 3) {
11             case 1:     // 01
12                 P += A;
13                 break;
14             case 2:     // 10
15                 P += S;
16                 break;
17             default:    // 11 or 00
18                 break;
19         }
20         P >>= 1;
21     }
22
23     return P>>1;
24 }
```

Fig. 2.8 The Booth algorithm for multiplication of two's complement 8-bit integers

$16 \times (8-2)$ but since we are in binary we can always write any string of ones as the next higher bit minus one. So, we have,

`00011100 = 00100000 + 000000-10`

where we have written a `-1` for a specific bit position to indicate subtraction. This means, if we scan across the multiplicand and see that at bit position i and $i-1$ there is a `0` and `1`, respectively, we can add the multiplier. Similarly, when we see a `1` and `0`, respectively, we can subtract the multiplier. At other times, we neither add nor subtract the multiplier. After each pair of bits, we shift to the right.

In Fig. 2.8 we initialize the algorithm by setting `A` to the multiplier, `m`, shifted nine places to the left, and similarly set `S` to the negative of the multiplier (two's complement form), also shifted nine positions to the left. The product, `P`, is initialized to the multiplicand in `r` but shifted one position to the left. This is done in lines 5 through 7. Why all the shifting? We are multiplying two eight bit signed numbers so the result may have as many as 16 bits, hence using `signed int` for `A`, `S` and `P`. This is the origin of eight of the nine bit positions. The extra bit position for `A` and `S`, and the single extra bit for `P` (line 7), is so that we can look at the last bit position and the one that would come after which we always set to zero. This means that the last bit position, bit 0, and the next, bit -1, could be `1` and `0` to signal the end of a string of ones. We mask the multiplicand, `r`, with `0xFF` to ensure that the sign is not extended when we move from the `signed char` to `signed int` data type. See the next section for a description of sign extension.

The loop in lines 9 through 21 examines the first two bits of `P`, which are the two we are currently considering, and decides what to do based on their values. Our options are,

bit i	bit $i-1$	operation
0	0	*do nothing*
0	1	add multiplier to product
1	0	subtract multiplier from product
1	1	*do nothing*

which is reflected in the `switch` statement of line 10. The phrase `P & 3` masks off the first two bits of `P`, which is what we want to examine. After the operation, we shift the product (`P`) to the right to examine the next pair of bits. When the loop finishes, we shift `P` once more to the right to remove the extra bit we added at the beginning in line 7. This completes the algorithm and we have the product in `P`, which we return. This algorithm substitutes a starting add and ending subtraction for what might be a long string of additions for each `1` bit in a string of `1` bits. Also, when not adding or subtracting, we simply shift bits. This makes the algorithm particularly efficient.

Sign Extension and Signed Division Just as we did for multiplication above, we can modify the unsigned integer algorithm for division in Fig. 2.5 to work with signed integers by determining the proper sign of the output, then making all arguments positive and dividing, negating the answer if necessary. However, before we do that, let's take a quick look at sign extension.

Sign Extension What happens if we take an 8-bit number and make it a 16-bit number? If the number is positive, we simply set the upper eight bits of the new 16-bit number to zero and the lower eight bits to our original number like so,

`00010110` → `0000000000010110`

which is pretty straightforward. Now, if we have a negative 8-bit signed integer in two's complement notation we know that the leading bit will be a `1`. If we simply add leading zeros we will get,

`11011101` → `0000000011011101`

which is no longer a negative number because the leading bit is now a `0`. To avoid this problem and preserve the value we extend the sign when we form the 16-bit number by making all the new higher bits `1` instead of `0`,

`11011101` → `1111111111011101`

which we know is the same value numerically and we can check it by looking at the magnitude of the number. Recall, we convert between positive and negative two's complement by flipping the bits and adding one. So, the 8-bit version becomes,

$$\texttt{11011101} \rightarrow \texttt{00100010 + 1} \rightarrow \texttt{00100011} = 35_{10}$$

and the 16-bit version becomes,

$$\texttt{1111111111011101} \rightarrow \texttt{0000000000100010+1} \rightarrow \texttt{0000000000100011} = 35_{10}$$

which means that both bit patterns represent −35 as desired. We intentionally frustrated sign extension in Fig. 2.8 by masking `r` with `0xFF` before assigning it to `P` which was a 32-bit integer.

Signed Division Figure 2.5 implements unsigned division. If we track the signs properly we can modify it to work with signed integers. Division actually returns two results. The first is the quotient and the second is any remainder. The sign we should use for the quotient is straightforward enough,

Dividend	*Divisor*	*Quotient*
+	+	+
+	−	−
−	+	−
−	−	+

ambiguity arises when we think about what sign to apply to the remainder. It turns out that different programming languages have adopted different conventions. For example, C chooses to make the remainder have the same sign as the dividend while Python gives the remainder the sign of the divisor. Unfortunately, the situation is more complicated still. When dividing negative numbers we are often returning an approximate quotient (unless the remainder is zero) and that approximate quotient has to be rounded in a particular direction. All division algorithms in programming languages satisfy $d = nq + r$ which means that the quotient, q, times the divisor, n, plus the remainder, r, equals the dividend, d. However, we have choices regarding how to set the signs and values of q and r. There are three options,

1. *Round towards zero.* In this case, we select q to be the integer closest to zero that when multiplied by n is less than or equal to d. In this case, if d is negative, r will also be negative. For example, $-33/5 = -6$ with a remainder of -3 so that $-33 = 5(-6) + (-3)$. This is the option used by C.
2. *Round towards negative infinity.* Here we round the quotient down and end up with a remainder that has the same sign as the divisor. In this case, $-33/5 = -7$ with a remainder of 2 giving $-33 = 5(-7)+2$. This is the option used by Python.
3. *Euclidean definition.* This definition makes the remainder always positive. If $n > 0$ then $q = floor(d/n)$ and if $n < 0$ then $q = ceil(d/n)$. In either case, r is always positive, $0 \leq r < |n|$.

Let's make these definitions more concrete. The table below shows the quotient and remainder for several examples in both C and Python. These will give us an intuitive feel for how these operations work.

Dividend	*Divisor*	*Quotient*	*Remainder*	
33	7	4	5	*C*
		4	5	*Python*
−33	7	−4	−5	*C*
		−5	2	*Python*
33	−7	−4	5	*C*
		−5	−2	*Python*
−33	−7	4	−5	*C*
		4	−5	*Python*

We see that differences only arise when the signs of the dividend and divisor are opposite. It is here that the C choice of rounding towards zero and the Python choice of rounding towards negative infinity come into play. The C choice seems more consistent at first because it always results in quotients and remainders with the same magnitude, only the signs change, but from a mathematical point of view it is less desirable because certain expected operations do not give valid results. For example, to test whether or not an integer is even it is common to check if the remainder is zero or one when dividing by two. If the remainder is zero, the number is even, if one, it is odd. This works in Python for negative integers since `-43 % 2 = 1` as expected, but in C this fails since we get `-43 % 2 = -1` because of the convention to give the remainder the sign of the dividend.

With all of the above in mind, we can now update our `div2` algorithm in Fig. 2.5 to handle signed integers. We show the updated algorithm, now called `signed_div2`, in Fig. 2.9. Let's look at what has changed.

The actual division algorithm in lines 20 through 30 is the same as in Fig. 2.5 since we are still performing unsigned division. Lines 6 through 18 check whether any of the arguments, dividend (`n`) or divisor (`m`) or both, are negative. We use two auxiliary variables, `sign` and `rsign`, to track how we deal with the sign of the answer. If the dividend is negative but the divisor is positive the quotient should be negative so we set `sign` to `1`. If the dividend is positive but the divisor is negative, we also need to make the quotient negative. If both the dividend and divisor are negative, the quotient is positive. In all cases we make the dividend and divisor positive after we know how to account for the sign of the quotient. For the remainder, we use the variable `rsign` to decide when to make it negative. The division algorithm itself will make the remainder, in `*r`, positive but in order for our answer to be sound we must sometimes make `*r` negative. When to do this? A sound answer will always satisfy,

$$n = m \times q + r$$

so if the dividend in `n` was negative, we will require the remainder to be negative as well. In this case we follow the C convention.

If we run `signed_div2` on `n = 123` and `m = 4` with all combinations of signs, we get the following output,

```
signed char signed_div2(signed char n,
                        signed char m,
                        signed char *r) {
    unsigned char i, sign=0, rsign=0;

    if ((n < 0) && (m > 0)) {
        sign = rsign = 1;
        n = -n;
    }
    if ((n > 0) && (m < 0)) {
        sign = 1;
        m = -m;
    }
    if ((n < 0) && (m < 0)) {
        rsign = 1;
        n = -n;
        m = -m;
    }

    *r = 0;

    for(i=0; i<8; i++) {
        *r = (*r << 1) + ((n & 0x80) != 0);
        n <<= 1;

        if ((*r-m) >= 0) {
            n |= 1;
            *r -= m;
        }
    }

    if (sign) n = -n;
    if (rsign) *r = -*r;

    return n;
}
```

Fig. 2.9 Shift, test, and restore unsigned integer division updated to handle signed integers

n	m	q	r	$m \times q + r$
123	4	30	3	4(30)+3
-123	4	-30	-3	4(-30)-3
123	-4	-30	3	-4(-30)+3
-123	-4	30	-3	-4(30)-3

where the column on the right shows that our choice of sign for the remainder is correct.

In this section we have reviewed implementations of signed arithmetic on integers. In some cases we were able to build directly on existing algorithms for unsigned arithmetic while in some we worked directly in two's complement format.

There is no need to talk about signed integer square root since the square root of a negative number is imaginary and we are not yet ready to work with complex numbers.

2.6 Binary-Coded Decimal

Binary-Coded Decimal (BCD) numbers make use of specific bit patterns corresponding to the digits $0 \ldots 9$ in order to store one or two decimal digits in each byte. Storing numbers in this format allows for decimal operations in place of binary and, indeed, some early microprocessors such as the Western Digital 6502 had BCD modes. In this section we describe how to encode numbers in BCD and how to do simple arithmetic with those numbers.

2.6.1 *Introduction*

As we saw earlier in this chapter, working with numbers expressed in binary can be cumbersome. Binary is the natural base for a computer to use given its construction but humans, with ten fingers and ten toes, generally prefer to work in decimal. One possibility is to encode decimal digits in binary and let the computer work with data in this format. Binary-coded decimal does just this. For our purposes we will work with what is generally called *packed BCD* where we use each nibble of a byte to represent exactly one decimal digit. Historically, other sizes were used, one digit per byte (unpacked) for example, or even other values, typically with early computers. We will examine one of these formats below (zoned decimal). In addition, while any set of distinct bit patterns can be used to represent decimal digits we will use the obvious choice, $0 = 0000, 1 = 0001, \ldots, 9 = 1001$, so there is a direct conversion between a decimal digit and its BCD bit pattern. The remaining six bit patterns, $1010 \ldots 1111$ are not used or allowed in properly formatted BCD numbers. In the previous sections we ignored the difficulties of converting human entered decimal data to binary and vice versa. BCD simplifies this process. BCD is no longer frequently used but it will turn up again when we consider some of the more specialized ways in which computers represent numbers, particularly Decimal Floating-Point numbers, Chap. 7.

If we consider only positive numbers, a single byte can represent decimal numbers from 0 through 99 as so,

BCD	*Decimal*
0 0	0
0 1	1
0 2	2
1 0	10
1 1	11
...	
9 8	98
9 9	99

where the two digits in the *BCD* column on the left represent the two 4-bit nibbles of the byte. This works but does not allow for negative numbers. Historically, several variations were used for storing a sign with a BCD number. The form we will use is directly analogous to signed binary integers but instead of two's complement we will use *ten's complement* with a leading nibble to represent the sign. If the leading nibble is `0000` the number is positive, if `1001` the number is negative and the value is in *ten's complement format*. We use the leading sign nibble because unlike two's complement where the leading bit is always `1` if the number is negative, ten's complement has no such quick test.

In two's complement we negate a number by flipping all the digits making `0` → `1` and `1` → `0` and then add one. For the ten's complement of a decimal number we first find the nine's complement which involves subtracting the number from 99, for a single byte, and then add one. So, we see that the BCD number 51 can be thought of as −49 because,

$$99 - 51 = 48, \ \ 48 + 1 = 49$$

A multidigit BCD number is represented with multiple bytes. Endian issues arise again in this case and we chose to use big-endian so that it is easier to see the decimal numbers in the binary bit patterns. With this convention, including the sign nibble, the decimal number 123 is represented in BCD as,

+	0	1	2	3
`0000`	`0000`	`0001`	`0010`	`0011`

where we need the leading `0` digit to fill out the byte representing the leading two digits. Similarly, using the negative sign nibble we can represent −732 as,

−	0	2	6	8
`1001`	`0000`	`0010`	`0110`	`1000`

because $999 - 732 = 267, \ \ 267 + 1 = 268$ is the ten's complement of 732.

2.6.2 *Arithmetic with BCD*

Let's look at arithmetic using BCD numbers. We only consider addition and subtraction as multiplication and division are rarely performed in BCD. For performance reasons in software it is faster to convert from BCD to binary, do the multiplication or division in binary, and then convert the answer back to BCD.

Addition of two BCD numbers is straightforward at first glance. We simply add nibble by nibble since each nibble is a digit,

	123		`0000`	`0001`	`0010`	`0011`
+	732	+	`0000`	`0111`	`0011`	`0010`
	855		`0000`	`1000`	`0101`	`0101`

where for each nibble we simply add in binary as if the numbers were unsigned binary integers. For this example, everything works out nicely. But consider this,

	123		`0000`	`0001`	`0010`	`0011`
+	738	+	`0000`	`0111`	`0011`	`1000`
	861		`0000`	`1000`	`0101`	`1011`

here we have a small problem, the sum of the first digits gives us `1011` which is 11, not an allowed bit pattern in BCD. We see the source of the problem which is that $3 + 8 = 11$ so we have produced a carry. A simple way to deal with any carries is to take the resulting bit pattern, `1011`, and add six, `0110`, which will give us the correct bit pattern for the decimal digit and a carry of `1` for the next digit. This correction gives,

$$1011 + 0110 = 1\ \ 0001$$

with the carry separated from the remaining digit which is now the BCD representation of `1`. Applying this correction to the addition leads to the correct result,

	123		`0000`	`0001`	`0010`	`0011`
+	738	+	`0000`	`0111`	`0011`	`1000`
			`0000`	`1000`	`0101`	`1011`
	861		`0000`	`1000`	`0110`	`0001`

where we have added the carry into the next digit to change the `5` to a `6`. Why add six? There are 16 possible 4-bit nibbles but we are only using the first 10 of them for decimal digits. If we exceed 10 for any single digit adding six will wrap the bit pattern around to the digit we would get if we, in decimal, subtracted ten. It also leaves the carry set for the next digit. If adding the carry to the next higher digit results in 10 or greater the "add six" trick can be used again and repeated as many times as necessary until the BCD number is in the proper form, each time moving to the next higher digit position.

We have chosen to represent negative BCD numbers using ten's complement. This format works in the same way as two's complement so that in order to subtract we simply add. For example, consider,

	123	0000	0001	0010	0011
−	732	+ 1001	0010	0110	1000
		1001	0011	1000	1011
	−609	1001	0011	1001	0001

where the overflow in the first digit was addressed by adding six and then adding the carry to the next higher digit. The sign is negative and the BCD number reads as 391. Since the number is negative, we expect that 391 is the ten's complement form of −609 which is the actual answer. To see that it is, we negate it,

$$999 - 391 = 608, \quad 608 + 1 = 609$$

so we know that we have, in fact, reached the correct result.

2.6.3 Conversion Routines

Binary to BCD and BCD to binary conversion is straightforward. First, let's consider binary to BCD. This routine could be used during multiplication or division of BCD numbers or might be used on its own to prepare for output of a binary value as a decimal number. Once the value is in BCD conversion to ASCII for output is straightforward, simply examine each nibble, add it to 48 which is the ASCII code for "0" and output the resulting byte as an ASCII character.

Our binary to BCD conversion routine will convert an 8-bit unsigned binary number to a three digit BCD number. It goes by the amusing name of the "double dabble" algorithm and we present our implementation in Fig. 2.10. The routine itself is referred to briefly in [7].

What is this code doing? We have an unsigned 8-bit binary number passed in as `b`. We will return the three digit BCD number in `p`. Since a single byte can hold any value from 0 to 255 we need exactly 12 bits to store the equivalent BCD number. Using an `unsigned short` as a return value gives us 16 bits, the top four of which we leave as zero. Inside the algorithm we work with `p` as an `unsigned int` so that we have 32-bits to work with. The algorithm is going to perform eight shifts to the left so that it can examine each of the eight bits in `b`. For example, if `b` is `123` $= 01111011_2$ we initialize `p` with `b` so that at the start `p` looks like,

			p				
0000	0000	0000	0000	0000	0000	0111	1011
			100s	*10s*	*1s*		

where the bit positions of the hundreds, tens and ones digits for our BCD representation are indicated.

The loop in lines 7 through 24 runs eight times and shifts `p` one bit position to the left each time. However, before the shift we look at each of the BCD digit locations to see if the value there is five or greater. If it is, we add three before shifting to the left. Line 8 pulls out the four bits representing the ones value in the

```
1  unsigned short bin2bcd8(unsigned char b) {
2      unsigned int p=0;
3      unsigned char i,t;
4
5      p = (unsigned int)b;
6
7      for(i=0; i<8; i++) {
8          t = (p & 0xf00) >> 8;
9          if (t >= 5) {
10             t += 3;
11             p = ((p>>12)<<12)|(t<<8)|(p&0xff);
12         }
13         t = (p & 0xf000) >> 12;
14         if (t >= 5) {
15             t += 3;
16             p = ((p>>16)<<16)|(t<<12)|(p&0xfff);
17         }
18         t = (p & 0xf0000) >> 16;
19         if (t >= 5) {
20             t += 3;
21             p = ((p>>20)>>20)|(t<<16)|(p^0xffff);
22         }
23         p <<= 1;
24     }
25
26     return (unsigned short)(p>>8);
27 }
```

Fig. 2.10 Unsigned 8-bit binary to three digit BCD conversion

BCD representation. Specifically, it first masks `p` with `0xf00` which leaves only the four bits in the ones place nonzero. It then shifts to the right eight bits and assigns this value to `t`. This sets `t` to the ones value. Line 9 asks if this value is greater than or equal to five. If so, line 10 increments `t` by three. Line 11 then updates the proper position in `p` with the new value of `t`. Let's look at this line more closely. Recall that OR combines bits and here three pieces are combined to replace the proper four bits of `p` with the value of `t`. First, `((p>>12)<<12)` loses the lower 12 bits of `p`, which includes the four bits of the BCD ones digit along with the eight bits originally set to `b`. It then shifts back up so that the net result is a value that has the lowest 12 bits set to zero and the remaining bits set to whatever they were previously. This is OR-ed with `t` shifted up eight bits so that the lower four bits of `t`, which are the only ones to have a nonzero value, are in the proper position to be the BCD ones digit. Lastly, `p & 0xff` keeps the lowest eight bits of `p` and OR's them in as well. When each of these operations is complete `p` is the same as it was previously with the exception that the four bits in the BCD ones position have been increased by three. We will see below why three was added and why five was the cutoff value. Lines 13 through 22 perform the same check for the BCD tens and BCD hundreds digits. The only change is the mask and shift values to isolate and work with the correct four bits of `p`. Lastly, line 23 shifts all of `p` one position to the left to move the next bit of `b` (in the lowest eight bits of p) into position. After all bits of b have been examined the final BCD result is in bits 9 through 20 so we shift down eight positions to get the final BCD value.

The operation of the entire algorithm is shown below for b = 123. For convenience we ignore the top three nibbles of p, a 32-bit unsigned integer. This leaves us with,

100s	*10s*	*1s*	b		*comment*
0000	0000	0000	0111	1011	initial
0000	0000	0000	1111	0110	shift 1
0000	0000	0001	1110	1100	shift 2
0000	0000	0011	1101	1000	shift 3
0000	0000	0111	1011	0000	shift 4
0000	0000	1010	1011	0000	add 3, ones
0000	0001	0101	0110	0000	shift 5
0000	0001	1000	0110	0000	add 3, ones
0000	0011	0000	1100	0000	shift 6
0000	0110	0001	1000	0000	shift 7
0000	1001	0001	1000	0000	add 3, tens
0001	0010	0011	0000	0000	shift 8

which after the last left shift has 123 as a BCD number in the indicated columns. The return statement shifts this value to the right eight bits to return the final three digit BCD number.

This shows us that the algorithm works but not why it works. The key to understanding the algorithm is to recall that to convert a BCD digit that is not a proper digit, say 1011 which is 11, to a proper digit is to add six, 0110. This will produce a carry bit to the next BCD digit position while preserving the proper value in the existing digit. So, one way to think of how to convert binary to BCD is to look at each digit position and if it is greater than ten subtract ten before doubling. This is to say, calculate,

$$2x + 6$$

for cases where the digit, x, is ten or greater before looking at the next bit of our binary number. This is equivalent to considering if the value is greater than five and if so adding three and then doubling because,

$$2x + 6 = 2(x + 3)$$

which has the added advantage of not requiring an extra bit since $2x + 6$ may be greater than 15 while $x+3$ never will be for x a valid BCD digit. This is the approach of the double-dabble algorithm shown in Fig. 2.10.

Conversion from a three digit BCD number to binary is especially straightforward. We simply examine each nibble, multiply it by the proper power of 10, and accumulate the result,

```
unsigned char bcd2bin8(unsigned short b) {
    unsigned char n;

    n = (b & 0x0f);
    b >>= 4;
    n += 10*(b & 0x0f);
    b >>= 4;
    n += 100*(b & 0x0f);

    return n;
}
```

The phrase `b & 0x0f` masks off the lowest nibble of `b`. This is the ones digit, so we simply set the output value in `n` to it. We then shift `b` to the right by four bits to make the lowest nibble the tens digit. We add this to the running total after first multiplying it by ten. We then shift one last time to make the hundreds digit the first nibble, multiply it by 100 and add to `n` to get the final value of the three digit BCD number.

2.6.4 Other BCD Encodings

So far in this section we have used packed BCD encoding. Here we explore two others: zoned decimal and densely packed decimal (DPD). We will only briefly discuss densely packed decimal because we cover it in detail in Sect. 7.2 as it is a critical part of the IEEE decimal floating-point format.

Densely Packed Decimal The DPD format encodes three decimal digits using ten bits. It is a refined version of Chen-Ho encoding [8] and is used by the IEEE 754-2008 Decimal Floating-Point Format [9] to encode the significand in base ten. There is a two bit savings in storage over packed BCD which requires 12 bits to encode three decimal digits.

Packing three BCD numbers into a DPD declet is straightforward. We label the bits of each BCD digit as *abcd*, *efgh*, and *ijkm*. Using these labels for the bits we can take any set of three BCD digits and encode them into a DPD declet using Table 2.3. The ten bits of the declet are labeled *pqr stu v wxy* which can be read from the table by finding the row matching the *aei* bit values from the input BCD numbers.

For example, to create a declet encoding the number 359 we first encode the digits in packed BCD as `0011`, `0101`, and `1001`, and then look for the row in Table 2.3 that starts with `001` (the first bits of each packed BCD digit). This directs us to line 2 so that we can map the remaining input bits to the output DPD bits like so,

Table 2.3 Packing three BCD numbers (*abcd efgh ijkm*) into a declet (*pqr stu v wxy*)

aei	*pqr*	*stu*	*v*	*wxy*
`000`	bcd	fgh	0	jkm
`001`	bcd	fgh	1	00m
`010`	bcd	jkh	1	01m
`011`	bcd	10h	1	11m
`100`	jkd	fgh	1	10m
`101`	fgd	01h	1	11m
`110`	jkd	00h	1	11m
`111`	00d	11h	1	11m

Locate the row by matching the first bit of each BCD number to the value under the *aei* column, then, the bits of the declet are read from the remaining columns of that row. After [11]

Table 2.4 Unpacking a declet (*pqr stu v wxy*) by translating it into three groups of binary digits representing three decimal digits (*abcd efgh ijkm*)

vxwst	*abcd*	*efgh*	*ijkm*
`0----`	0pqr	0stu	0wxy
`100--`	0pqr	0stu	100y
`101--`	0pqr	100u	0sty
`110--`	100r	0stu	0pqy
`11100`	100r	0pqu	100y
`11110`	0pqr	100u	100y
`11111`	100r	100u	100y

After [11]

$$\begin{array}{rcccl} pqr & \rightarrow & bcd & = & \texttt{011} \\ stu & \rightarrow & fgh & = & \texttt{101} \\ v & \rightarrow & 1 & = & \texttt{1} \\ wxy & \rightarrow & 00m & = & \texttt{001} \end{array}$$

so that the final output declet is `011 101 1 001`.

A declet can be unpacked using Table 2.4 by forming the value *vxwst* and locating the matching row. For example, to undo the example above where 359 → `011 101 1 001` (*pqr stu v wxy*), we form *vxwst* = `10010` which matches line 2 of Table 2.4 so that the bits of the unencoded declet become,

$$\begin{array}{rcccccl} abcd & \rightarrow & \text{0pqr} & = & \texttt{0011} & = & 3 \\ efgh & \rightarrow & \text{0stu} & = & \texttt{0101} & = & 5 \\ ijkm & \rightarrow & \text{100y} & = & \texttt{1001} & = & 9 \end{array}$$

as we expect.

Zoned Decimal Zoned decimal is a storage format that was supported on early IBM mainframes starting with the IBM 360, if not earlier. The format is still used, in fact, and is supported by certain programming languages, for example, RPG IV [10].

Zoned decimal is a BCD format used primarily during input and output. While packed BCD stores two decimal digits per byte, zoned decimal stores only one in the lower nibble. The upper nibble, the "zone", is used to store a value that makes the entire byte a valid EBCDIC or ASCII character code for the number itself. In this way, output is made simple. The only twist is storing the sign of the decimal number. In this case, the upper nibble of the lowest order digit is altered to indicate that the entire number is negative. We will implement zoned decimal routines that store up to eight digits in an unsigned 64-bit integer. If we do this we see that,

$$\begin{aligned} 8 &\rightarrow \text{F0F0F0F0F0F0F0F8} \\ -8 &\rightarrow \text{F0F0F0F0F0F0F0D8} \\ 8675309 &\rightarrow \text{F0F8F6F7F5F3F0F9} \\ -8675309 &\rightarrow \text{F0F8F6F7F5F3F0D9} \end{aligned}$$

where we are using the EBCDIC representation which uses `F` for the high nibble and `D` if the number is negative. For ASCII the conversion is similar but `3` is used for the high nibble and `7` if the number is negative.

For example, the single decimal digit, 8, maps to `F0F0F0F0F0F0F0F8` where the number is stored in the 64-bit unsigned integer with leading zeros, seven of them, as `F0`, the EBCDIC code for "0", followed by `F8` for the digit, "8". For −8 the conversion is the much the same but the lowest order byte is now `D8` to indicate that the entire number is negative. Notice that an unsigned 64-bit integer can store eight bytes and that each zoned decimal digit is a byte so the range of representable integers is [−99999999, +99999999], which fits in a standard C signed integer. The code for converting a signed integer to zoned decimal is,

```
1 #include <inttypes.h>
2 #define PNIB 0xF0
3 #define NNIB 0xD0
4
5 uint64_t bin2zoned(int b) {
6     uint8_t i,s = (b<0) ? 1:0;
7     uint64_t d,z=0,p=10000000;
8
9     b = abs(b);
10    for(i=7; i>0; i--) {
11        d = b / p;
12        b -= d * p;
13        p /= 10;
14        z += (PNIB+d) << (uint64_t)(i*8);
15    }
16    z += (s) ? (NNIB+b) : (PNIB+b);
17    return z;
18 }
```

where we include the `inttypes` standard C library to access the `uint64_t` data type. Lines 2 and 3 define the upper nibbles for the EBCDIC representation. Replace these with `0x30` and `0x70` respectively for ASCII. The `bin2zoned` function (line 5) accepts a standard signed integer value (b) and returns the eight digit zoned decimal representation. Line 6 sets the sign of the input (s) and then works with the positive version (line 9). The loop starting in line 10 sets the highest seven digits of the output. The last digit is a special case because the upper nibble changes if the input is negative. The output value is in z.

Line 11 sets d to the integer part of the input divided by the power of ten the current zoned digit represents (p, initialized in line 7). The loop counts down so that we process the input from most-significant to least-significant digit. Line 12 subtracts the digit value from the input. Line 13 drops p to the next lower power of ten. Finally, line 14 adds the digit value to the positive upper nibble value, PNIB+d, and shifts it into the proper byte position, i*8.

The lowest-order digit in handled in line 16. At this point the value in b is the digit we need so we check the sign in s to decide which upper nibble to add before adding the final digit value to z. Line 17 returns the zoned decimal representation.

Converting a zoned decimal back to binary is similarly straightforward,

```
1  int zoned2bin(uint64_t z) {
2      int32_t b=0,p=1;
3      int8_t i,s=1;
4  
5      for (i=0; i<8; i++) {
6          b += (z & 0xF) * p;
7          p *= 10;
8          if ((z & 0xF0) == NNIB) s=-1;
9          z >>= 8;
10     }
11     return s*b;
12 }
```

where we loop over each digit in the input, lowest to highest, starting in line 5. The output is stored in b and we mask off the lowest nibble of the input, z, and multiply it by the current power of ten represented by that digit, p (line 6). Line 7 moves to the next higher power of ten. Line 8 checks if the current zoned digit's high nibble is negative. If so, it sets the sign in s. Line 9 shifts the input down so that the highest zoned digit is now in the first byte. The binary value is multiplied by the sign and returned (line 11). Conversion from zoned decimal to packed BCD is left as an exercise for the reader (see Problem 2.9).

2.7 Chapter Summary

In this lengthy but essential chapter we covered a lot of important topics. We learned key terminology regarding bits, bytes, and words. We explored unsigned integers at length learning how to perform such low-level functions as manipulation of individual bits to the basic logic operations of AND, OR, NOT and XOR.

We completed our tour of low-level manipulation by studying shifts and rotates and how they affect the value of an unsigned integer. We then examined how to compare the magnitude of two unsigned integers.

Unsigned integer arithmetic was our next topic and with it we learned how to implement the basic operations of addition, subtraction, multiplication and division with remainder. To round out the unsigned integers we reviewed one approach for calculating the square root of a number.

Signed integers were our next target. We built upon what we learned with unsigned integers and delved into the main ways signed data is stored: sign-magnitude, one's complement and two's complement. We learned how to compare signed integers and how to either extend existing unsigned arithmetic routines to handle signed arguments or how to implement algorithms that operate on two's complement numbers directly.

To round out our examination we took a look at binary-coded decimal numbers, how to add and subtract them, in ten's complement, and how to convert between binary-coded decimal and pure binary as well as vice versa.

It would not be too much to say that integers are the heart of computers. We now know how to work with them at any level. There are more specialized representations of integers, and we will examine them in other chapters, but for the vast majority of computer work involving data that does not need fractions, including characters which are represented as integer codes, fixed-width integers are our main tools and understanding them in some detail is well worth the effort.

Exercises

2.1 Interpret the following bit patterns first as little-endian then as big-endian unsigned short integers. Convert your answer to decimal.

- `1101011100101110`
- `1101010101100101`
- `0010101011010101`
- `1010101011010111`

2.2 Given `A` = $\texttt{6B}_{16}$ and `B` = $\texttt{D6}_{16}$, write the result of each expression in binary and hexadecimal.

- `(A AND B) OR (NOT A)`
- `(A XOR B) OR (A AND (NOT B))`
- `((NOT A) OR (NOT B)) XOR (A AND B)`

2.3 Using C syntax write an expression that achieves each of the following. Assume that the unsigned char `A` = `0x8A`.

- Set bits 3 and 6 of `A`.
- Keep only the low nibble of `A`.

- Set unsigned char `B` to `100` if bit 3 of `A` is not set.
- Clear bit 1 of `A`.
- Toggle bits 4 and 5 of `A`.

2.4 Using C syntax write an expression for each of the following.

- Swap the upper and lower nibbles of unsigned char `v = 0xC4`.
- Multiply `v` by `5` using at least one shift operation.

2.5 Express each of the following numbers in one's and two's complement notation. Write your answer in binary and hexadecimal. Assume 8-bit signed integers.

- -14
- -127
- -1

2.6 Write a function to reverse the bits of an unsigned 8-bit integer. *

2.7 Write a function to count the number of `1` bits in an unsigned 32-bit integer. *

2.8 Modify the cube root algorithm in Sect. 2.3.6 to work properly with *signed* integers.

2.9 Modify the `zoned2bin` function of Sect. 2.6.4 to return a *packed BCD* representation of the input. Call this new function `zoned2bcd`. Note, the input is an unsigned 64-bit integer representing 8 digits so the packed BCD representation will require 32-bits and will fit in an unsigned 32-bit integer.

2.10 The *Hamming distance* between two integers is the number of places where their corresponding bits differ. For example, the Hamming distance between `1011` and `0010` is `2` because the numbers differ in bits 0 and 3. Write a function to calculate the Hamming distance between two unsigned short integers. *

2.11 A *Gray code* is a sequence of bit patterns in which any two adjacent patterns in the sequence differ by only one bit. Gray codes were developed to aid in error correction of mechanical switches but are used more generally in digital systems. For example, for four bits, the first six values in the Gray code are: `0000, 0001, 0011, 0010, 0110, 0111`. The rule to convert a binary number n into the corresponding Gray code g is to move bit by bit, from lowest bit to highest bit, setting the output bit to the original bit or inverted if the next higher bit is `1`. So, to change $n = 0011$ to $g = 0010$ we output a `0` for bit 0 of g since bit 1 of n is set and output the existing bit value of n for all other positions since the next higher bit is not set.

Using this rule, write a C function that calculates the Gray code g for any unsigned short n input. *(Hint: there are many ways to write a function that works but in the end it can be accomplished in one line of code. Think about how to check if the next higher bit of the input is set by shifting and how to invert the output bit if it is.)* **

References

1. Booth, A.D., A signed multiplication technique. The Quarterly Journal of Mechanics and Applied Mathematics, Vol IV, Part 2 (1951)
2. Shannon, C., "A mathematical theory of communication." ACM SIGMOBILE Mobile Computing and Communications Review 5.1 (2001): 3–55.
3. As mentioned in *Anecdotes*, Annals of the History of Computing, Vol 6, No 2, April pp 152–156 (1984)
4. Swift, J., Gulliver's Travels (1726). Reprinted Dover Publications; Unabridged edition (1996)
5. The Internet Protocol. RFC 791 (1981)
6. Boole, G., An Investigation of the Laws of Thought (1854). Reprinted with corrections, Dover Publications, New York, NY, 1958. (reissued by Cambridge University Press, 2009)
7. Gao, S., Al-Khalili, D., Chabini, N., "An improved BCD adder using 6-LUT FPGAs", IEEE 10th International New Circuits and Systems Conference (NEWCAS 2012), pp. 13–16 (2012)
8. Chen, T. C., "Decimal Number Compression (Internal memo to Irving T. Ho)", San Jose Research Laboratory: IBM: 1–4 (1971).
9. IEEE Standards Association. Standard for Floating-Point Arithmetic. IEEE 754-2008 (2008).
10. IBM Rational Development Studio for i, ILE RPG Language Reference 7.1 (2010).
11. Duale, A., *et al.* Decimal floating-point in z9: An implementation and testing perspective. IBM Journal of Research and Development 51.1.2 (2007): 217–227.

References

Chapter 3
Floating Point

Abstract We live in a world of floating-point numbers and we make frequent use of floating-point numbers when working with computers. In this chapter we dive deeply into how floating-point numbers are represented in a computer. We briefly review the distinction between real numbers and floating-point numbers. Then comes a brief historical look at the development of floating-point numbers. After this we compare two popular floating-point representations and then focus exclusively on the IEEE 754 standard. For IEEE 754 we consider representation, rounding, arithmetic, exception handling, and hardware implementations. We conclude the chapter with some comments about binary-coded decimal floating-point numbers.

3.1 Floating-Point Numbers

This chapter is about floating-point numbers. This is a highly complex area with active research even today. As such, we can only scratch the surface of representation and computation with floating-point numbers in a single chapter. Therefore, we will, after some preliminaries, focus on the current Institute of Electrical and Electronics Engineers (IEEE) 754-2008 standard which we will refer to simply as "IEEE 754". For a more in-depth treatment of floating-point numbers and arithmetic please see Muller et al. [1].

It would be good to have an idea of what, exactly, a "floating-point" number actually is and how it differs from a "real" number. Let's start with the latter concept. This one we know from school: a real number is a number that can be found on the real number line. We have an intuitive feel for what a real number is, and we know many of them because we also know that all integers ($\mathbb{Z}$) and rationals ($\mathbb{Q}$) are also real numbers. But, of course there are still many more real numbers besides these. For example, π is transcendental and therefore a real number but it is neither an integer nor a rational (capable of being written as p/q with $p, q \in \mathbb{Z}$).

How, then, to define the real numbers, $\mathbb{R}$? Formally, there are several ways to define $\mathbb{R}$. The one we will consider is the synthetic approach with specifies $\mathbb{R}$ by defining several axioms which hold only for $\mathbb{R}$. The set of numbers which satisfy these axioms is, by definition, $\mathbb{R}$. Specifically, $\mathbb{R}$ may be defined by considering a set ($\mathbb{R}$), two special elements of $\mathbb{R}$, 0 and 1, two operations called addition ($+$) and multiplication ($\times$), and finally a relation on $\mathbb{R}$ called $\leq$ which orders the elements

R.T. Kneusel, *Numbers and Computers*, DOI 10.1007/978-3-319-50508-4_3

of $\mathbb{R}$. With this and the following four axioms, we have arrived at a specification of the real numbers. The axioms are,

1. $(\mathbb{R}, +, \times)$ forms a field.
2. $(\mathbb{R}, \leq)$ forms a totally ordered set.
3. $+$ and $\times$ maintain $\leq$.
4. The order $\leq$ is complete in a formal sense.

where a full description of their meaning and implications are beyond the scope of this text. The interested reader is directed to [2] for a presentation on fields and rings. For the present, in this sense, $+$ is simply normal addition and $\times$ is normal multiplication. The order $\leq$ also acts as expected. A *floating-point number* is an approximation of a real number that is representable on a computer. It is "close enough" (we'll see below what that means) to the actual real number we wish to represent to be useful and lead to meaningful results.

To be specific, in this chapter a floating-point number is a number which is represented in exponential form as,

$$\pm d_0.d_1d_2d_3\ldots d_{p-1} \times \beta^e, \quad 0 \leq d_i < \beta$$

where $d_0.d_1d_2d_3\ldots d_{p-1}$ is the p digit *significand* (or *mantissa*), β is the base and e is the integer exponent. For the above we follow the notation given in [3]. While it is possible for the base β to be any integer for this chapter we restrict ourselves to $\beta = 2$ so that our numbers are binary and the only allowed digits are 0 and 1.

If there are p digits in the significand then this part of the floating-point number can have β^p unique values. This defines the precision with which values can be represented for a particular exponent. To complete the format we need to also specify the range for the exponent from e_{min} to e_{max}. Before moving on to a brief history of floating-point numbers in computers let's take a quick look at working with floating-point numbers in base 2.

The significand, $d_0.d_1d_2d_3\ldots d_{p-1}$ is a number with the value of

$$d_0 + d_1\beta^{-1} + d_2\beta^{-2} + d_3\beta^{-3} + \cdots + d_{p-1}\beta^{-(p-1)}$$

which is then scaled by the exponent β^e, so that the final value of the number is,

$$(d_0 + d_1\beta^{-1} + d_2\beta^{-2} + d_3\beta^{-3} + \cdots + d_{p-1}\beta^{-(p-1)})\beta^e$$

where we have ignored a possible minus sign to make the entire number negative.

The above means that the distribution of floating-point numbers on the real number line is not uniform as shown in Fig. 3.1.

This is the case because as we saw above the digits of the significand become multipliers on the exponent value β^e. So, for a fixed exponent, the smallest interval between two floating-point numbers is $\beta^{-(p-1)}\beta^e$ which is of fixed size as long as the

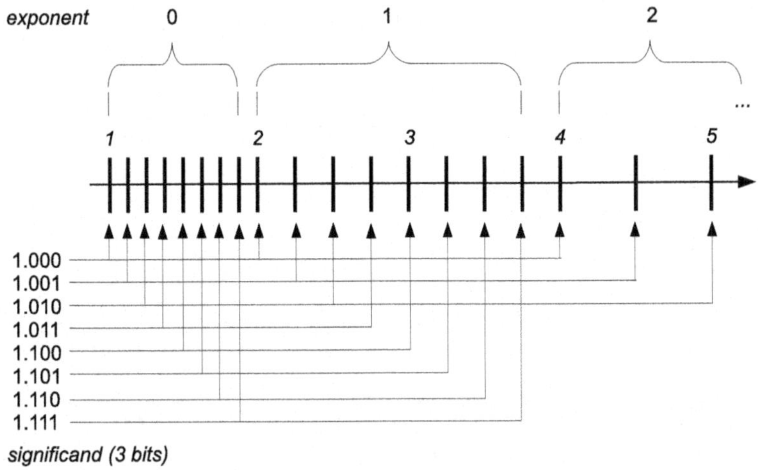

Fig. 3.1 The distribution of floating-point numbers. For this figure the number of significand bits is three with an implied leading bit of one: $1.m_0m_1m_2 \times 2^e$. The location of representable numbers for exponents of $e = 0, 1, 2$ are shown. For each increase in the exponent the number of representable numbers is halved. So, between $[1, 2)$ there are eight values while between $[2, 3)$ there are only four and between $[4, 5)$ there are only two

exponent does not change. Once the exponent changes from e to $e+1$ the minimum interval becomes,

$$\beta^{-(p-1)}\beta^e \rightarrow \beta^{-(p-1)}\beta^{e+1} = \left(\beta^{-(p-1)}\beta^e\right)\beta$$

which is β times larger than the previous smallest interval.

Above we said that there are β^p possible significands. However, not all of them are useful if we want to make floating-point numbers unique. For example, in decimal 1.0×10^2 is the same as 0.1×10^3 and the same holds true for any base β. Therefore, to make floating-point numbers uniquely representable we add a condition which states that the first digit of the significand must be nonzero. For binary floating-point this means that the first digit of the significand must be a 1. Floating-point numbers with this convention are said to be *normalized*. This requirement also buys us one extra digit of precision when the base is 2 because the only nonzero digit is 1 which means all normal binary floating-point numbers have a known leading digit for the significand. This known digit need not be stored in memory leaving room for a $p+1$-th digit in the significand.

To make all of this concrete, let's look at expressing some binary floating-point numbers ($\beta = 2$). We will calculate the actual decimal values for the floating-point numbers shown in Fig. 3.1. They are,

Significand	Exponent	Expanded	Decimal value
1.000	0	$(1) \times 1$	1.0000
1.001	0	$(1 + \frac{1}{8}) \times 1$	1.1250
1.010	0	$(1 + \frac{1}{4}) \times 1$	1.2500
1.011	0	$(1 + \frac{1}{4} + \frac{1}{8}) \times 1$	1.3750
1.100	0	$(1 + \frac{1}{2}) \times 1$	1.5000
1.101	0	$(1 + \frac{1}{2} + \frac{1}{8}) \times 1$	1.6250
1.110	0	$(1 + \frac{1}{2} + \frac{1}{4}) \times 1$	1.7500
1.111	0	$(1 + \frac{1}{2} + \frac{1}{4} + \frac{1}{8}) \times 1$	1.8750
1.000	1	$(1) \times 2$	2.0000
1.001	1	$(1 + \frac{1}{8}) \times 2$	2.2500
1.010	1	$(1 + \frac{1}{4}) \times 2$	2.5000
1.011	1	$(1 + \frac{1}{4} + \frac{1}{8}) \times 2$	2.7500
1.100	1	$(1 + \frac{1}{2}) \times 2$	3.0000
1.101	1	$(1 + \frac{1}{2} + \frac{1}{8}) \times 2$	3.2500
1.110	1	$(1 + \frac{1}{2} + \frac{1}{4}) \times 2$	3.5000
1.111	1	$(1 + \frac{1}{2} + \frac{1}{4} + \frac{1}{8}) \times 2$	3.7500
1.000	2	$(1) \times 4$	4.0000
1.001	2	$(1 + \frac{1}{8}) \times 4$	4.5000
1.010	2	$(1 + \frac{1}{4}) \times 4$	5.0000

where we see that the three bits of the significand has indeed set the precision of each number for a fixed exponent but that the larger the exponent the greater the delta between successive floating-point numbers.

3.2 An Exceedingly Brief History of Floating-Point Numbers

Perhaps the first description of floating-point numbers, meaning a method for storing and manipulating real numbers with a machine, is that of Leonardo Torres y Quevedo in his paper "Essays on Automatics" (1914). This paper is included in Randell's book "The Origins of Digital Computers" [4]. Torres y Quevedo implemented an electromechanical calculating machine, the "electromechanical arithmometer", inspired by the work of Charles Babbage, which was capable of accepting typed statements such as *365* × *256* via a typewriter and would respond, by typing on the same line, with = *93440* and then advance to the next line. This machine was demonstrated at a conference in Paris in 1920.

As for floating-point, Torres describes a way to store floating-point numbers in a consistent manner in his "Essays" paper. First he states that numbers will be stored in exponential format as $n \times 10^m$. He then offers three rules by which consistent manipulation of floating-point numbers by a machine could be implemented. They are, in his own words (as translated into English),

1. *n will always be the same number of digits (six for example).*
2. *The first digit of n will be of order of tenths, the second of hundredths, etc.*
3. *One will write each quantity in the form: n; m.*

On the whole this is quite remarkable for 1914. The format he proposed shows that he understood the need for a fixed-sized significand as is presently used for floating-point data. He also fixed the location of the decimal point in the significand so that each representation was unique. Finally, he even showed concern for how to format such numbers by specifying a syntax to be used that could be entered through a typewriter, as was the case for his "electromechanical arithmometer" some years later. We use essentially the same syntax for most programming languages today; simply replace the ";" with an "E" or "e" so that where he would have written *0.314159;1* we now write *0.314159E1*. It is unfortunate that Torres y Quevedo's work is not more widely known outside his native Spain.

Konrad Zuse is typically credited with being first to implement practical floating-point numbers for his *Z1* and *Z3* computers [5]. The *Z1* was developed in 1938 with the *Z3* completed in 1941. These computers used binary floating-point numbers with one bit for the sign, 14 bits for the significand, and a signed exponent of 7 bits. Zuse also recognized the value of an implicit "1" bit in the significand and stored the numbers in normalized format. This format is in many ways identical to the modern floating-point formats we will be investigating in this chapter and is a testament to Zuse's often unrecognized brilliance.

According to [5] Zuse called any number with an exponent of -64 "zero" and any number with an exponent of 63 "infinity". This means that the smallest number possible on the *Z3* was $2^{-63} = 1.08 \times 10^{-19}$ while the largest was $1.999 \times 2^{62} = 9.2 \times 10^{18}$. Numbers were entered in decimal and converted to binary format. Similarly, output was converted from binary to decimal. However, this conversion did not cover the entire range of the machine and the user was limited to a much smaller floating-point range for input and output.

Computers in the 1950s through the 1970s used a wide variety of floating-point formats. For example, the Burroughs B6700 used a base 8 ($\beta = 8$) format while the IBM System/360 used a base 16 ($\beta = 16$) format. Many other computers copied the IBM format including the Data General Nova and Eclipse computers.

In 1985 IEEE released a standard for floating-point arithmetic. This standard has since dominated the industry and has been implemented in hardware by the Intel x86 series of microprocessors. This standard is the main focus of this chapter and we begin by comparing its floating-point representation to the older IBM format.

3.3 Comparing Floating-Point Representations

In this section we examine the 32-bit IBM S/360 format for floating-point numbers and compare it to the 32-bit IEEE 754 format. We then focus exclusively on the IEEE format for the remainder of the section and chapter. By this comparison we will see that the choice of $\beta = 16$ was not a particularly good one and why the $\beta = 2$ of IEEE is a better one.

A single-precision floating-point number in the IBM S/360 format was stored in memory as,

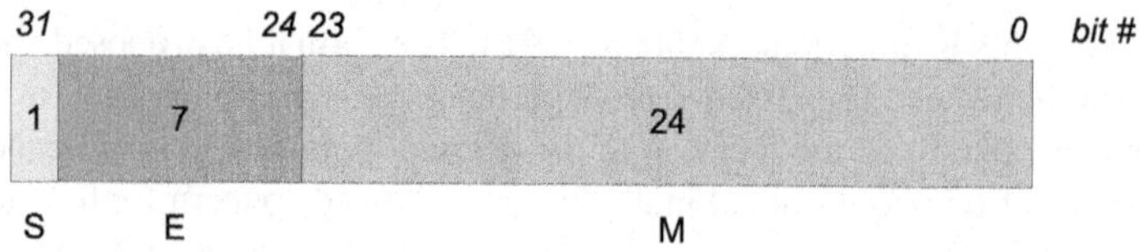

with 1 bit for the sign (S), 7 bits for the exponent (E), and 24 bits for the significand (M). The exponent was stored in excess-64 format meaning an exponent of 0 represented −64 and an exponent of 127 represented 63. The significand was in binary but manipulated in groups of four bits since the base was 16 and $2^4 = 16$. Because of this, one can think of the significand as representing six hexadecimal digits for the first six negative powers of 16.

In order to normalize a floating-point number at least one of the first four high bits of the significand had to be nonzero. Because of this the smallest positive floating-point number that could be represented was,

which is, in base 16, $0.1_{16} \times 16^{0-64} = 1 \times 16^{-1} \times 16^{-64} = 5.397605 \times 10^{-79}$.

Similarly, the largest positive number was,

equal to $0.\text{FFFFFF}_{16} \times 16^{127-64} = 15\times16^{-1}+15\times16^{-2}+15\times16^{-3}+15\times16^{-4}+15\times16^{-5}+15\times16^{-6}\times16^{63} = (1-16^{-6})\times16^{63} = 7.237005\times10^{75}$ since the significand of $0.\text{FFFFFF}_{16}$ is only one 16^{-6} away from 1, hence $0.\text{FFFFFF}_{16} = 1 - 16^{-6}$. The number zero was represented by all bits set to zero.

The IEEE 754 standard defines a number of different floating-point sizes. For the moment we consider only the *binary32* single-precision format which fits in a 32-bit number. On modern x86-based hardware this is the format used when one declares a variable to be of type `float` in C, for example. This format is stored in memory as,

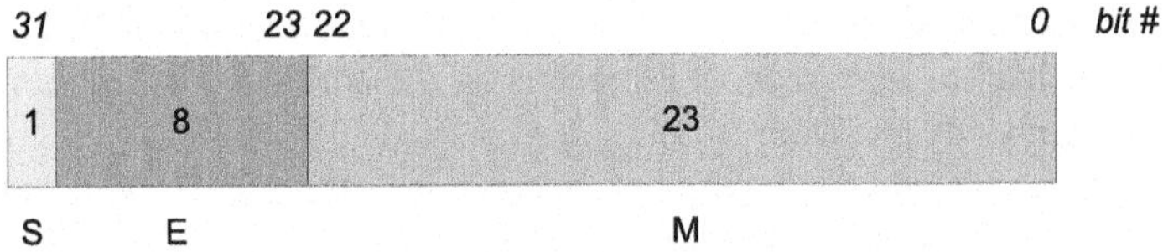

with 1 bit for the sign (S), 8 bits for the exponent (E), and 23 bits for the significand (M). The exponent is stored in excess-127 format with two special exponent values of 0 and FF_{16} reserved. The base is 2 ($\beta = 2$) and the significand has an implied leading 1 bit always for a normalized number. So, in this format, a number is stored as,

$$\pm 1.b_{22}b_{21}b_{20}\ldots b_0 \times 2^{E-127}$$

where b_n is the value of that bit of the significand. Since the base is 2 the bits are interpreted as individual bits as opposed to groups of four bits for the base 16 IBM format.

Because of the implied 1 bit, the smallest allowed *binary32* number is,

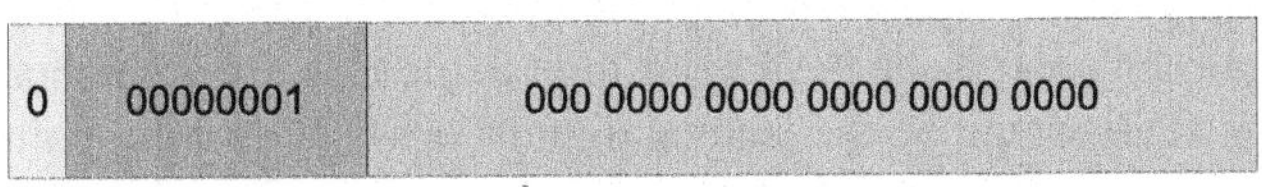

which is $1 \times 2^{1-127} = 1.1754944 \times 10^{-38}$. Similarly, the largest positive number is,

which, using the reasoning above for the significand of the largest IBM S/360 format number, is $(2 - 2^{-23}) \times 2^{254-127} = 3.4028235 \times 10^{38}$. The mantissa is $(2 - 2^{-23})$ instead of $(1 - 2^{-23})$ because of the implied leading 1 bit of the *binary32* significand.

Immediately we see that the IBM format has a much larger range than the IEEE format, approximately $[10^{-79}, 10^{75}]$ versus $[10^{-38}, 10^{38}]$. This is encouraging in terms of expressing large numbers using fewer bits of memory but there is a problem hidden in the significand. Namely, for the IBM format, a normalized number is one in which the first hexadecimal digit of the six hexadecimal digit significand is non-zero. This means that any of the possible bit patterns for 1 through 15 are allowed. If the bit pattern is 0001_2 then there are three fewer significant bits in the significand than if the leading digit is 1000_2 up to 1111_2. This represents a potential lost of precision of up to one decimal digit and is due entirely to the way in which the number is represented in base 16. This loss of precision as a function of the representation of the number in the selected base is known as *wobbling precision* and is more severe when the base is larger, as it is for the IBM format. This wobbling precision can lead to very inaccurate results as the following example illustrates (borrowed from Cody and Waite [7]).

The approximate value of $\pi/2$ in hexadecimal is $1.922_{16} \times 16^0$. We get this from Python using the `float.hex()` class method:

$$\begin{aligned}\texttt{float.hex(pi/2.0)} &= \texttt{0x1.921fb54442d18p} + \texttt{0}\\ &= 1.921\text{fb}54442\text{d}18_{16} \times 2^0\end{aligned}$$

where we must be careful because Python expresses floats in hexadecimal with a base of 16, not 2. The conversion to binary is straightforward, however, simply replace each hexadecimal digit with four binary digits equal to the hexadecimal digit. In this case, the exponent is zero so the actual value is just the mantissa. Similarly, the approximate value of $2/\pi$ in hexadecimal is $\text{a}.2\text{f}9_{16} \times 16^{-1}$ since,

$$\begin{aligned}\texttt{float.hex(2.0/pi)} &= \texttt{0x1.45f306dc9c883p} - \texttt{1}\\ &= 1.45\text{f}306\text{dc}9\text{c}883_{16} \times 2^{-1}\\ &= 1.45\text{f}306\text{dc}9\text{c}883_{16}/2\\ &\approx \text{a}.2\text{f}9_{16} \times 16^{-1}\end{aligned}$$

with the important point being the leading digit of the base 16 representation of these two numbers. This would be the leading digit of the IBM floating-point format for these numbers and in the case of $\pi/2$ we see that it is a 0001_2 while for $2/\pi$ we get 1010_2. The former has three leading zeros meaning only 21 significant bits in the IBM representation while the latter has no leading zeros meaning all 24 significand bits are important. This difference of three bits is a loss of about one decimal digit of accuracy.

What is the case for the IEEE base 2 format? In this case there is no loss of precision at all which we can see immediately from the Python output. For $\pi/2$ we get $1.921\text{fb}54442\text{d}18_{16} \times 2^0$ while for $2/\pi$ we have $1.45\text{f}306\text{dc}9\text{c}883_{16} \times 2^{-1}$ both of which have a leading 1 digit as the IEEE format specifies. This means that all the bits of the significand are meaningful and there is no wobbling precision effect when $\beta = 2$. This is the reason why the IEEE 754 format is to be preferred to the older IBM S/360 floating-point format.

As an aside, before we leave this section, we describe a small C function for converting 32-bit IBM S/360 floating point numbers to IEEE double precision numbers. Double precision is necessary as we have seen that the single precision IEEE format does not have the numeric range necessary to hold all 32-bit IBM floats. The IBM format is assumed to be stored in an unsigned 32-bit integer which allows all bit patterns. The function is,

```
1  double ibm_to_ieee_64(uint32_t n) {
2      int s, e, m;
3      double ans;
4
5      s = (n & (1<<31)) != 0;
6      e = (n >> 24) & 0x7F;
7      m = n & 0xFFFFFF;
8
9      ans = m * pow(16,-6) * pow(16,e-64);
10     return (s == 1) ? -ans : ans;
11 }
```

This function may be useful if one encounters an old data file in IBM format. The sign of the IBM float is extracted in line 5, the exponent in line 6, and the significand in line 7 using simple masking and shifting operations. The resulting double precision floating point number is calculated in line 9. The significand of the IBM format is of the form $0.h_0h_1h_2h_3h_4h_5$ so the value in m must be multiplied by 16^{-6} before the final multiplication by 16^{e-64} where we subtract 64 because the exponent is in excess-64 format. Line 10 simply sets the sign of the output and returns it. This function assumes that it is running on a system which uses IEEE floating-point by default.

3.4 IEEE 754 Floating-Point Representations

The IEEE 754 standard defines multiple floating-point formats for binary ($\beta = 2$) and decimal ($\beta = 10$) numbers. We are only concerned with the binary formats here. These are, with their ranges,

Name	Minimum	Maximum
binary16	$2^{-14} \approx 6.104 \times 10^{-5}$	$(2 - 2^{-10}) \times 2^{15} = 65504.0$
binary32	$2^{-126} \approx 1.175 \times 10^{-38}$	$(2 - 2^{-23}) \times 2^{127} \approx 3.403 \times 10^{38}$
binary64	$2^{-1022} \approx 2.225 \times 10^{-308}$	$(2 - 2^{-52}) \times 2^{1023} \approx 1.798 \times 10^{308}$

where *binary16* is half precision, *binary32* is single precision (C `float`), *binary64* is double precision (C `double`), and *binary128* (not shown) is quadruple precision, sometimes called "quad precision". The *binary128* format uses 128-bits and some C compilers implement it in software. More often, however, compilers for x86 architectures will use the 80-bit extended precision hardware type instead. Because of this we will largely ignore *binary128* in this chapter. If you are familiar with Fortran, the *binary32* format is roughly equivalent to a `REAL`, while the *binary64* format is similar to `REAL*8` and the *binary128* format is like `REAL*16`.

The *binary32* format was illustrated above in Sect. 3.3. For completeness we illustrate *binary16*,

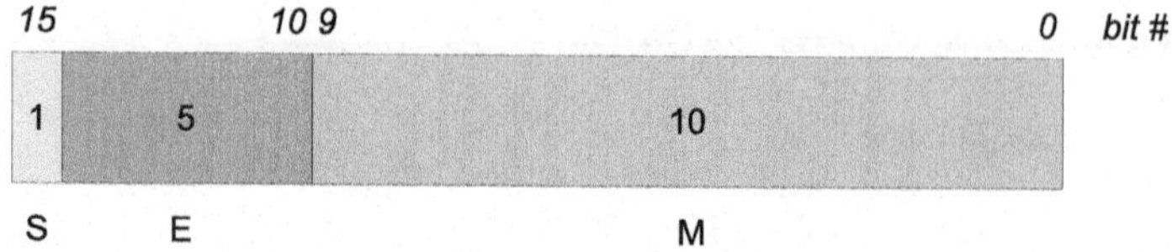

and *binary64*,

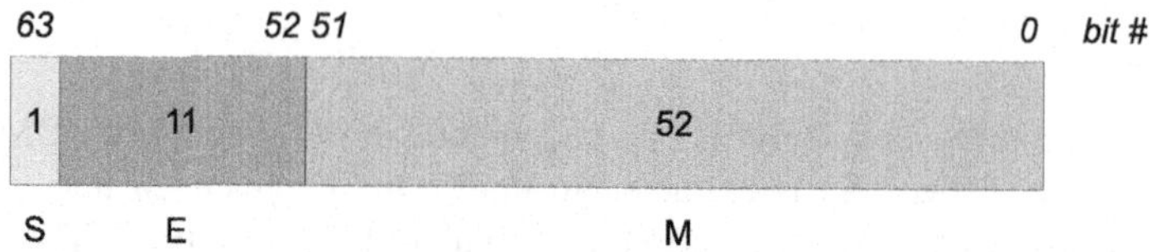

which follow the same format as *binary32* but differ in the exponent and significand bits. For *binary16* the exponent is 5 bits and is stored in excess-15 format. The exponent for *binary32* is 8 bits and uses excess-127 format. Lastly, *binary64* uses an exponent of 11 bits stored in excess-1023 format.

The observant reader will see that the exponent ranges in the table above do not cover the full range of bit values possible. For example, the *binary32* format uses 8 bits for the exponent which means any exponent from $0 - 127 = -127$ to $255 - 127 = 128$ should be possible but the actual exponent range is from -126 to 127 instead. This is because IEEE 754 uses the smallest and largest exponent values for specific purposes. The smallest exponent value, the bit pattern of all zeros, designates either zero itself, or a *subnormal* number. We will deal with subnormal numbers shortly. The largest exponent bit pattern, either 11111_2, 11111111_2, or 11111111111_2 for *binary16*, *binary32*, or *binary64*, respectively, is reserved for infinity or NaN (not-a-number). The selection between zero and a subnormal number or infinity or NaN depends upon the value of the significand (M). Specifically,

Exponent	$M = 0$	$M \neq 0$
largest	$\pm\infty$	NaN
smallest (=0)	± 0	*subnormal*

where the sign bit sets positive or negative. Note carefully that this convention implies that zero itself is signed but the standard requires $+0 = -0$, i.e., zero is interpreted as zero, regardless of the sign bit.

Most readers are likely familiar with the single and double precision floating point numbers, *binary32* and *binary64*, respectively. These generally map to the well used C data types of `float` and `double`. Note, however, that the Python data type `float` is really a *binary64* number. The addition of *binary16* to the IEEE standard seems a bit odd as it does not map to any primitive data type typically used in modern programming languages.

Half precision floats, another common name for *binary16* numbers, are supported in recent versions of `gcc` for ARM targets. Newer Intel microprocessors also have instructions for working with *binary16* numbers. In both cases the storage in memory uses half precision but once read from memory the numbers are converted to 32-bit floats. The main use for these numbers is to increase the precision of values that are stored on graphics processors. Indeed, *OpenGL* supports *binary16* numbers. Naturally, however, this comes with a price, which is a rapidly increasing loss of precision as the value stored gets further and further from zero. As illustrated in Fig. 3.1 the difference between floating-point values increases as one moves further away from zero. On the plus side, this means that if one needs to work with numbers in the range [0, 1] then unless very high precision is required *binary16* is a reasonable way to keep relatively high precision (about 6 or 7 decimals) while using half the memory of a 32-bit float and one quarter the memory of a 64-bit float. This might be significant if there are large arrays of these values to work with as the bandwidth to move that memory is much less than it might otherwise be. It also may help by allowing more data to fit in a processor cache thereby reducing the number of expensive and slow accesses from main memory. Recall that images are typically stored as large arrays so it makes sense to want to be as efficient as possible with memory while not sacrificing precision or dynamic range. It is worth noting that NVIDIA CUDA, starting with version 7.5, supports half precision floating-point storage and arithmetic. CUDA is heavily used for implementing modern neural networks, see Sect. 6.4, and these networks typically do not require extreme precision.

Infinity For any of the IEEE formats, if the exponent is maximum, i.e., all bits set to 1, and the significand is zero, then the number is infinite (∞) and positive or negative depending upon the sign bit. Infinity is produced when the result of an operation would exceed the range of numbers that can be expressed in the given floating-point format. For example,

$$\texttt{exp(1000000.0)} = \texttt{Inf} = +\infty$$
$$\texttt{-exp(1000000.0)} = \texttt{-Inf} = -\infty$$

Infinity is also returned for certain other operations such as,

$$\texttt{1.0/0.0} = \texttt{Inf} = +\infty$$
$$\texttt{log(0.0)} = \texttt{-Inf} = -\infty$$

which follow the usual rules of arithmetic. The $-\infty$ result for `log(0.0)` makes sense as a limit as $x \to 0$ from the right.

As mentioned in the IEEE standard, the following operations are valid and work as expected when using ∞ as an argument and x finite (and not zero),

$$\begin{aligned}
\infty + x = \infty\ ,\ \ x + \infty &= \ \ \infty \\
\infty - x = \infty\ ,\ \ x - \infty &= -\infty \\
\infty \times x = \infty\ ,\ \ x \times \infty &= \ \ \infty \\
\infty \div x = \infty\ ,\ \ x \div \infty &= \ \ 0
\end{aligned}$$

Additionally, the square root of infinity is also defined to be infinity.

Not-A-Number While infinity is straightforward, Not-a-Number or NaN, gets a little messy. A NaN is stored as a floating-point number with the largest possible exponent (all bits 1) and a significand that is *not* zero. There are two kinds of NaNs, quiet and signaling. We will discuss this distinction more fully below when we talk about floating-point exceptions. For now we are concerned with how these two kinds of NaNs are stored so that we can distinguish between them.

According to the standard, the first bit of the significand, which is d_1 in the representation $\pm d_0.d_1d_2d_3\ldots d_{p-1}\times\beta^e$ since d_0 is always implied and 1, determines the NaN type. If this bit is zero the NaN is a signaling NaN, otherwise it is a quiet NaN. Note that this implies that a signaling NaN must have at least one other bit of the significand be nonzero otherwise the number is not a NaN but infinity. The remaining $p-2$ digits of the significand are available for a "payload", to use the standard's term, which may contain diagnostic information about what caused the NaN in the first place.

The standard is intentionally fuzzy about what, if anything, should constitute the payload of a NaN. However, the standard is explicit that NaNs, when used as an argument in an operation, should preserve the payload of at least one of the NaNs, or the NaN, if it is the only NaN in the operation. Operations on NaNs produce NaNs as output. The vague nature of the NaN payload invites creative use. Indeed, some groups and companies have used NaNs for their own purposes. Since it is not specified, beyond indicating a bit for quiet or signaling, users are free to use NaNs as they see fit. One possibility might be to use NaNs as symbols in a symbolic expression parser. Another would be to use NaNs as missing data values and the payload to indicate a source for the missing data or its class.

The following set of C functions can be used to check for NaN and to set and get a payload from a NaN of up to 22 bits,

```
typedef union {
    float f;
    unsigned int d;
} fp_t;

char nan_isnan(float nan) {
    fp_t x;
    x.f = nan;
    return (((x.d >> 23) & 0xFF) == 0xFF) &&
        ((x.d & 0x7FFFFF) != 0);
}

```

```
12 float nan_set_payload(float nan, unsigned int payload) {
13     fp_t x;
14     x.f = nan;
15     x.d |= (payload & 0x3FFFFF);
16     return x.f;
17 }
18
19 unsigned int nan_get_payload(float nan) {
20     fp_t x;
21     x.f = nan;
22     return (x.d & 0x3FFFFF);
23 }
```

where `nan_isnan` checks to see if a float is actually a NaN, `nan_set_payload` sets the free 22-bits of a NaN to a user-specified value and `nan_get_payload` returns this value from a NaN. Most systems already have a `isnan` function, we include it here for completeness. The functions make use of a trick to work with the same piece of memory as both a float and an unsigned integer, both of which occupy 32-bits. This is the `fp_t` data type defined in lines 1 through 4. The function `nan_isnan` accepts a *binary32* number and puts it into `x` (line 8). We then interpret the memory of the float as an unsigned integer by accessing the `d` field. Line 9 is the magic in two parts both of which come directly from the definition of a NaN. The first part, `(((x.d >> 23) & 0xFF) == 0xFF)`, examines the exponent field of the number. Recall that a *binary32* number has 23 bits in the significand. We shift 23 bits down to move the exponent field into bit position 0. We then mask this to keep the first eight bits (the AND with `0xFF`). This removes any sign bit which the standard ignores for NaNs. If this is the exponent of an actual NaN then, by definition, the exponent is all 1 bits, which is `0xFF` so we check if the value is equal to `0xFF`. An exponent of all 1 bits is not enough to define a NaN, however, as the number could still be infinity. We must consider the significand itself. This is the second part of line 9, `((x.d & 0x7FFFFF) != 0)`, which masks off the 23-bits of the significand by the AND with `0x7FFFFF` and then checks to see if this value is zero. If the number is a NaN, this value will not be zero. If it is a quiet NaN the highest bit will be set, at a minimum, and if it is a signal NaN, at least one of the other 22-bits will need to be set. If both of these parts are true, the number is a NaN and the result of the final logical AND is returned.

The `nan_set_payload` function accepts a NaN and a payload which must fit in 22-bits. A `fp_t` is used to hold the NaN. The payload is masked with `0x3FFFFF` to keep it in range and then OR'ed into the bits of the NaN in line 15. The updated NaN is returned in line 16. To access the NaN value use `nan_get_payload` which accepts a NaN and after setting `x` to it, returns the lowest 22-bits by again masking with `0x3FFFFF`.

To illustrate the idea of NaN propagation consider the following code snippet, compiled on a 32-bit Linux system using `gcc`,

```
1  const char *pp(unsigned int x) {
2      static char b[33];
3      unsigned int z;
4      b[0] = '\0';
5      for (z = (1<<31); z > 0; z >>= 1)
6          strcat(b, ((x & z) == z) ? "1" : "0");
7      return b;
8  }
9
10 int main() {
11     fp_t x;
12     float n = log(-1);
13     unsigned int d;
14
15     x.f = n;
16     printf("%d  original     = %s\n", nan_isnan(n),pp(x.d));
17     n = nan_set_payload(n, 12345);
18
19     x.f = n;
20     printf("%d  set payload = %s\n", nan_isnan(n),pp(x.d));
21     d = nan_get_payload(n);
22     printf("payload is %d (%s)\n", d, pp(d));
23
24     n = 123 * n;
25
26     d = nan_get_payload(n);
27     printf("payload is still %d (%s)\n", d, pp(d));
28 }
```

where the function pp simply prints the given 32-bit unsigned integer in binary as a string. In line 12 we define n and set it so `log(-1)` which is a NaN. In line 15 we put this NaN into x and print the result of asking whether n is a NaN (it is) and showing the bits of n,

```
1  original    = 01111111110000000000000000000000
```

Line 17 sets the payload for the NaN to the unimaginative air-shield combination number of 12345. Lines 19 through 22 show that the NaN is still a NaN but now has a payload,

```
1  set payload = 01111111110000000011000000111001
payload is 12345 (00000000000000000011000000111001)
```

where $11000000111001_2 = 12345$. So far, this isn't anything particularly impressive. Line 24, however, multiplies n by 123 and assigns back to n. This is to illustrate NaN propagation, which the IEEE standard strongly encourages. The output of line 27 is,

```
payload is still 12345 (00000000000000000011000000111001)
```

which clearly demonstrates that the payload was propagated.

Subnormal Numbers As we previously stated, a normal number is one in which the leading bit of the significand is an implied 1 and the exponent, e, is $0 < e < m$, for m the maximum exponent value which depends upon the precision of the number. For *binary32*, $m = 255$ since eight bits are used to store the exponent. Therefore, the smallest normal *binary32* number is,

$$0\ 00000001\ 00000000000000000000000_2 = 2^{-126} = 1.1754944 \times 10^{-38}$$

where we have separated the binary representation into the sign, exponent and significand. If we insist on using an implied leading 1 bit for the significand the next smallest number we can represent is immediately zero. However, if we are willing to suffer a loss of precision in the significant we can drop the exponent to zero and change the implied leading digit of the significand from 1 to 0 to represent numbers as,

$$\pm 0.d_1 d_2 \dots d_{23} \times 2^{-126}$$

where numbers of this form are called *subnormal* or *denormal*. Subnormal numbers allow us to work with very small numbers around zero but at the small cost of a loss of precision in the significand. However, there is a very large cost in terms of performance as modern floating-point hardware does not work with subnormal numbers implying calculations must be done in software.

The program in Fig. 3.2 illustrates the transition between normal and subnormal numbers for 32-bit floats. It initializes a float to 1.0 and then repeatedly divides it by 2 until the result is zero. The function ppx prints the bits of the floating point number with a space between the sign and exponent and the exponent and the significand. If we run this program the tail end of the output is,

```
x = (0 00000010 00000000000000000000000) 2.3509887e-38
x = (0 00000001 00000000000000000000000) 1.1754944e-38
x = (0 00000000 10000000000000000000000) 5.8774718e-39
x = (0 00000000 01000000000000000000000) 2.9387359e-39
x = (0 00000000 00100000000000000000000) 1.4693679e-39
x = (0 00000000 00010000000000000000000) 7.3468397e-40
x = (0 00000000 00001000000000000000000) 3.6734198e-40
x = (0 00000000 00000100000000000000000) 1.8367099e-40
x = (0 00000000 00000010000000000000000) 9.1835496e-41
x = (0 00000000 00000001000000000000000) 4.5917748e-41
x = (0 00000000 00000000100000000000000) 2.2958874e-41
x = (0 00000000 00000000010000000000000) 1.1479437e-41
x = (0 00000000 00000000001000000000000) 5.7397185e-42
x = (0 00000000 00000000000100000000000) 2.8698593e-42
x = (0 00000000 00000000000010000000000) 1.4349296e-42
x = (0 00000000 00000000000001000000000) 7.1746481e-43
x = (0 00000000 00000000000000100000000) 3.5873241e-43
```

```
x = (0 00000000 00000000000000010000000) 1.793662e-43
x = (0 00000000 00000000000000001000000) 8.9683102e-44
x = (0 00000000 00000000000000000100000) 4.4841551e-44
x = (0 00000000 00000000000000000010000) 2.2420775e-44
x = (0 00000000 00000000000000000001000) 1.1210388e-44
x = (0 00000000 00000000000000000000100) 5.6051939e-45
x = (0 00000000 00000000000000000000010) 2.8025969e-45
x = (0 00000000 00000000000000000000001) 1.4012985e-45
x = (0 00000000 00000000000000000000000) 0
```

where the first two numbers are normalized, the exponent is nonzero. After these start the subnormalized numbers the first of which is,

$$0\ 00000000\ 10000000000000000000000_2 = 2^{-1} \times 2^{-126} = 5.8774718 \times 10^{-39}$$

and the smallest positive subnormal number is therefore $2^{-23} \times 2^{-126} = 1.4012985 \times 10^{-45}$.

```c
#include <stdio.h>
#include <string.h>

const char *ppx(unsigned int x) {
    static char b[36];
    unsigned int z, i=0;
    b[0] = '\0';
    for (z = (1<<31); z > 0; z >>= 1) {
        strcat(b, ((x & z) == z) ? "1" : "0");
        if ((i == 0) || (i == 8))
            strcat(b, " ");
        i++;
    }
    return b;
}

typedef union {
    float f;
    unsigned int d;
} fp_t;

int main() {
    fp_t x;
    int i;
    x.f = 1.0;
    for(i=0; i < 151; i++) {
        printf("x = (%s) %0.8g\n", ppx(x.d), x.f);
        x.f /= 2.0;
    }
}
```

Fig. 3.2 A program to display the transition from normal to subnormal floating-point numbers

The penalty in performance for using subnormal numbers can be found by timing how long it takes to run 10 million iterations of a simple multiplication of x by 1 where x is first the smallest normal number and then the largest subnormal number. If we do this, being careful to make sure the loop is actually performed and not optimized away by the compiler, we see that subnormal numbers are about 23 times slower than normal numbers on the machine used. This reflects the difference between hardware and software support.

3.5 Rounding Floating-Point Numbers (IEEE 754)

Often when a real number is expressed in floating-point format it is necessary to round it so that it can be stored as a valid floating-point number. The IEEE standard defines four rounding modes, or rules, for binary floating-point. These are available to programmers in C on Linux if using the `gcc` compiler (among others). The rounding modes are listed in Fig. 3.3.

The default mode for `gcc`, and the one recommended by the IEEE standard, is to round towards the nearest value with ties rounding towards the even value. The even value is the one with a zero in the lowest order bit of the significand. For example, using decimal and rounding to an integer value "round to nearest" would act in this way,

$$
\begin{aligned}
+123.7 &\rightarrow +124\\
+123.4 &\rightarrow +123\\
+123.5 &\rightarrow +124\\
-123.7 &\rightarrow -124\\
-123.4 &\rightarrow -123\\
-123.5 &\rightarrow -124
\end{aligned}
$$

which is more or less the way we would expect rounding to work. The stipulation of rounding ties to even helps prevent the accumulation of rounding errors as in binary this will happen about 50% of the time.

The other, directed, rounding modes work in this way,

Round to Nearest	Round to the nearest floating-point number. Ties round to even.
Round Towards Zero	Round down, if positive, or up, if negative, always towards zero.
Round Towards $+\infty$	Round up, always towards positive infinity.
Round Towards $-\infty$	Round down, always towards negative infinity.

Fig. 3.3 The four IEEE 754 binary floating-point rounding modes

Value	Round towards zero	Round towards $+\infty$	Round towards $-\infty$
+123.7	+123	+124	+123
+123.4	+123	+124	+123
+123.5	+123	+124	+123
-123.7	-123	-123	-124
-123.4	-123	-123	-124
-123.5	-123	-123	-124

The directed rounding modes always shift results, consistently up or down for rounding towards $\pm\infty$, regardless of sign, or up or down for negative and positive numbers in the case of "round towards zero". The default mode is used so often that many software libraries do not function properly if one of the directed modes is used [8].

It is possible to alter the rounding mode programmatically. Figure 3.4 is a Linux program that changes the rounding mode while displaying the floating-point representation of `exp(1.1)` and `-exp(1.1)`.

This program produces the following output,

```
FE_TONEAREST:
   exp(1.1) = (0 10000000 10000000100010001000010)  3.00416613
  -exp(1.1) = (1 10000000 10000000100010001000010) -3.00416613

FE_UPWARD:
   exp(1.1) = (0 10000000 10000000100010001000010)  3.00416613
  -exp(1.1) = (1 10000000 10000000100010001000001) -3.00416589

FE_DOWNWARD:
   exp(1.1) = (0 10000000 10000000100010001000001)  3.00416589
  -exp(1.1) = (1 10000000 10000000100010001000010) -3.00416613

FE_TOWARDZERO:
   exp(1.1) = (0 10000000 10000000100010001000001)  3.00416589
  -exp(1.1) = (1 10000000 10000000100010001000001) -3.00416589
```

which shows the floating-point number in decimal as well as the actual bits used to store the value. Note that Fig. 3.4 includes a comment line describing the actual command used to compile the program. The `-frounding-math` option is required otherwise `fesetround()` will not actually change the rounding mode.

If we look at the output we see that `FE_TONEAREST`, which is the default rounding mode, produces results which match what we expect since negating the value is not expected to alter the result other than the sign bit. We see this clearly by looking at the bits of the results; they are all the same except for the sign bit.

The remaining directed modes act as expected. For `FE_UPWARD` we see that the negative result is smaller than the positive, in magnitude, because it has been rounded towards positive infinity. Similarly, we see the magnitudes reversed for `FE_DOWNWARD` which is rounding towards negative infinity. Lastly, `FE_TOWARDZERO` rounds towards zero giving the smaller magnitude result for both positive and negative cases. The floating-point bit patterns for `FE_TONEAREST` and `FE_TOWARDZERO`

```
#include <stdio.h>
#include <fenv.h>
#include <math.h>
#include <string.h>

// compile: gcc rounding.c -o rounding -lm -frounding-math

int main() {
    fp_t fp;

    printf("\nFE_TONEAREST:\n");
    fesetround(FE_TONEAREST);
    fp.f = exp(1.1);
    printf("    exp(1.1) = (%s) %0.8f\n", ppx(fp.d), fp.f);
    fp.f = -exp(1.1);
    printf("   -exp(1.1) = (%s) %0.8f\n\n", ppx(fp.d), fp.f);

    printf("FE_UPWARD:\n");
    fesetround(FE_UPWARD);
    fp.f = exp(1.1);
    printf("    exp(1.1) = (%s) %0.8f\n", ppx(fp.d), fp.f);
    fp.f = -exp(1.1);
    printf("   -exp(1.1) = (%s) %0.8f\n\n", ppx(fp.d), fp.f);

    printf("FE_DOWNWARD:\n");
    fesetround(FE_DOWNWARD);
    fp.f = exp(1.1);
    printf("    exp(1.1) = (%s) %0.8f\n", ppx(fp.d), fp.f);
    fp.f = -exp(1.1);
    printf("   -exp(1.1) = (%s) %0.8f\n\n", ppx(fp.d), fp.f);

    printf("FE_TOWARDZERO:\n");
    fesetround(FE_TOWARDZERO);
    fp.f = exp(1.1);
    printf("    exp(1.1) = (%s) %0.8f\n", ppx(fp.d), fp.f);
    fp.f = -exp(1.1);
    printf("   -exp(1.1) = (%s) %0.8f\n\n", ppx(fp.d), fp.f);
}
```

Fig. 3.4 A program to display the effect of different rounding modes. Note that `fp_t` and `ppx()` are defined in Fig. 3.2

are the same in each case except for the sign bit while the other two cases are similar but the positive and negative bit patters are flipped (for the significand).

If the default of "round to nearest" is so widely used, why are the other rounding modes present in the IEEE standard? One example of why can be found in [9] where directed rounding significantly improved the accuracy of the calculations. We will make use of rounding modes in Chap. 8 when we implement interval arithmetic. However, specialized uses aside, in most cases it is completely unnecessary to modify the rounding mode from the default.

3.6 Comparing Floating-Point Numbers (IEEE 754)

The IEEE standard requires that floating-point numbers maintain an explicit ordering for comparisons. The simplest way to compare two floating-point numbers is to look at their bit patterns. For example, the following two numbers are as close as two 32-bit floating-point numbers can be without being equal. This means there are no representable floating point numbers between them. The numbers are,

```
0.500000000 = 0 01111110 00000000000000000000000 = 1056964608
0.500000060 = 0 01111110 00000000000000000000001 = 1056964609
```

where the floating-point value is on the left, the actual bit patterns are in the center and the equivalent 32-bit signed integer is on the right. Note that when viewed as an integer the smaller number, 0.5, is also smaller. This is no accident. The floating-point format is designed to make comparison of floating-point numbers as straightforward as comparing two signed integers where negative floating-point numbers are interpreted in two's-complement format. The only exception is zero since it can be signed or unsigned. But it is easy to check for this single case.

The C code for a floating-point comparison function that checks for the special case of signed zero is,

```c
typedef union {
    float f;
    int d;
} sp_t;

int fp_compare(float a, float b) {
    sp_t x,y;

    x.f=a;
    y.f=b;

    if ((x.d == (int)0x80000000) &&
        (y.d == 0)) return 0;
    if ((y.d == (int)0x80000000) &&
        (x.d == 0)) return 0;

    if (x.d == y.d) return 0;
    if (x.d < y.d) return -1;
    return 1;
}
```

where we introduce `sp_t` instead of `fp_t` which we used previously. The difference is we now use a signed integer instead of an unsigned integer. The signed integer is necessary so that the integer comparisons will treat negative floating-point numbers properly.

The `fp_compare` function accepts two floats, a and b, as input and returns -1 if $a < b$, 0 if $a = b$, and +1 if $a > b$. In lines 9 and 10 the input floats are placed into two `sp_t` variables, x and y. This is so we can get the equivalent signed integer from the floating-point bit pattern. Lines 12 and 13 are the special check for negative zero compared to positive zero. In this case we want a return value of 0 since the standard says these values are equal. The cryptic `(int)0x80000000` is the bit pattern for a negative zero number cast to a signed integer. If this is the first value and the second is positive zero, consider them the same. Lines 14 and 15 make the same comparison with the arguments reversed. Line 17 checks if the bit patterns are the same and returns zero if they are to declare the values equal. Line 18 checks to see if $a < b$ and returns -1 if they are. The only option left is that $a > b$ so we return +1 in line 19. This completes the predicate function to compare two floating point numbers using their integer bit patterns.

For completeness, let's define functions for the basic comparison operators. We can then check these against the intrinsic operators in C for floating-point numbers. The functions are,

```
int fp_eq(float a, float b) {
    return fp_compare(a,b) == 0;
}

int fp_lt(float a, float b) {
    return fp_compare(a,b) == -1;
}

int fp_gt(float a, float b) {
    return fp_compare(a,b) == 1;
}
```

And we can use them with the following code,

```
int main() {
    float a,b;

    a = 3.13;  b = 3.13;
    printf("3.13 == 3.13: %d %d\n", fp_eq(a,b), a==b);

    a = 3.1;   b = 3.13;
    printf("3.1 < 3.13: %d %d\n", fp_lt(a,b), a<b);

    a = 3.2;   b = 3.13;
    printf("3.2 > 3.13: %d %d\n", fp_gt(a,b), a>b);

    a = -1;  b = 1;
    printf("-1 < 1: %d %d\n", fp_lt(a,b), a<b);
}
```

which shows that comparing the integer representations gives the results we would expect,

```
 3.13 = 3.13 : 11
 3.10 < 3.13 : 11
 3.20 > 3.13 : 11
-1.00 < 1.00 : 11
```

In this discussion we have ignored NaNs and infinity. NaNs, according to the standard, are unordered even when compared to themselves. Infinities of the same sign compare as equal while those of opposite signs should compare as expected. This is not the case with the simple example above in `fp_compare` which would need to be updated to compare infinities properly.

3.7 Basic Arithmetic (IEEE 754)

The standard mandates that conforming implementations supply the expected addition (+), subtraction (−), multiplication (×) and division (/) operators. It also requires square root of a positive number and a fused multiply-add which implements $(xy) + z$ with no exceptions thrown during the multiplication. This combined operation only involves one rounding step, not the two that would happen if the multiplication was done separately from the addition.

Addition and Subtraction The addition of two floating-point numbers is straightforward. Let's consider the steps necessary to calculate 8.76 + 1.34. First, we represent each number in binary as:

```
8.76 → 8.760000229 = 0 10000010 00011000010100011110110
1.34 → 1.340000033 = 0 01111111 01010111000010100011111
```

where we immediately see that neither 1.34 nor 8.76 can be exactly represented using a 32-bit float and that we have again shown the binary representation separating the sign, exponent and significand. If we convert the bit pattern to actual binary we get,

$$8.76 \rightarrow 1.00011000010100011110110 \times 2^3$$
$$1.34 \rightarrow 1.01010111000010100011111 \times 2^0$$

where we see that the exponents are not the same. In order to add the significands we need to make the exponents the same first. So, we rewrite the smaller number so that it has the same exponent as the larger. This means we add the difference of the exponents to the exponent of the smaller number, $3 - 0 = 3 \rightarrow 3 + 0 = 3$, and shift the significand of the smaller number that many bit positions to the left to get,

$$8.76 \rightarrow 1.00011000010100011110110\ 000 \times 2^3$$
$$1.34 \rightarrow 0.00101010111000010100011\ 111 \times 2^3$$

with the bits that would have been shifted out of the 32-bit float separated from the rest of the significand by a space. Normally, the calculation is done with a higher precision than the storage format and rounding is applied. The top value is filled in by three zeros for the extended precision bits.

With the exponents the same the addition is simple, add the significand, bit by bit, just as one would add 2.5×10^3 and 4.6×10^3 to get 7.1×10^3,

$$\begin{array}{rl} 8.76 \rightarrow & 1.00011000010100011110110\ 000 \times 2^3 \\ +\ 1.34 \rightarrow & 0.00101010111000010100011\ 111 \times 2^3 \\ \hline & 1.01000011001100110011001\ 111 \times 2^3 \end{array}$$

If the answer does not have a single 1 to the left of the decimal point we shift the significand to the right until it does so that our answer is a normalized number. Naturally, if we shift the significand n bits to the right we need to raise the exponent by adding n to it. In the example above the answer is already normalized since the first, and only, bit to the left of the decimal point is 1.

There is one step remaining, which is to account for rounding. Assuming that the rounding mode is the default of "round to nearest" we need to consider the first 1 in the three lowest order bits of the answer, the three bits on the far right of the significand. In order to round up to the nearest floating-point number we add one to the right-most bit on the significand, before the space, to make $001 \rightarrow 010$. In this case, then, the final answer is,

$$10.100000381 \rightarrow 1.01000011001100110011010 \times 2^3$$

Subtraction works in the same way. Specifically, the steps to add or subtract two floating-point numbers x and y are,

1. Raise the exponent of the number with the smaller exponent by the difference between the larger and smaller exponent. Also, shift the significand of the smaller number to the right by the same number of bits. This causes both numbers to have the same exponent. If the smaller number was y the shifted number is y'.
2. Add or subtract the significands of x and y' as they now have the same exponent. Put the answer in z.
3. If there is not a single 1 to the left of the decimal point in z, shift the significand of z to the right until there is. Increase the exponent of z by the number of bits shifted.

Multiplication If we want to multiply two numbers in scientific notation we multiply the significands and add the exponents,

$$(a \times 10^x)(b \times 10^y) = ab \times 10^{x+y}$$

similarly, for two floating-point numbers expressed in binary, we multiply the significands and add exponents then normalize. So, to multiply 3.141592 by 2.1, we have,

$$
\begin{array}{lll}
3.141592 & \rightarrow 3.141592026 & \rightarrow \texttt{0 10000000 10010010000111111011000} \\
2.1 & \rightarrow 2.099999905 & \rightarrow \texttt{0 10000000 00001100110011001100110}
\end{array}
$$

where the first column is the real number we want to multiply, the second column is the actual floating-point number we get when using *binary32* and the last column is the actual bit pattern in memory. Using the bit pattern we can write these values in binary as,

$$
\begin{array}{ll}
3.141592 & \rightarrow 1.10010010000111111011000 \times 2^1 \\
2.1 & \rightarrow 1.00001100110011001100110 \times 2^1
\end{array}
$$

which means that our product will have an exponent of $1 + 1 = 2$. We can multiply the significands, bit by bit, which is tedious on paper so instead we approximate the product using only seven bits after the radix point,

```
          1.1001001
×         1.0000110
-------------------
                  0
         1110010010
        11100100100
                ...
    110010010000000
-------------------
+   1.1010010011011 0
```

where we have already moved the radix point to the proper position in the final sum. The ... represents a series of all zero places corresponding to the string of four zeros in the multiplier. The last line shows the last partial product. The full product of the significands is `1.101001100011101011011 11`.

Since the significand product already has a leading value of one we do not need to shift in order to get a leading bit of one. For example, if the significand product started with `10.00101...` we would shift the radix point one position to the left so make the significand `1.000101...` and increase the exponent by one position.

So, finally, our now approximate product (because we dropped many significand bits) can be represented as,

$$1.10100100110110 \times 2^2 \rightarrow 6.5756835938$$

while the full precision product is $3.141592 \times 2.1 = 6.5973429545$.

In summary, then, the steps to multiply two floating-point numbers x and y are,

1. Add the exponents of x and y.
2. Multiply the significands of x and y.
3. Shift the product of the significands to get a single leading one bit to the left of the radix point. Adjust the exponent as well by adding one for every bit position the product of the significands is shifted to the right. This is equivalent to shifting the radix point to the left.
4. Reconcile the signs so that final product has the expected sign.

This concludes our look at floating-point arithmetic. Next we consider an important part of the IEEE 754 standard we have neglected until now, exceptions.

3.8 Handling Exceptions (IEEE 754)

Not all floating-point operations produce meaningful results for all operands. An important part of the IEEE 754 standard addresses what to do when exceptions to normal floating-point processing arise. The standard recommends a default set of actions for these cases and then allows implementations to give users the power to trap and respond to exceptions. In this section we learn what these exceptions are, the default responses the standard recommends for them, and how to trap exceptions to bring them under user control in C. It is important to remember that floating-point exception handling is distinct from the concept of exceptions in programming languages, though they are, of course, related and could be intermingled. Modern processors, including the x86 line, process floating-point exceptions in hardware and as such are programming language independent.

Exceptions and Default Handling The standard defines five types of exceptions: *inexact, invalid, underflow, overflow*and *divide-by-zero*:

Exception type	Description	Examples
inexact	When the result is not exact due to rounding.	2.0/3.0
invalid	When the operands are not valid for the operation.	$0/0, 0 \times \infty, \sqrt{x}, x < 0$
underflow	When the result is too small to be accurate or when subnormal.	$(1.17549435 \times 10^{-38})/3.0$
overflow	When the result is too large to represent	$(3.40282347 \times 10^{38})^2$
divide-by-zero	When a finite number is divided by zero	1/0

The standard also defines default behavior for these exceptions. The goal of the default behavior is to allow the computation to proceed by substituting a particular value for the exception or to just ignore the exception altogether. For example, the default behavior for an *inexact* exception is to ignore it and continue with the calculation. The default behavior for an *invalid* exception is to generate a NaN and then propagate the NaN through the remainder of the calculation. If *divide-by-zero* happens the result is passed on as $\pm\infty$ depending upon the signs of the operands.

The *overflow* exception is raised when the result does not fit. What the default behavior is depends upon the rounding mode in effect at the time as well as the sign of the number. Specifically,

Round to Nearest	Return $\pm\infty$ depending upon the sign of the result.
Round Towards Zero	Return the largest number for the format with the sign of the result.
Round Towards $+\infty$	Return $+\infty$ if positive, largest negative number if negative.
Round Towards $-\infty$	Return $-\infty$ if negative, largest positive number if positive.

Lastly, *underflow* is raised when the number is too small. The implementer is allowed to determine what "too small" means. The default result returned will be the subnormal number or zero.

Trapping Exceptions For many programs the default behavior is perfectly acceptable. However, for total control over floating-point operations the standard allows for traps which run user code when an exception happens. For x86 architectures, traps are in hardware but can be used from programming languages such as C. When a trap executes, or is used by a higher-level library, the programmer is able to control what happens when exceptions occur. Let's take a look at how one can do this using `gcc`.

The first thing to be aware of is that by default, the `gcc` compiler is not 100% compliant with IEEE 754. So, to get the performance we expect we need to be sure to use `-frounding-math` and `-fsignaling-nans` when we compile. Also, we need to include the `fenv.h` library. With these caveats, the set of functions we can use depends upon the way in which we want to control floating-point exceptions. We can throw a signal when they happen (`SIGFPE`) or we can test individual operations for individual exceptions. Let's look at using a signal first.

A signal is a Unix concept, recall we assume a 32-bit Linux implementation, which is a message to a program to indicate that a particular event has happened. There are many possible signals, but the one we want is `SIGFPE` to signal a floating-point error. We need to enable floating-point exceptions, then arrange a signal handler like so,

```
1  #include <stdlib.h>
2  #include <stdio.h>
3  #include <signal.h>
4  #include <fenv.h>
5
6  void fpe_handler(int sig) {
7      if (sig != SIGFPE) return;
8      printf("Floating point exception!\n");
9      exit(1);
10 }
11
12 int main() {
13     float a=1, b=0, c;
14
15     feenableexcept(FE_INVALID   | FE_DIVBYZERO |
16                    FE_OVERFLOW  | FE_UNDERFLOW);
17     signal(SIGFPE,  fpe_handler);
18
19     c = a/b;
20 }
```

where lines 1 through 4 include necessary libraries. Lines 6 through 10 define our signal handler. We check to see if the signal is actually a floating-point exception (line 7) and then report it and exit. In `main` we define some floats and then enable floating-point exception traps by calling `feenableexcept()` with the bit masks for the floating-point exceptions we want to trap. Line 17 ties the `SIGFPE` signal to our handler, `fpe_handler()`. Finally, line 19 tries to divide by zero which will trigger the exception. This code should be compiled with,

```
$ gcc fpe.c -o fpe -lm -frounding-math -fsignaling-nans
```

assuming `fpe.c` to be the name of the file.

Another way to work with floating-point exceptions is to explicitly check if any were raised during a calculation. In this case we need the `feclearexcept()` and `fetestexcept()` functions to clear any exceptions prior to the calculation and then to test for particular exceptions after. For example, this program,

```
#include <stdio.h>
#include <signal.h>
#include <fenv.h>

int main() {
    float c;

    feclearexcept(FE_ALL_EXCEPT);
    c = 0.0/0.0;
    if (fetestexcept(FE_INVALID) != 0)
        printf("FE_INVALID happened\n");

```

```
    feclearexcept(FE_ALL_EXCEPT);
    c = 2.0/3.0;
    if (fetestexcept(FE_INEXACT) != 0)
        printf("FE_INEXACT happened\n");

    feclearexcept(FE_ALL_EXCEPT);
    c = 1.17549435e-38 / 3.0;
    if (fetestexcept(FE_UNDERFLOW) != 0)
        printf("FE_UNDERFLOW happened\n");

    feclearexcept(FE_ALL_EXCEPT);
    c = 3.40282347e+38 * 3.40282347e+38;
    if (fetestexcept(FE_OVERFLOW) != 0)
        printf("FE_OVERFLOW happened\n");

    feclearexcept(FE_ALL_EXCEPT);
    c = 1.0 / 0.0;
    if (fetestexcept(FE_DIVBYZERO) != 0)
        printf("FE_DIVBYZERO happened\n");
}
```

will produce this output,

```
FE_INVALID happened
FE_INEXACT happened
FE_UNDERFLOW happened
FE_OVERFLOW happened
FE_DIVBYZERO happened
```

when compiled with,

```
$ gcc fpe_traps.c -o fpe_traps -lm -frounding-math -fsignaling-nans
```

thereby showing how to explicitly test for floating-point exceptions. Note, just as `feenableexcept()` took an argument that was the bitwise `OR` of one or more exceptions, so too will `fetestexcept()` so that in order to test for *invalid* and *divide-by-zero* one would call,

```
int i = fetestexcept(FE_INVALID | FE_DIVBYZERO)
```

and then check for either with `i != 0` or one explicitly with `(i & FE_INVALID) != 0`.

3.9 Floating-Point Hardware (IEEE 754)

For this section we assume a system using modern x86 processors. These processors implement an 80-bit extended precision version of floating-point in hardware. This is why floating-point calculations are so fast when in the past floating-point was generally, especially for smaller systems, done in software. This is still true in the large embedded processor world which we will of necessity neglect entirely.

An x86 extended precision float is stored in registers as,

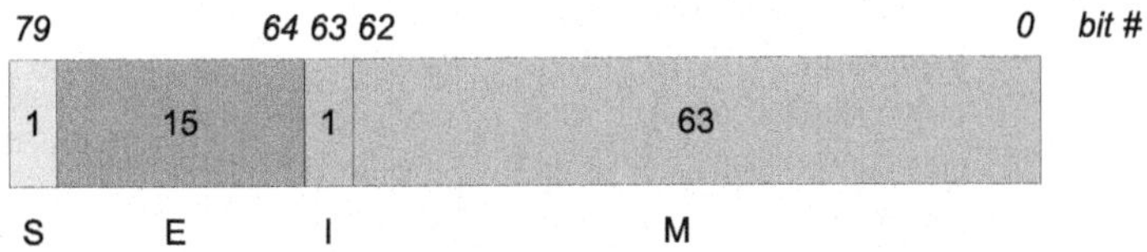

where it is very important to note that unlike the IEEE storage formats, there is no implied leading 1 bit for the significand. Instead, it is given explicitly as the first bit marked a "I" above. The sign (S) is 0 for positive and 1 for negative. The exponent (E) is stored as excess-16383 and the fractional part of the significand (M) is 63-bits long. This means that a floating-point number (n) is,

$$n = (-1)^s i_0.d_{62}d_{61} \ldots d_0 \times 2^{e-16383}$$

where there is now an explicit integer part (i_0) and the binary significand ($d_{62} \ldots d_0$). How many bits of the significand are actually used in a particular calculation is a function of the processor (32-bit vs 64-bit), the compiler, and various compiler settings. We ignore these important but intricate details here and instead focus on the representation.

Just as an IEEE *binary32* or *binary64* float uses certain bit combinations to indicate NaN and infinity so does the hardware version. The explicit integer part helps in quickly determining the type of number that is being worked with. Table 3.1 illustrates how to decide whether or not the float is normal, subnormal, zero, infinity, NaN or invalid.

The top part of the table illustrates normal, subnormal and zero. As is the case for IEEE, there are two possible representations for zero, signed and unsigned. The sign is ignored in calculations. An integer part of zero indicates a subnormal number. If the integer part is 1 and the exponent is any value except all zeros or all ones the number is a normalized floating-point number. In the bottom part of the table are the exceptional values. All of these have the integer part set to 1 and the exponent set to all ones. These values are distinguished based on the value of the leading fractional part of the significand and the remaining bits of the significand. If the leading bit, M_{62}, is zero the number is infinity (all remaining significand bits zero) or a signaling NaN. Recall that a signaling NaN raises an exception. If M_{62} is set the number is either a quiet NaN (remainder of the significand bits are not zero) or an invalid number (all zero bits).

The bit patterns covered in Table 3.1 are not exhaustive. There are other combinations which could be present. In the past some of these were used but now all are considered invalid and will not be generated by the floating-point hardware.

The large number of bits in the significand means that 32-bit and 64-bit floating point operations can be performed with full precision and proper rounding of results. Above we discussed the basic arithmetic operations called out by the IEEE 754

Table 3.1 How to interpret the bit patterns of an x86 extended precision floating-point number

(a) Normal, subnormal and zero

Exponent	I	M	Value
000000000000000	0	All 0	Zero (sign ignored)
000000000000000	0	Non-0	Subnormal
Any but all 1's	1	Any	Normal

(b) Infinity, not-a-number, and invalid (exponent = 111111111111111)

I	M_{62}	$M_{61} \ldots M_0$	Value
1	0	0	$\pm\infty$ (uses sign)
1	0	Non-0	Signaling NaN
1	1	0	Invalid result (e.g., $0/0$)
1	1	Non-0	Quiet NaN

Only the meaningful combinations of bits are shown, all others should be interpreted as *invalid*. When the CPU performs floating-point calculations it uses this format converting to and from IEEE 754 formats like *binary32* and *binary64* as needed. In (a) the exponent and integer part are used to decide between zero, subnormal numbers and normal numbers. All normal numbers have an integer part of 1 matching the implied leading 1 bit of the IEEE formats. For (b) the exponent is the special case of all ones. The integer part and leading bit of the significand determine the number type

standard. There is one we intentionally ignored at the time. This is the fused multiply-add operation, $xy + z$, which first calculates xy and then adds z *without* rounding after the multiplication. If this were not done the expression would involve two roundings, one for the multiplication and another for the addition. This is unnecessary and with the extended precision of the x86 floating-point format will result in fewer rounding artifacts. The $xy + z$ form was chosen because it appears frequently in expressions and a smart compiler would look for these expressions so as to use the fused multiply-add capability whenever possible.

3.10 Binary Coded Decimal Floating-Point Numbers

In this section we take a look at a seemingly obscure representation which is, in fact, one of the most widely used representations in the world: binary coded decimal floating-point or "BCD float". BCD floats use packed BCD for the significand and ten as the exponent to represent numbers. The integer BCD format is described in detail in Chap. 2, Sect. 2.6. The floating-point version described here is found in a similar form in virtually all desktop calculators making it one of the most widely used number formats in the world. Specifically, here we will describe two BCD float formats and provide conversion routines for one of them.

The first format is the one used by the popular line of TI calculators (Texas Instruments). This format uses nine bytes to store a floating-point number. This seems excessive but desktop calculators only operate on a few numbers at a time so the price in processing time is well-worth the precision gained. The format is described in more detail in [10]. As a C structure the format is, ignoring guard digits used to minimize rounding error during calculations,

```
struct FP {
    uint8_t sign;
    uint8_t exponent;
    uint8_t significand[7];
};
```

where the overall sign of the number is stored as 0x00 for positive and 0x80 for negative. Other values are used for complex numbers but we'll ignore those here. The exponent is stored with a bias of 0x80 giving a range of $[-128, +127]$. I.e., the actual exponent is found by subtracting 0x80 from the value stored in the exponent field. The significand is stored in packed BCD format in seven bytes giving 14 decimal digits of precision.

The second BCD float format is one found in compilers for microcontrollers. For example, the Raisonance RC-51 C compiler for the 8051 microcontroller supports three sizes of BCD floats: 32-bits (6 decimal digits), 48-bits (10 decimal digits), and 56-bits (14 decimal digits). We will describe the 32-bit format in detail and develop conversion routines to and from IEEE doubles.

A BCD float-32 number is stored in four bytes as:

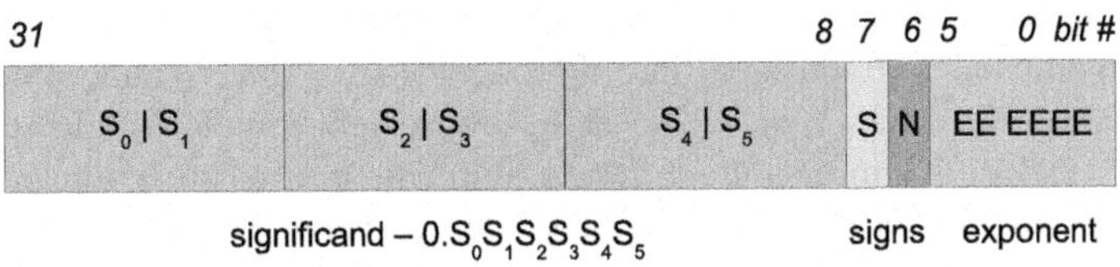

where the significand is 3 bytes long (6 decimal digits) and one byte is used to store the signs and exponent. The first sign ("S" above) is the overall sign of the number where S=1 means the number is negative. The exponent, instead of being stored as a two's-complement number, uses a sign bit as well ("N" above). However, in this case the sense of the sign bit is reversed so that N=0 means that the exponent is negative. The magnitude of the exponent is stored in 6 bits which means that the BCD float-32 format can store numbers with exponents in the range $[-63, +63]$. This same format is used for the other BCD floats, only the number of digits in the significand changes.

For example, the number -3.14159×10^6 would be stored in BCD float-32 format as `0x314159C7` $= 00110001010000010101100111000111_2$,

$$-3.14159 \times 10^6 \rightarrow \underset{3}{0011}\ \underset{1}{0001}\ \underset{4}{0100}\ \underset{1}{0001}\ \underset{5}{0101}\ \underset{9}{1001}\ \underset{S}{1}\ \underset{N}{1}\ \underset{7}{000111}$$

where the significand is stored so that there is no integer part: 0.314159 instead of 3.14159. Therefore the number is stored as 0.314159×10^7 with S=1 because it is negative and N=1 because the exponent is positive.

Conversion from BCD float-32 to double precision (to allow for exponents > 38) is straightforward,

```
1  double bcdfloat32_to_double(uint32_t bcd) {
2      uint32_t sig = bcd >> 8;
3      uint8_t i, sexp = bcd & 0xFF;
4      double ans=0, exp;
5  
6      for(i=6; i > 0; i--) {
7          ans += (sig & 0xF) * pow(10,-i);
8          sig >>= 4;
9      }
10 
11     exp = (sexp & 0x40) ? (sexp & 0x3F) : -(sexp & 0x3F);
12     ans *= pow(10,exp);
13 
14     return (sexp & 0x80) ? -ans : ans;
15 }
```

where `sig` stores the significand (all but lowest order byte, line 2). We store the sign and exponent byte in `sexp` (line 3). Lines 6 through 9 process the nibbles of the significand and add them into the output (`ans`) after scaling them by the appropriate power of ten. Line 7 looks at the lowest significand nibble and multiplies it by 10^{-i} and adds it to the output. Line 8 shifts the next most-significand digit of the significand into place. At the end of the loop, `ans` contains the significand so we only need now look at the exponent. Line 11 extracts the exponent by masking off the lowest-order bits (AND with 0x3F) and also sets the exponent sign by looking to see if bit 6 is set (`sexp & 0x40`). Line 12 then scales the significand by the exponent. Line 14 checks the overall sign of the number and returns it negating if necessary.

Conversion from a double to a BCD float-32 is also straightforward. For simplicity, we ignore range checks and implement the conversion as,

```
uint32_t double_to_bcdfloat32(double d) {
    char s[20];
    uint32_t ans=0, nsgn=1, sgn=0, sexp=0;
    int32_t exp;

    sprintf(s, "%+0.5e", d);
    if (s[0] == '-') sgn=1;
    if (s[9] == '-') nsgn=0;
    exp = 10*(s[10]-'0') + (s[11]-'0');
    exp = (s[9]=='-') ? -exp : exp;
    exp++;
    sexp = (abs(exp) & 0x3F) + (sgn*0x80) + (nsgn*0x40);

```

```
14      ans += (s[1]-'0') << 7*4;
15      ans += (s[3]-'0') << 6*4;
16      ans += (s[4]-'0') << 5*4;
17      ans += (s[5]-'0') << 4*4;
18      ans += (s[6]-'0') << 3*4;
19      ans += (s[7]-'0') << 2*4;
20      ans += sexp;
21
22      return ans;
23  }
```

where we cheat a bit by using a call to `sprintf` to convert the double input to a formatted base-10 character string (line 6). Once we have the string representation we need only pull it apart to set the fields of the BCD float-32 number. Lines 7 and 8 set the bits for the signs. Lines 9 through 11 extract the exponent. We add one to the exponent (line 11) because the conversion expresses the significand as "3.14159" and not "0.314159". Line 12 sets the bits of the lowest-order byte of the output using the absolute value of the exponent and the signs. Lines 14 through 19 extract the digits of the significand from the string representation. The numeric value is the digit character minus the ASCII code for zero. Each of these is then shifted to the proper nibble position and added into the output. Finally, the signs and exponent are added (line 20) and the resulting value is returned.

So, why do calculators use BCD floating-point? The best answer to that question is found in Chap. 7 which discusses the IEEE decimal floating-point format which uses a variation on packed BCD. Here we will give an example which shows that decimal calculations in a calculator are generally more precise, if slower, than binary floating-point calculations.

Consider this expression,

$$(1001 \times 0.01 - 10) \times 100 \rightarrow 1$$

which is exactly the answer we would expect to get mathematically. However, as we now understand from this chapter, when we use IEEE floating-point for the calculation we do not get 1 but instead get 0.9999999999999787 (double precision). If we try this same calculation using an inexpensive Rockwell 9TR calculator from 1976 we do get exactly 1 for the answer because the calculator is using BCD floating-point.

3.11 Chapter Summary

In this chapter we took a thorough look at floating-point numbers and how they are represented in a computer. We discussed just what we mean by "floating-point" and talked about some of the history surrounding floating-point and computers.

We compared two popular (though one is largely disused now) floating-point representations. We then explored the main parts of the IEEE 754 standard covering representations, rounding modes, comparisons, basic arithmetic, and exceptions. Next, we discussed the relationship between the standard and common floating-point hardware. Lastly, we explored packed BCD decimal floating-point which is often implemented in desktop and pocket calculators.

Exercises

3.1 Convert the following 32-bit IEEE bit patterns to their corresponding floating-point values:

(a) `0 10000000 10010010000011111011011`
(b) `0 10000000 01011011111100001010100`
(c) `0 01111111 10011110001101110111101`
(d) `0 01111110 01100010111001000011000`
(e) `0 01111111 01101010000010011110011`

Where the first number is the sign, the second group of 8 bits is the exponent and the remaining bits are the significand. It will help to write a computer program to handle the powers of two in the significand.

3.2 Assume a simple floating-point representation that uses four bits in the significand, three bits for the exponent, and one for the sign. The exponent is stored in excess-3 format and all exponent values are valid. In this case we have the following representations,

```
 4.25 = 0 101 0100
  1.5 = 0 011 1000
 -1.5 = 1 011 1000
-4.25 = 1 101 0100
```

Work out the following sums and express the answer in bits. Use a floating-point intermediate with eight bits for the significand and then round your answer to the nearest expressible value. Find `4.25 + 1.5` and `-4.25 + -1.5`. **

3.3 Modify `fp_compare` to handle infinities properly.

3.4 Summing the values of an array is a common operation. If the naive implementation is used, simply adding each new array element to the accumulated sum of all the previous, there will be an overall round-off error proportional to the number of terms added. For a small array this is negligible but for a large array it may matter. One way to avoid the accumulated round-off error is to use *pairwise summation* [6]. This is a recursive divide-and-conquer algorithm which checks to see if the input array length is below a cutoff value, say five elements, and if so it returns their sum by simply adding them together. Otherwise, it takes the length of the input array, divides it by two using integer division, and calls itself on each of those subarrays adding the return values together to form the final output value. In pseudo-code,

```
function pairwise(x) {
    if (length(x) <= 5) {
        s = x[0] + x[1] + ...
    } else {
        n = length(x) / 2
        s = pairwise(x[0..n-1]) + pairwise(x[n..])
    }
    return s
}
```

Implement this function in C and Python. Demonstrate the effect of round-off by comparing the output of this function with the naive summation for input arrays of 10,000 elements or more. Use a pseudo-random number generator to generate values in the range $[0, 1)$. **

3.5 Extend the BCD float-32 routines in Sect. 3.10 to process BCD float-48 and BCD float-56. For simplicity, store these in 64-bit unsigned integers and convert to and from C doubles.

References

1. Muller, J., Brisebarre, N., *et al.* Handbook of Floating-Point Arithmetic, Birkhäuser Boston (2010).
2. Saracino D., Abstract Algebra: A First Course, Addison-Wesley (1980).
3. Goldberg, D., "What every computer scientist should know about floating-point arithmetic." ACM Computing Surveys (CSUR) 23.1 (1991): 5–48.
4. Randell, B., The Origins of Digital Computers-Selected Papers. Springer-Verlag, (1982).
5. Rojas, R., "Konrad Zuse's legacy: the architecture of the Z1 and Z3." Annals of the History of Computing, IEEE 19.2 (1997): 5–16.
6. Higham, N. Accuracy and stability of numerical algorithms, Siam, (2002).
7. Cody, W., Waite, W. Software Manual for the Elementary Functions, Prentice-Hall, (1980).
8. Monniaux, D., "The pitfalls of verifying floating-point computations." ACM Transactions on Programming Languages and Systems (TOPLAS) 30.3 (2008): 12.
9. Rump, SM., "Accurate solution of dense linear systems, Part II: Algorithms using directed rounding." Journal of Computational and Applied Mathematics 242 (2013): 185–212.
10. McLaughlin, S., Learn TI-83 Plus Assembly In 28 Days, http://tutorials.eeems.ca/ASMin28Days/lesson/day18.html. (retrieved 09 Sep 2016).

Chapter 4
Pitfalls of Floating-Point Numbers (and How to Avoid Them)

Abstract Floating-point numbers can be hazardous. In this chapter we take a look at some of the pitfalls of floating-point computation and develop rules of thumb through which we can (hopefully) avoid most of them. We first examine some of the catastrophic and even deadly consequences of using floating-point computation poorly. We then bring things down a bit and examine expressions and processes that can lead to a loss of precision. Next we enumerate some rules of thumb which can help us design code less susceptible to these pitfalls thereby leading to more robust software. Finally, we end with a description of a software tool that can help refactor floating-point code to be more reliable.

4.1 What Pitfalls?

Programmers use floating-point numbers very frequently and sometimes without careful consideration of the consequences of the imprecision which may arise. In this chapter we will look at several examples of where floating-point numbers have led to disaster, often with the loss of human life. After these cautionary tales we will run some experiments illustrating the effects of casual floating-point use. We end by offering some general (and hopefully helpful) advice on how to improve our chances of creating a correctly functioning program involving floating-point values and describe a software tool that can help improve the reliability of floating-point expressions.

Throughout this chapter it will be good to keep this quote from Kahan as given by Krämer [1] in mind,

> Significant discrepancies [between the computed and the true result] are very rare, too rare to worry about all the time, yet not rare enough to ignore. [2]

meaning that in general floating-point calculations "just work" but that it is important to pay at least a little bit of attention to what you are doing. Naturally, if the application is a homework assignment the level of concern about the reliability of the floating-point results is considerably less than if the application is a radiation therapy machine with the potential to seriously injure or kill a patient.

R.T. Kneusel, *Numbers and Computers*, DOI 10.1007/978-3-319-50508-4_4

With this in mind, let's look at some instances of catastrophic failures of floating-point computation.

Patriot Missile Failure During the Gulf War (August 1990 through February 1991) the Iraqi army employed Soviet-built Scud missiles. The United States responded by installing Patriot anti-missile batteries to shoot down approaching Scuds. On February 25, 1991 a Scud missile hit a US army barracks in Dharan, Saudi Arabia, killing 28 solders. The base was protected by Patriot missile batteries yet the battery in range of the Scud failed to fire. The reason why is due to round-off error in floating-point calculations. The full report of the incident is available from the United States General Accounting Office (GAO) [4] but the essence of the problem has to do with trying to represent a simple value, in this case 0.1, in a floating-point format which was itself then truncated to fit into a fixed-point number.

The Patriot missile software tracked the position of incoming Scuds using radar. It tracked time by counting tenths of a second and used a 24-bit fixed-point representation of the floating-point value for 0.1 to convert to seconds. The only problem is that in binary the value of 0.1 is an infinitely repeating decimal,

$$0.1 = 0.0001100110011001100110011001100\ldots_2$$

that simply truncating at 24-bits gives, $0.00011001100110011001100_2$ meaning an error of $0.0000000000000000000000000110011001100\ldots_2 \approx 0.000000095$ s was in introduced for each truncation. This is not much, but if the missile battery was left in operation for a long period of time the error would accumulate.

The GAO report indicates that Israeli forces, also operating Patriot batteries, noticed that after approximately 8 h of operation targeting was off by about 20%. A software fix was created and scheduled to be sent to the batteries. However, no long running tests were performed to see how long the battery could be in operation before the timing error was catastrophic. Battery operators were warned that long runtimes might cause inaccuracy in targeting but were not given any notion of what "long runtime" might mean. On February 25, 1991 the Patriot battery in Dharan had been in continuous operation for about 100 h. This means the small timing error introduced by the truncation of 0.1 to 24-bits was now multiplied by the number of tenths of a second in 100 h or 3,600,000 causing a discrepancy of about 0.34 s. During this time a Scud missile would cover about half of a kilometer putting it out of range of the battery. Since the battery thought the missile was out of range it did not fire and the missile went on to kill 28 soldiers. The software patch arrived at Dharan on February 26th.

Ariane 5 Explosion The European Space Agency has a successful track record when it comes to launching sophisticated spacecraft. However, the maiden flight of its Ariane 5 rocket on June 4, 1996, was a total loss, to the tune of $500 million USD, all because one of the software modules attempted to put a 64-bit floating-point value into a 16-bit signed integer.

Just under 40 s after its launch from Kourou, French Guiana, the rocket self-destructed. The software putting a 64-bit floating-point value into the signed 16-bit integer had caused an overflow (ie, the value exceeded 32,767) which caused the rocket to believe it was off-course. It therefore fired thrusters to "correct" its course which put so much stress on the rocket that its self-destruct was triggered. The final report of the inquiry board summarizes the events [5].

Schleswig-Holstein Parliament Election This story does not involve loss of life or property but was politically troubling. In the German state of Schleswig-Holstein no party receiving less than 5% of the vote is seated in parliament. In the election of 1992 it looked like the Green party had precisely 5% of the vote allowing them a seat. This was according to the software that prints the results. However, closer examination showed that the Green party only had 4.97% of the vote, less than the required 5%, and that the software had simply rounded to one decimal so that 4.97% $\rightarrow$ 5.0%. Because of this, the Green party votes were dropped and the net result was a one seat majority for the Social Democrats [6].

Vancouver Stock Exchange This story does not involve loss of life or property, either, but was rather embarrassing all the same. In 1982 the Vancouver Stock Exchange created a new index set to an initial value of 1000.000 [7]. The "000" is meaningful, the index was calculated to three decimals. However, a programming error simply truncated the index to three decimals without rounding so, to use the example from the original Wall Street Journal article, if the actual index value was 540.32567 the software simply truncated the index to 540.325 dropping the final "67" digits. Of course, this is not a good idea as the proper three decimal value is 540.326 rounding the "5" up to "6" based on the value of the fourth decimal. The error is small but cumulative and as the index was updated some 2800 times per day it didn't take long before the index value was completely meaningless. Recalculation of the index by hand, when the error was realized, put the value at over 1098 instead of the 574 or so that was being reported.

4.2 Some Experiments

The examples above show that floating-point can sometimes lead to fatal and costly disasters. More practically, floating-point, in combination with compilers, hardware, and even the form of the equations used, can lead to wildly inaccurate results as the experiments in this section illustrate.

The Logistic Map A favorite equation used to illustrate the onset of chaotic behavior is the logistic map, $x_{i+1} = rx_i(1 - x_i)$ with r a constant between 1 and < 4 and x a value from $(0, 1)$. This map exhibits very interesting behavior as an initial x_0 is iterated repeatedly and as r varies from say 2 up to just under 4. If one were to

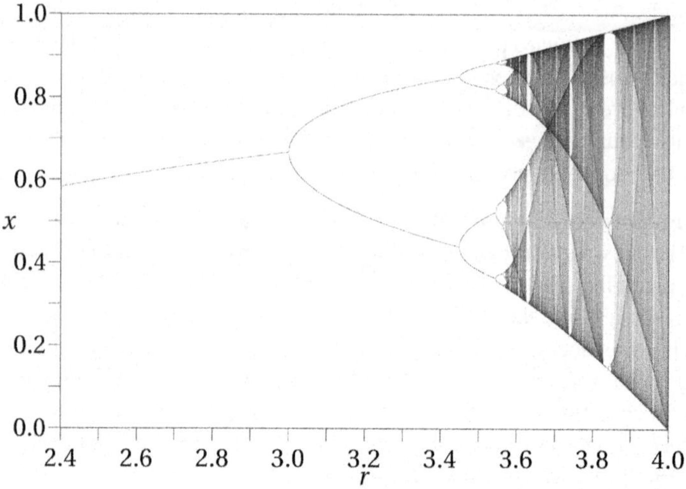

Fig. 4.1 The bifurcation plot of the logistic map. The plot shows several hundred x values from the iterated logistic map for each r value as a function of r. This map illustrates the period-doubling route to chaotic behavior. For the experiment $r = 3.8$ which is in the chaotic region ("Logistic Map" by Jordan Pierce, Creative Commons CC0 1.0 Universal Public Domain Dedication License)

select a starting value of 0.1 (or any other) and run the iteration forward a thousand times and then plot the next thousand as a function of r one would get Fig. 4.1 which is known as a bifurcation plot. This plot shows the period-doubling route to chaotic behavior as r increases. There is a deep correlation between the structure of this plot and the famous Mandelbrot set. Of interest to us is the sequence of values one gets if r is in the chaotic regime for different but algebraically equivalent forms of the logistic equation. This experiment is inspired by previous work by Colonna [8].

We can rewrite the logistic equation as,

$$x_{i+1} = rx_i(1 - x_i) = rx_i - rx_ix_i = x_i(r - rx_i) = rx_i - rx_i^2$$

where each of the four forms will be computed in our test program. The program itself is straightforward,

```
#include <stdio.h>
#include <math.h>

int main() {
    double x0,x1,x2,x3;
    double r = 3.8;
    int i;

    x0 = x1 = x2 = x3 = 0.25;

    for(i=0; i < 100000; i++) {
        x0 = r*x0*(1.0 - x0);
        x1 = r*x1 - r*x1*x1;
        x2 = x2*(r - r*x2);
        x3 = r*x3 - r*pow(x3,2);
    }

    printf("     x0           x1           x2           x3\n");

    for(i=0; i < 8; i++) {
        x0 = r*x0*(1.0 - x0);
        x1 = r*x1 - r*x1*x1;
        x2 = x2*(r - r*x2);
        x3 = r*x3 - r*pow(x3,2);
        printf("%0.8f  %0.8f  %0.8f  %0.8f\n", x0,x1,x2,x3);
    }
}
```

This program iterates each of the four versions of the logistic map and outputs the final eight iterates after an initial burn in of 100,000 iterations. Note that the program uses double precision values and that the initial value is 0.25 which is not a repeating binary. The program produces this as output,

```
     x0          x1          x2          x3
0.76670670  0.65123778  0.80925715  0.38542453
0.67969664  0.86308311  0.58656806  0.90011535
0.82729465  0.44904848  0.92152269  0.34164928
0.54293721  0.94013498  0.27481077  0.85471519
0.94299431  0.21386855  0.75730128  0.47187310
0.20427297  0.63888941  0.69842500  0.94699374
0.61767300  0.87669699  0.80038457  0.19074708
0.89738165  0.41077765  0.60712262  0.58657800
```

when compiled with,

```
$ gcc logistic.c -o logistic -lm
```

where we already see the effect of the different way the compiler has decided to evaluate the logistic expressions even though algebraically they are identical. If we compile again with optimization,

```
$ gcc logistic.c -o logistic -lm -O2
```

we get,

x0	x1	x2	x3
0.18402020	0.67978266	0.32056222	0.92875529
0.57059572	0.82717714	0.82764792	0.25144182
0.93106173	0.54322944	0.54205800	0.71523156
0.24390598	0.94289862	0.94327827	0.77396643
0.70078024	0.20459509	0.20331661	0.66478109
0.79681171	0.61839658	0.61552007	0.84681933
0.61523066	0.89673255	0.89928943	0.49292215
0.89954320	0.35189247	0.34415822	0.94980964

which is different still. Clearly, the way in which an expression is coded matters significantly. Which value, if any, is most correct? How can one tell? One way would be to use arbitrary precision floating-point which we discuss in detail in Chap. 9. Alternatively, one could use rational arithmetic and convert back to floating-point at the end. For an explanation of rational arithmetic see Chap. 5 but be warned that the iteration will quickly lead to fractions with very large numerators and denominators and require a great deal of memory.

Taylor Series for e^x When $x < 0$ Let's consider another experiment. In this case we want to compute e^x for any x using the Taylor series expansion,

$$e^x = 1 + x + \frac{x^2}{2!} + \frac{x^3}{3!} + \cdots = 1 + \sum_{i=1}^{\infty} \frac{x^i}{i!}$$

which we translate into Python as,

```
def fact(n):
    if (n == 1):
        return 1.0
    return n * fact(n-1)

def exponential(x, tol=1e-10):
    ans = 0.0
    term= 1.0
    p = 1.0

    while (abs(term) > tol):
        ans += term
        term = x**p / fact(p)
        p += 1

    return ans
```

where we accumulate terms of the series until one of the terms is smaller than a given threshold. If we run this program for positive values of x and compare them with the output of a gold standard `exp(x)` function (recall that Python floats are double precision), we get,

```
20   4.8516519540979016e+08   4.8516519540979028e+08
21   1.3188157344832141e+09   1.3188157344832146e+09
22   3.5849128461315928e+09   3.5849128461315918e+09
23   9.7448034462489052e+09   9.7448034462489033e+09
24   2.6489122129843472e+10   2.6489122129843472e+10
25   7.2004899337385880e+10   7.2004899337385880e+10
```

where the first column is the argument, the second is the output of `exponential()` and the third is the output of our gold standard function. Clearly, our Taylor series approximation for e^x is working quite well.

Now, let's run again but this time make $x < 0$. In this case we get,

```
-20   4.9926392547479328e-09   2.0611536224385579e-09
-21   2.7819487708032662e-08   7.5825604279119066e-10
-22  -2.4369901527907362e-08   2.7894680928689246e-10
-23   3.1518333727167153e-09   1.0261879631701890e-10
-24   3.7814382919759864e-07   3.7751345442790977e-11
-25   1.1662837734562158e-06   1.3887943864964021e-11
```

Not so good. What happened and how can we fix it? What happened is that the Taylor series for $e^x, x < 0$ is made up of terms that are alternately even and odd powers of x. This means that the sign of the term changes from term to term. In other words, the series sum is made up of values that are wildly different from each other. Consider the approximation for e^{-20},

$$
\begin{aligned}
e^{-20} &= 1 - 20 + \tfrac{20^2}{2} - \tfrac{20^3}{6} + \tfrac{20^4}{24} - \tfrac{20^5}{120} + \tfrac{20^6}{720} - \cdots \\
&= 1 - 20 + 200 - 1333.3 + 6666.7 - 26666.7 + 88888.9 - \cdots
\end{aligned}
$$

where the effect of each term changing sign is that the sum to term n is always the opposite sign as term $n + 1$. This addition of two very different numbers leads to a rapid loss of precision making the approximation functionally useless even though mathematically it is valid.

The fix in this case is to use only positive x and to recall that $e^{-x} = 1/e^x$. If we run our Python program again with only positive x and report `1.0/exponential(x)` we greatly improve our final results,

```
-20   2.0611536224385583e-09   2.0611536224385579e-09
-21   7.5825604279119097e-10   7.5825604279119066e-10
-22   2.7894680928689241e-10   2.7894680928689246e-10
-23   1.0261879631701888e-10   1.0261879631701890e-10
-24   3.7751345442790977e-11   3.7751345442790977e-11
-25   1.3887943864964019e-11   1.3887943864964021e-11
```

thereby proving that thought is always required when expecting meaningful results from floating-point operations.

Summing an Array of Numbers Summing the elements of an array is a frequent operation with floating-point numbers. For example, calculation of the mean of a collection of data points typically involves storing the data in an array and then processing the elements of the array to calculate the mean by summing the array elements and then dividing by the number of elements.

The straightforward way to sum an array is to simply add the elements one after the other to an overall sum. As we will see, this simple approach suffers quite a bit from round-off error. We can do better and compensate for the error by using Kahan summation [3] which is a form of compensated summation [9]. For example, this code will generate a vector of random values and sum them using simple summation and Kahan summation,

```
1  #include <stdio.h>
2  #include <time.h>
3  #include <stdlib.h>
4
5  #define N 10001
6  double A[N];
7
8  int main() {
9      double sum, c, y, t;
10     int i;
11
12     srand(time(NULL));
13     for(i=0; i<N; i++)
14         A[i] = 1001.0*((double)rand()/(double)RAND_MAX);
15
16     sum = 0.0;
17     for(i=0; i<N; i++)
18         sum += A[i];
19     printf("simple sum = %0.16f\n", sum);
20
21     sum = c = 0.0;
22     for(i=0; i<N; i++) {
23         y = A[i] - c;
24         t = sum + y;
25         c = (t - sum) - y;
26         sum = t;
27     }
28     printf("Kahan sum  = %0.16f\n", sum);
29     return 0;
30 }
```

where lines 12 through 14 fill the array `A` with random numbers in the range $[0, 1001]$. The simple sum code is in lines 16 through 18. It is self-explanatory. We print the result in line 19. Kahan summation is implemented in lines 21 through 27. These lines need some explanation. The main goal of Kahan summation is to preserve some notion of the error (`c`) and compensate for it. Line 23 adjusts the to-be-added value (`A[i]`) by the current value of `c`. This subtracts off the estimated error from the previous value which may actually be negative. Line 24 stores the new sum adding in the current value (`t`). Line 25 would, in the absence of error, be exactly zero. The difference `t-sum` should be exactly `A[i]-c` as is `y` so the value of the expression should be zero. However, round-off error will (sometimes) make this value nonzero. This is the estimated error introduced by adding `A[i]` and we compensate for it when the next term is added (line 23). Line 26 updates the actual returned sum.

How well does Kahan summation work? We can calculate it. If we sum an array of 10001 random values, each in the range [0,1001], using simple summation and Kahan summation we can compare the result with a high-precision sum of the same array using 100 decimal digits. Taking the 100 decimal digit answer as the gold standard we can calculate the deviation for the simple and Kahan sums. If we repeat this process, say 300 times, we can calculate an average deviation. Doing just this gives,

$$\begin{aligned}\text{simple sum}|\bar{\Delta}| &= 0.0000000103093065730\\ \text{Kahan sum}|\bar{\Delta}| &= 0.0000000002381457745\end{aligned}$$

where running the test again will give slightly different results but it will consistently show that Kahan summation is approximately two orders of magnitude more precise than the simple summation code. The high-precision floating-point library used is the one developed in Chap. 9.

Comparing Floating-Point Numbers Our final example involves one of the most common of activities with floating-point numbers: testing for equality.

To illustrate the problem, consider the simple expression $0.1+0.2 == 0.3$ where $==$ implies a test for equality. This expression fails when using 32-bit or 64-bit IEEE floating-point. However, this expression does not fail for IEEE floating-point: $0.1 \times 10^1 + 0.2 \times 10^1 == 0.3 \times 10^1$. So, for what values of x in 10^x will the test pass? If we test empirically using only positive exponents to avoid cluttering the final plot, we get Fig. 4.2 where we plot 1 if the test passes for that particular exponent x and plot 0 if not. The results for negative exponents are similar. Looking at the figure we plainly see islands where the expression works as expected but also many places where it fails. Comfortingly, there is a large set of exponents covering what is probably a very common range of values encountered in practice where the expression is evaluated correctly with the notable exception of $x = 0$.

Why is the expression not always evaluated properly? Round-off error is to blame. We have seen previously that 0.1 is infinitely repeating in binary so the requirement to fit it in a finite number of bits will inevitably result in error that can make a mathematically true expression appear false on the computer.

Fig. 4.2 Comparing $0.1 \times 10^x + 0.2 \times 10^x = 0.3 \times 10^x, x \in [0, 200]$ using double precision floating-point numbers. If the comparison passes plot 1, otherwise plot 0 for exponent x. It is clear that there are islands where the comparison works as expected but also many places where it fails

Another area where round-off error in comparisons comes into play is inside of a loop. It is not uncommon to want to iterate over a range of floating-point values,

```
for(d=start; d < end; d += inc)
    ...
```

where `start` is a starting value, `end` is an ending value and `inc` is an increment. For example, we can set `inc` using `inc = (end - start) / n` in order to calculate `n` values in the loop. This seems straightforward but round-off error and the test `d < end` will not let us easily predict whether the loop will execute n or $n + 1$ times. This is bad if we hope in the loop body to fill an array of exactly n elements.

We can determine the frequency of this error with a simple program that counts the number of times the body of a loop defined in this way actually executes for random `start` and `end` values fixing `n`,

```
int main() {
    double start, end, inc, d, s=0.0;
    int i=0,j=0,n,m;

    srand(time(NULL));
    n = 100;
    m = 10000000;

    for(j=0; j<m; j++) {
        start = 100001.0*((double)rand()/(double)RAND_MAX);
        end = start + 1001.0*((double)rand()/(double)RAND_MAX);
        inc = (end - start) / n;
        i = 0;

        for(d=start; d < end; d += inc)
            i++;
        s += i-n;
    }

    printf("fraction with extra loop = %0.9f\n", s/(double)m);
    return 0;
}
```

where we set `start` to a random value in the range [0, 100001] and `end` to `start` plus a value in the range [0, 1001]. We fix n at 100 meaning we expect the loop in lines 15 and 16 to execute exactly n times. We count the actual number of loop iterations in `i` and accumulate a running total in `s` (line 17). When we've run `m` tests we report the fraction of loops that actually executed $n + 1$ times (line 20). Running this program many times always reports a fraction very close to 49.5% meaning for any arbitrary start and end conditions, expecting exactly 100 iterations, we are actually getting 101 iterations nearly half the time. So, this issue is a real one and must be considered when developing code. In this case we can correct the problem by rewriting the loop,

```
for(j=0; j < n; j++) {
    d = start + j*inc;
    ...
```

where we are now assured of executing exactly n times and we set `d` at the beginning of each iteration so we can use it in the body of the loop.

What about comparing for equality in general? Can we be accurate when using floating-point values? The answer is "yes" with some attention paid to what we mean by equality. We saw in Chap. 3 that the bit representation of IEEE floating-point values are ordered and when two values are the same their bit values are the same, with one notable exception. So, a simple, highly strict, test for floating-point equality is to look at the bits of the representation. If they are the same the numbers must be the same. The exception is zero which may be signed or unsigned according to IEEE 754. In this case the numbers are the same but the bit values are different. We should also be careful in the case of NaNs. Recall that a NaN has a payload which is intended to convey application-specific information about the source of the NaN. So, while we might want to say that NaNs are equal they might have different payloads. Infinity, positive or negative, is not an issue because, by definition, there is only one bit pattern that represents each of them.

With this in mind, we can define a very strict equality test as,

```
1 uint8_t equal_strict(float a, float b) {
2     uint32_t A = *(uint32_t *)&a;
3     uint32_t B = *(uint32_t *)&b;
4 
5     if ((a == 0.0) && (b == 0.0)) return 1;
6     if (isnan(a) && isnan(b)) return 1;
7     return (A == B);
8 }
```

where we take the inputs and re-interpret them as unsigned integers (lines 2 and 3). This will let us compare the bits directly. Line 5 checks for zero since the IEEE standard demands that $0.0 == -0.0$. Line 6 looks at NaNs, regardless of the payload. Lastly, line 7 compares the bits directly now that we know the inputs are not NaNs or zero. This is the strictest possible test for floating-point equality and is probably not what we want since it requires *exact* equality for all bits.

A common technique is to use a tolerance value. If we know the range of values we are working with, for example, the result of actual physical measurements, we can usually set a tolerance below which we are willing to call the values the same. In this case we can use an expression like,

```
if (fabs(a-b) < t) ...
```

to check if the difference between two floating-point values, here doubles because we are using `fabs`, is less than a tolerance value, `t`. In many cases this is a perfectly reasonable thing to do. However, this is not a general solution because in the general case the magnitudes of `a` and `b` may be way out of concordance with the given tolerance. To be more accurate and still allow a tolerance over which we are willing to call two floating-point values equal we can return to the integer representation and measure the difference between the two values in integer space and as long as that is within a certain tolerance we can call the numbers "equal" (with the tests above for zero and NaN cases).

We introduce a new term, "ulps", which means "units in the last place". An ulp is the smallest difference between two consecutive floating-point numbers. In integer space, this is the difference between a floating-point number with integer representation n and the floating-point number with representation $n + 1$. We can use this to test for equality by passing, along with the two numbers, the maximum number of ulps difference we are willing to tolerate and still call the numbers equal. This difference may be quite large if we are in a situation where we are willing to call two very different numbers equal. However, in that case, we may be able to intelligently use the tolerance approach above.

Our ulps equality routine is a simple modification of the strict equality routine above,

```
1 uint8_t equal_ulps(float a, float b, uint32_t ulps) {
2     uint32_t A = *(uint32_t *)&a;
3     uint32_t B = *(uint32_t *)&b;
4
5     if ((a == 0.0) && (b == 0.0)) return 1;
6     if (isnan(a) && isnan(b)) return 1;
7     return abs(A-B) < ulps;
8 }
```

where the only change, besides passing in a tolerance value (`ulps`) is line 7 which checks if the difference between the two inputs, as integers, is less than the max tolerance. Note that `equal_ulps` will not be correct for numbers that are very close to zero. This is because the high bit of an IEEE float is the sign and even if the numbers are both very close together, if one is negative and the other positive, then the raw difference in ulps will be very large indeed. We leave correcting `equal_ulps` as an exercise for the reader, see Problem 4.1.

We can test our new function with this program,

```
1  int main() {
2      float a,b;
3      uint32_t A,B,i;
4      uint32_t S=0, UMAX=0, UMIN=80000000;
5      uint32_t p0=0,f0=0,p1=0,f1=0, M=50000000;
6
7      srand(time(NULL));
8
9      for(i=0; i < M; i++) {
10         a = (float)(1001.0*(double)rand()/
                 (double)RAND_MAX);
11         b = 2.1*a;
12         b = b - 2.0*a;
13         b = 10.0*b;
14         A = *(uint32_t*)&a;
15         B = *(uint32_t*)&b;
16         S += abs(A-B);
17         if (abs(A-B) > UMAX) UMAX = abs(A-B);
18         if (abs(A-B) < UMIN) UMIN = abs(A-B);
19         if (equal_ulps(a,b,30)) p0++; else f0++;
20         if (a==b) p1++; else f1++;
21     }
22
23     printf("p0=%8d, f0=%8d, passed=%0.8f\n", p0, f0,
               (double)p0/(double)M);
24     printf("p1=%8d, f1=%8d, passed=%0.8f\n", p1, f1,
               (double)p1/(double)M);
25     printf("mean difference=%0.8f, [%u,%u]\n", (double)S/
               (double)M,UMIN,UMAX);
26
27     return 0;
28 }
```

This program performs many tests of the difference between a random floating-point value (a) and b which is set to $(2.1a - 2.0a) \times 10$ which, mathematically, should be exactly equal to a but which we know will not be because of round-off error. We split the formula up among several lines to stymie the compiler lest it do any advanced optimizations. The variable S accumulates the difference between a and b in ulps. The largest and smallest difference encountered is stored in UMAX and UMIN respectively. Line 19 calls the new equality routine and increments the counters for passed and failed. Line 20 does the same for the straightforward equality test.

When the loop completes the program outputs the totals for each case along with the fraction of tests that passed, the mean difference, minimum difference and maximum difference in ulps. A typical run produces output like,

```
p0=50000000, f0=        0, passed=1.00000000
p1= 2408450, f1=47591550, passed=0.04816900
mean difference=5.37257440, [0,20]
```

where all the calls to `equal_ulps` have passed because we set the tolerance to 30 ulps and in this case the maximum difference encountered was 20 ulps. Notice that only 4.8% of the direct comparisons were actually successful. These percentages are maintained when the range of `a` is changed across many orders of magnitude.

Note that the routines above work for double precision floating-point as well. Simply map `float` → `double` and `uint32_t` → `uint64_t`. The results found are still valid in the double precision case.

4.3 Avoiding the Pitfalls

The previous section illustrated through empirical experiments some of the pitfalls of floating-point computation. The topics addressed are not exhaustive by any means. However, we can now put together a short list of "rules of thumb" that are worth keeping in mind when writing floating-point code. After this list we take a quick look at a software tool that can automatically improve floating-point expressions as another approach to avoiding pitfalls.

Rules of Thumb We end this chapter with some "rules of thumb" regarding the use of floating-point numbers. These rules are not absolute, of course, and others could be given, but they will hopefully be found helpful and cause you to pause and consider carefully when writing floating-point programs.

1. Do not use floating-point numbers if integers will suffice. As a corollary, do not use floating-point numbers if exact computation is required, rather, scale if a fixed number of decimals will always be used. For example, if the values are dollars do the computations in cents and divide by 100 when done.
2. Do not represent floating-point numbers in text form in files or as strings unless absolutely necessary. If necessary, use scientific notation and be sure to have enough digits after the decimal point of the mantissa to represent the significand accurately. For a 32-bit float use at least 8 digits. For a 64-bit float use at least 16. For example, in Python,

```
>>> from math import pi
>>> float("%0.15g" % pi) == pi
False
>>> float("%0.16g" % pi) == pi
True
```

where the `0.15g` format specifier uses 15 digits after the decimal which is not enough, 16 are required so that the text representation of π, when converted back to a floating-point number, matches the original. Floating-point numbers are often found in XML and JSON files and frequently without enough precision.

3. The spacing between floating-point numbers increases as the number increases. If working with large numbers, rescale, if possible, to map to values closer to one.
4. Comparing two floating-point values for equality can be problematic if they were computed using a different code sequence. If possible, replace expressions like `a == b` with `abs(a-b) < e` where `e` is a tolerance value or use the equality functions defined above. Even the tolerance value is dangerous in that the choice of `e` depends upon the scale of the values being worked with.
5. Subtracting two nearly equal numbers results in a large loss of precision as the result is made up of only the lowest order bits of the significand. Generally, this should be avoided. Naturally, the rule above does just this which is why it is best to compare integer representations directly when possible and absolute precision is needed.
6. For real numbers we know that $(ab)c = a(bc)$ however this is sometimes not the case for floating-point numbers due to rounding. Use care when assuming associativity.

The rules above are ad hoc examples of a large research area known as *numerical methods*. An important part of numerical methods involves the analysis of the equations being implemented by the computer and a reworking of those equations to minimize floating-point errors to improve the stability of the program. The stability of numerical algorithms is a vast field as might be expected. For a scholarly introduction, especially its early chapters, see Higham [9]. For a more popular treatment see Acton [10].

Using a Tool: Herbie One emerging area of research in numerical methods is the automatic rewriting of floating-point expressions to improve their reliability. Here we will take a look at one such tool, Herbie. The description of how Herbie works is beyond the scope of this book, see [11] for more information. We will describe how to install and test Herbie and how to use it to simplify expressions. There is also a web demo of Herbie that uses a more familiar syntax, see http://herbie.uwplse.org/demo/ (retrieved 14-Sep-2016).

Herbie is written in Racket. Racket is a programming language derived from Scheme which itself is derived from Lisp. In a previous life, Racket was called MzScheme, though Racket goes well beyond MzScheme. In order to use Herbie we need to install Racket. We can install Racket by downloading it from here: http://download.racket-lang.org/racket-v6.5.html (retrieved 14-Sep-2016). After we install Racket we can get Herbie from github with:

```
git clone https://github.com/uwplse/herbie
```

where can test the installation with,

```
racket src/reports/run.rkt bench/hamming/
```

If 24 or more tests pass we can assume that Herbie is working properly (patience, it takes a while). Compile Herbie with,

```
raco make src/reports/run.rkt
```

to improve performance. We are now ready to use Herbie to improve some floating-point expressions.

In order to use Herbie we need to enter the expressions we wish to optimize in an external file using Lisp syntax. This syntax is prefix-order meaning operators come before the operands using parentheses to group operators and operands. For example, consider the following translations from standard infix to prefix format,

Infix	Prefix
$x+y$	`(+ x y)`
$x-y$	`(- x y)`
$2x^2-3x+4$	`(+ (- (* 2 (sqr x)) (* 3 x)) 4)`
$100(10.01x-10)$	`(* 100 (- (* 10.01 x) 10))`
$\sqrt{x^2+y^2}$	`(sqrt (+ (sqr x) (sqr y)))`
$\cos^2 x+\sin^2 x$	`(+ (sqr (cos x)) (sqr (sin x)))`

The set of intrinsic functions Herbie understands is: +, −, *, /, abs, sqrt, sqr, exp, log, expt, sin, cos, tan, cot, asin, acos, atan, sinh, cosh, and tanh where the functions have their usual meaning. Use `(expt x y)` for x^y. Herbie also knows PI for π and E for e.

The expressions need to be entered into the file as follows,

```
(FPCore (x)
    (* 100 (- (* 10.01 x) 10)) )
```

where the expression is, using Lisp syntax, the second line. The expression is wrapped with

```
(FPCore (x) ... )
```

where `(x)` is a list of the variables in the expression and ... is the expression.

So, to use Herbie to process the expression $x(\sqrt{x-1}-\sqrt{x})$ we enter,

```
(FPCore (x) (* x (- (sqrt (- x 1)) (sqrt x))) )
```

into a file named `expr.fpcore` (the `fpcore` extension is suggested) and evaluate it with,

```
racket src/herbie.rkt expr.fpcore
```

to produce Herbie's rewritten expression,

```
(FPCore (x) (/ (- x) (+ (sqrt x) (sqrt (- x 1)))))
```

indicating that, according to Herbie, we should change,

$$x(\sqrt{x-1}-\sqrt{x}) \rightarrow \frac{-x}{\sqrt{x}+\sqrt{x-1}}$$

to maximize our accuracy. Let's test Herbie's claims using a simple Python program,

```
1  import random
2  from APFP import *
3  from math import sqrt
4
5  def f(x):
6      return x*(sqrt(x-1) - sqrt(x))
7
8  def g(x):
9      return (-x) / (sqrt(x) + sqrt(x-1))
10
11 def main():
12     APFP.dprec = 100
13     M = 10000
14     SA = APFP(0)
15     SB = APFP(0)
16
17     for i in xrange(M):
18         x = 1001.0*random.random() + 1.0
19         a = f(x)
20         b = g(x)
21         X0 = (APFP(x) - APFP(1)).sqrt()
22         X1 = APFP(x).sqrt()
23         B = APFP(x)*(X0 - X1)
24         DA = abs(B - APFP("%0.18f" % a))
25         DB = abs(B - APFP("%0.18f" % b))
26         SA += DA
27         SB += DB
28
29 print "Running %d tests:" % M
30 print " mean deviation using f() = %s" % (SA / APFP(M))
32 print " mean deviation using g() = %s" % (SB / APFP(M))
```

where the APFP library is the fixed-point arbitrary precision floating-point library developed in Chap. 9 which is capable of performing calculations with an arbitrary number of digits after the decimal point. Here we use 100 digits after the decimal point. There is no practical limit to the size of the integer portion. The functions `f()` and `g()` are the two forms of the expression we are testing, the original and the version Herbie suggests. The main loop (line 17) draws a random value, gets the output of the two forms of the expression and then calculates the original expression using arbitrary precision (lines 21 through 23). We determine the deviation from the arbitrary precision result for each expression (lines 24 and 25) and accumulate the deviation so we can calculate the mean (lines 30 and 32).

Running this program produces output similar to (truncating the 100 digits in the actual display),

```
Running 10000 tests:
    mean deviation using f() = 0.0000000000005792391185219435
    mean deviation using g() = 0.0000000000000006209725959118
```

where `f()` is $x(\sqrt{x-1}-\sqrt{x})$, the original expression, and `g()` is $\frac{-x}{\sqrt{x}+\sqrt{x-1}}$, the expression Herbie suggested we use instead. Indeed, Herbie is correct, the rewritten expression produces on average 1/1000-th the deviation from the arbitrary precision result when compared to the original expression. Note that the arbitrary precision result uses the original expression.

From this simple test is appears that tools like Herbie can be used profitably to improve floating-point accuracy in programs. In its present form Herbie is a tad tedious to use because of the requirement to convert expressions to prefix format or to enter them manually in the web interface. It would be nice if Herbie was infix-aware and multi-language aware so that we can mark expressions we want Herbie to automatically process directly in the source code. For example, a preprocessor using Herbie for C programs might look for special comments such as,

```
double f(double x) {
    double y;

    // herbie-input{
    y = x*(sqrt(x-1) - sqrt(x));
    // }herbie-input

    return y;
}
```

and produce new source code with,

```
double f(double x) {
    double y;

    // herbie-output{
    y = (-x) / (sqrt(x) + sqrt(x-1));
    // }herbie-output

    return y;
}
```

or something similar.

4.4 Chapter Summary

In this chapter we examined some real-world instances where floating-point errors caused severe damage and loss of life or property. We then used empirical experiments to understand some of the issues found in the application of floating-point numbers. These helped us formulate some rules of thumb for using floating-point. Lastly, we examined a powerful software tool for the automatic rewriting of floating-point expressions to improve their accuracy and demonstrated the efficacy of the rewritten expressions.

Exercises

4.1 Correct the deficiency in `equal_ulps` above by treating the input floating-point numbers as sign-magnitude integers. I.e., bit 31 of the integer representation is the sign and needs to be explicitly examined in some fashion.

4.2 The function `equal_ulps` uses a tolerance value to indicate equality. One could also use a tolerance value to help in deciding whether a floating-point number, `a`, is greater than or less than another number, `b`. Write a new function, `compare_ulps(a,b,u)`, which returns -1 if $a < b$, zero if $a = b$, and +1 if $a > b$. The third argument, `u`, is a tolerance in ulps. If `a` and `b` are within this tolerance they are considered equal. If not, and `a` is larger than `b`, return +1 otherwise return -1. (Hint: recall that bit 31 of an IEEE float is the sign bit. You will need to take this into account in your function.)

4.3 Experiment using Herbie to rewrite other expressions. Test them using the Python code in Sect. 4.3. In particular, consider expressions using $\sqrt{f(x)}$, $e^{f(x)}$, $\log f(x)$, and trig functions like $\sin f(x)$, etc. Is Herbie most effective with any specific transcendental function?

References

1. Krämer, Walter. "A priori worst case error bounds for floating-point computations." IEEE transactions on computers 47.7 (1998): 750–756.
2. Kahan, W. M., "The Regrettable Failure of Automated Error Analysis." mini-course, Conf. Computers and Mathematics, Massachusetts Inst. of Technology. 1989.
3. Kahan, W. M., "Further remarks on reducing truncation errors", Communications of the ACM, 8 (1): 40 (January 1965).
4. B-247094, Report to the House of Representatives. Washington, D.C.: GAO, Information Management and Technology Division, 4 Feb. 1992, url: www.fas.org/spp/starwars/gao/im92026.htm (retrieved 15 Oct 2014).
5. Lions, JL (chair), ARIANE 5 Flight 501 Failure Report by the Inquiry Board, Paris, 19 July 1996.
6. Weber-Wulff, D., The Risks Digest, Volume 13, Issue 37, 1992.
7. Quinn, K., Ever Had Problems Rounding Off Figures? This Stock Exchange Has, The Wall Street Journal, p 37, November 8, 1983.
8. Colonna, JF, The Subjectivity of Computers, http://www.lactamme.polytechnique.fr/descripteurs/subject.01..html. (retrieved 15 Oct 2014).
9. Higham, N. Accuracy and stability of numerical algorithms, Siam, (2002).
10. Acton, F., Real computing made real: Preventing Errors in Scientific and Engineering calculations, Courier Dover Publications, (2013).
11. Panchekha, P., et al. "Automatically improving accuracy for floating point expressions." ACM SIGPLAN Notices 50.6 (2015): 1–11.

Part II
Other Representations

Chapter 5
Big Integers and Rational Arithmetic

Abstract Big integers differ from standard integers in that they are of arbitrary size; the number of digits used is limited only by the memory available. In this chapter we look at how big integers are represented in memory and how to perform arithmetic with them. We also discuss some implementations which might be of use when using programming languages that do not support big integers natively. Next we examine rational arithmetic with big integers. Finally, we conclude with some advice on when it might be advantageous to use big integers and rational numbers.

5.1 What is a Big Integer?

Big integers, also known as bignums, bigints, multiple precision, arbitrary precision, or infinite precision integers, are integers that are not limited to a fixed number of bits. The standard representation for integers in computers involves a fixed number of bits, typically 8, 16, 32, or 64. These numbers therefore have a fixed range. A big integer, however, uses as much memory as is necessary to represent the digits and therefore is only limited in range by the memory available to it.

Several programming languages include big integers as part of their specification including Python. In Python we have an interpreter that seamlessly moves between fixed-width integers and big integers, of data type `long`, depending upon the result of a calculation. The transition point depends upon the version of the Python interpreter in use. The easy way to check this limit is to import the `sys` library and get the value for `sys.maxint`. This value is the largest positive integer which the interpreter will store in the standard representation. If this value is exceeded, the interpreter will switch to a big integer representation. For example, on a 32-bit system we have the following,

```
>>> import sys
>>> sys.maxint
    2147483647
>>> sys.maxint + 1
    2147483648L
>>> -sys.maxint
    -2147483647
```


R.T. Kneusel, *Numbers and Computers*, DOI 10.1007/978-3-319-50508-4_5

```
>>> -sys.maxint - 1
    -2147483648
>>> -sys.maxint - 2
    -2147483649L
```

where the Python prompt is given as >>> and replies from the interpreter are shown without the prompt and indented for clarity.

Several important things to note are happening in this example. First, we import the `sys` library and ask for the value of `maxint` to which Python replies it is `2,147,483,647` which we recognize as the largest positive two's complement 32-bit integer. This is to be expected on a 32-bit system. If we take this value and add one, which results in a value that cannot be represented by a 32-bit two's complement integer, we get the answer we expect with an extra "`L`" after it. This extra letter is Python's way of saying the data type is now `long`, meaning a big integer. If we negate the `maxint` value we get a two's complement integer. If we then subtract one from this value we still get a two's complement integer, which is expected because a two's complement integer gives us one extra negative number compared to one's complement notation. However, if we subtract two we get our expected answer but this time Python has moved to a big integer representation.

In this chapter we will look at ways in which big integers are represented in memory and some algorithms for dealing with big integer input and output when the base is a convenient one. We then look at arithmetic and explore some of the more common big integer algorithms. Next, we look at some more sophisticated big integer algorithms which are suitable when the numbers involved become quite large. We then look at rational arithmetic, a natural place to use big integers. We follow this with a, by necessity, time-sensitive view of key big integer implementations covering libraries for languages without big integers which we expect to be long-lived and certain languages which implement big integers natively. We conclude with some comments on when one might wish to resort to big integers or rational arithmetic.

5.2 Representing Big Integers

We know that place notation represents numbers using a given base, B, and the coefficients of the powers of that base which sum to the number we want to represent. Specifically, we can represent a number in any base with,

$$abcd.efg_B = a \times B^3 + b \times B^2 + c \times B^1 + d \times B^0 + e \times B^{-1} + f \times B^{-2} + g \times B^{-3}$$

where we typically use $B = 10$ as the base. When we represent a big integer in memory we would like to use a base which allows us to accomplish two things simultaneously,

1. Use a base which makes conversion between text input and output straightforward.
2. Use a base in which products of the "digits" of our base fit in a standard word.

If we have a 32-bit system and look at using signed 32-bit integers as the way to represent digits we need a base where any two digits, when multiplied, give us a value that fits in a 32-bit integer. We will consider representation of signs for big integers later. This will satisfy the second requirement above. What about the first? If one looks at the common number bases used in computer systems we see that they are all powers of two. Octal is base $2^3 = 8$ while hexadecimal is base $2^4 = 16$. In these cases we saw that it was particularly simple to change a binary number into an octal or hexadecimal number and vice versa. This is a clue we can use when choosing a base for our big integers. Since we would like to make things convenient for users we want to be able to input and output base ten numbers. This suggests using a power of ten for our big integer base, but which power? Since in this book we are assuming a 32-bit computer we want a base such that the base squared fits in a signed 32-bit integer. If we choose $B = 10{,}000$ we see that $B^2 = 100{,}000{,}000$ which does fit nicely in a signed 32-bit integer. However, if we go to the next power of ten, $B = 100{,}000$ we see that $B^2 = 10{,}000{,}000{,}000$ which is too large to fit in 32-bits. So, let's work with $B = 10{,}000$. Why choose such a large base? Why not simply use $B = 10$? The answer is that the larger the base the smaller the number of unique digits which are required to represent a large number. In binary we need eight bits to represent a number which in base ten is only three digits long and is only two digits in hexadecimal. So, the larger the base, the less memory we will need to store our numbers. As we are, by definition, working with big numbers we would like to be as economical as possible. If we were willing to work harder to handle input and output we could, naturally, use an even larger base. The largest value that fits in an unsigned 32-bit integer is 4294,967,295 and we see that $65{,}535^2 = 4294{,}836{,}225$ will fit meaning that is the largest base we could use. However, as it is not a power of ten, we would have to work harder in input and output and since the larger base does not change the algorithms we will be exploring in this chapter we are justified in working with the smaller but handier base of 10,000. An additional nice feature of our choice of base is that it fits in a *signed* 32-bit integer allowing us to make our internal representation quite compact as we will see.

We have selected to represent our big integers in memory as base 10,000 numbers with 10,000 digits. We will simply use standard binary to represent the digit with each word in memory being one digit of the number. When we want to output a big integer, then, we only need output each signed 32-bit integer as a group of four decimal digits, again, because our base is $10^4 = 10{,}000$. Similarly, when getting a number from a user, we will take the input, as a string of ASCII characters, and group it in sets of four characters to form the digits of our base 10,000 number in memory.

The previous paragraph contains a subtle piece of information. It implies that our big integers will be stored in memory in big-endian format. This means that if we use an array of signed integers to hold the digits that the first integer in that array will be the largest power of the base, not the smallest as it would be in little-endian format. This choice again simplifies our input and output operations.

At this point, we are able to write down how we will represent big integers for the examples that follow. We will use arrays of signed integers to represent the digits. Additionally, the very first element of the array will contain the number of digits in the number and the sign of this element will be the sign of the integer. Pictorially,

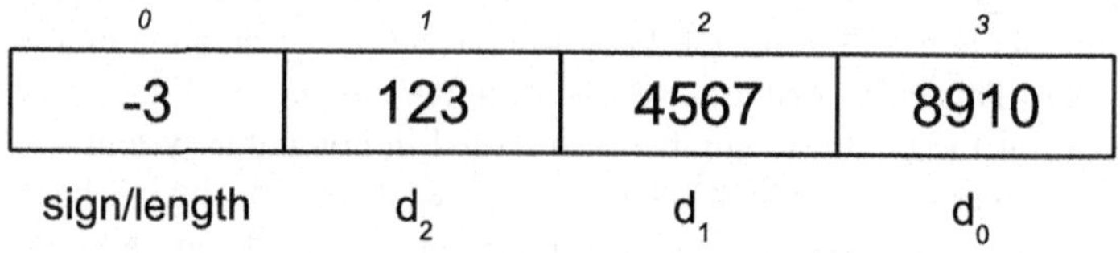

which illustrates how the number "−12,345,678,910" would be stored in memory using an array of four signed 32-bit integers. The first element of the array contains "−3" which says there are three digits in this number and that it is negative. The remaining three elements of the array are the digit values themselves in base 10,000. This simple format is what we will be using going forward in this chapter.

Input and Output With all of these preliminaries in place, Fig. 5.1 shows a simple C routine to convert a string into a base 10,000 number. Similarly, Fig. 5.2 shows a C routine for output of a base 10,000 number as a decimal number.

Figure 5.1 accepts a string in `s` and returns a pointer to a newly allocated big integer. Note that this memory must be released by a call to `free`. Let's look in detail at what this particular function is doing.

Line 10 checks to see if the input is negative. If it is we set the sign to `1` (default is `0`) and skip past the sign character by incrementing our character index `j`. Line 15 calculates the number of digits that will be in the big integer. The input string is in base 10 while the output integer is in base 10,000 so we divide the length of the input, accounting for the sign character, by four to see how many digits we need. We take the ceiling of this to avoid truncating the length. Line 16 allocates new memory on the heap for the number, using `n+1` to add in one element for the length and sign. For clarity we skip any check on the return value of `malloc` to see if the allocation succeeded. In line 17 the sign and length are set.

Lines 19 through 36 work through the big integer, digit by digit, using sets of four characters from `s` to set the value. The first digit, which is the highest order digit, is a special case and is handled by lines 21 through 29. This is because the highest order digit is the only digit which might have fewer than four digits in the element. The digit is set to zero, then the `switch` statement determines how many excess digits there are in the input string. These are digits that will be in the first output digit and there can be between one to four of them. Notice there are no `break` statements in the `switch`. This will cause the first matched condition to fall through to all the lower ones. This is what we want. Lines 31 through 34 handle all other digits of the output big integer. In this case, since we know there are actually four characters matched to this digit we do not need the `switch` statement and the logic is easier to see. We successively add in each character multiplied by the proper power of ten and converted to a digit value by subtracting the ASCII value for zero. When the loop over output digits completes the number is complete and a pointer to the memory is returned in line 38.

```c
#include <stdio.h>
#include <stdlib.h>
#include <math.h>
#include <string.h>

signed int *bigint_input(char *s) {
    int i, j=0, n, sign=0;
    signed int *b;

    if (s[0] == '-') {
        sign = 1;
        j++;
    }

    n = (int)ceil((strlen(s)-sign) / 4.0);
    b = (signed int *)malloc((n+1)*sizeof(signed int));
    b[0] = (sign) ? -n : n;

    for(i=1; i <= n; i++) {
        if ((i==1) && (((strlen(s)-sign)%4) != 0)) {
            b[i] = 0;
            switch ((strlen(s)-sign)%4) {
                case 3:
                    b[i] += (s[j++]-'0') * 100;
                case 2:
                    b[i] += (s[j++]-'0') * 10;
                case 1:
                    b[i] += (s[j++]-'0');
            }
        } else {
            b[i]  = (s[j++]-'0') * 1000;
            b[i] += (s[j++]-'0') * 100;
            b[i] += (s[j++]-'0') * 10;
            b[i] += (s[j++]-'0');
        }
    }

    return b;
}
```

Fig. 5.1 Big integer input in C. This routine takes a string representing a big integer in decimal and converts it to a base 10,000 number. The required standard C libraries are also listed

Figure 5.2 takes a pointer to a big integer in `b` and a pointer to an output string buffer in `s`. The big integer is then put into `s` as an ASCII string. This conversion is simpler than going the other way as in Fig. 5.1. If the first element of the big integer is negative the number is negative so we output the sign in line 8 bumping the output character index, `j`. Lines 10 through 16 pass over the big integer, digit by digit, and convert each digit to a set of up to four ASCII characters. To save space we use a temporary string, `t`, and the C library function `sprintf` which prints into a string. This changes the `i`-th big integer digit into a four character string. This

```
1  #include <stdio.h>
2
3  void bigint_output(signed int *b, char *s) {
4      int i, j=0;
5      char t[5];
6
7      if (b[0] < 0)
8          s[j++] = '-';
9
10     for(i=1; i <= abs(b[0]); i++) {
11         sprintf(t, "%04d", b[i]);
12         if (!((t[0] == '0') && (i == 1))) s[j++] = t[0];
13         if (!((t[1] == '0') && (i == 1))) s[j++] = t[1];
14         if (!((t[2] == '0') && (i == 1))) s[j++] = t[2];
15         s[j++] = t[3];
16     }
17
18     s[j] = '\0';
19 }
```

Fig. 5.2 Big integer output in C. This routine takes a base 10,000 big integer and outputs it as a decimal string

string, because of the `%04d` format statement may have leading zeros. If `i` is one, not zero because we skip the sign/length element, then we do not want leading zeros. This is the origin of the somewhat cryptic code in lines 12 through 14. Each of these `if` statements asks if we are working with the highest order digit, when `i` is one, and if that digit is zero. It then negates this to execute when we are not working with the highest order digit and the digit is zero. In that case, we want to copy the proper character from the temporary string, `t`, to the output string, `s`. Since we always have at least one character in the digit we are working with line 15 has no `if` statement, it is always executed. When the loop in lines 10 through 16 completes we have filled in all of the output string, character by character, with the decimal representation of the input big integer. All that remains is to add a null at the end of the output to make a valid C-style string out of it (line 18). With this the conversion is complete.

Comparisons Before we jump into arithmetic routines for big integers we need to take a quick look at comparisons using big integers.

Figure 5.3 shows basic big integer comparison in C. The function `bigint_compare` takes two big integer arguments, a and b, and returns -1 if $a < b$, 0 if $a = b$, and +1 if $a > b$. Let's take a look at what it is doing.

Lines 4 and 5 do a quick comparison to see if the signs of the numbers are different. If they are we already know the proper value to return and we do not need to look at the actual data values. Lines 7 through 10 do similar quick tests. We already know that the signs of a and b are the same so we look to see which has a larger number of digits. In line 7 we look to see if a has more digits than b and that they are both positive. If so, we know that a is larger so we return 1. If both are

```
1 int bigint_compare(signed int *a, signed int *b) {
2     signed int i;
3 
4     if ((a[0] > 0) && (b[0] < 0)) return  1;
5     if ((a[0] < 0) && (b[0] > 0)) return -1;
6 
7     if ((abs(a[0]) > abs(b[0])) && (a[0] > 0)) return  1;
8     if ((abs(b[0]) > abs(a[0])) && (a[0] > 0)) return -1;
9     if ((abs(a[0]) > abs(b[0])) && (a[0] < 0)) return -1;
10    if ((abs(b[0]) > abs(a[0])) && (a[0] < 0)) return  1;
11
12    if (a[0] > 0) {
13        for(i=1; i <= abs(a[0]); i++) {
14            if (a[i] > b[i]) return 1;
15            if (b[i] > a[i]) return -1;
16        }
17    } else {
18        for(i=1; i <= abs(a[0]); i++) {
19            if (a[i] > b[i]) return -1;
20            if (b[i] > a[i]) return  1;
21        }
22    }
23
24    return 0;
25 }
```

Fig. 5.3 Big integer comparison. The function compares two big integers, a and b, and returns -1 if $a < b$, 0 if $a = b$, and +1 if $a > b$. With this basic comparison function it is straightforward to define <, >, =, <=, >= and ! = via small wrapper functions

positive and b has more digits, the case in line 8, we return -1 because that means that a is the smaller value and we are determining the relationship between a and b (not b and a). Lines 9 and 10 do the same comparison but for both a and b negative. In this case, the return values are flipped since if a has more digits than b it must be larger in magnitude and negative making it smaller than b.

Line 12 looks at the sign of a. At this point we know a and b have the same sign and same number of digits. We must therefore look digit by digit, largest power of 10,000 first, to see where they might be different. This tells us which of the two is larger. The loop in lines 13 through 16 moves through a and b and uses digit values to decide if one is larger than the other. The loop in lines 19 through 21 does the same but in this case we know that a and b are both negative and have the same number of digits so the return values are reversed. Lastly, if we reach line 24 we know that the numbers have the same sign, same number of digits, and same digit values so we know that they must be equal and we return 0.

The function in Fig. 5.3 makes it easy to define wrapper functions to implement the actual comparison operations. If we were using C++ we would define a big integer class and overload <, >, =, etc to implement the comparisons. As we are using plain C we instead implement a function library. A possible definition for the equal function is,

```
int bigint_equal(signed int *a, signed int *b) {
    return bigint_compare(a,b) == 0;
}
```

We leave the definition of the remaining operators as an exercise.

Now that we know how to store big integers, how to read them from strings and how to write them to strings, and how to compare them, we are ready to look at basic arithmetic with big integers.

5.3 Arithmetic with Big Integers

Basic arithmetic with big integers generally follows the school book methods though as we will see there are some alternatives for multiplication that are effective if the number of digits is very large. We will continue with our implementation of big integers as above and develop routines for addition, subtraction, multiplication and division. We will not implement C versions of the alternative multiplication routines but we will discuss the operation of the algorithms. Note that the routines given here are bare-bones only. We are omitting necessary checks for memory allocation, etc., in order to make the routines as concise and easy to follow as possible. We proceed then with the understanding that a true big integer package would have all of these checks in place.

Addition and Subtraction Before we look at implementing addition and subtraction let's think a bit about how we will handle negative numbers. We are using a sign-magnitude representation of our big integers because using a complement notation might mean manipulating all the digits of the number when we can be more clever about handling operations with regard to the signs. For addition we have the following situations in Table 5.1 where we always work with the absolute value of the integers a and b. Similarly for subtraction we have Table 5.2 where it becomes apparent that for addition and subtraction we need the ability to compare the absolute value of two big integers, the ability to negate a big integer, and the ability to add and subtract two positive big integers. Therefore, to properly implement addition and subtraction we need two helper functions: `bigint_ucompare` to compare the absolute value of two big integers, and `bigint_negate` to negate a big integer. These are given in Fig. 5.4.

These functions are straightforward. Indeed, `bigint_negate` is about as simple as can be given the efficient way we have selected to represent our big integers in memory. To change the sign we need only negate the first element of the array. In addition we return a reference to the array so that we can compose the negate call with other big integer operations. The function `bigint_ucompare` is a reduced version of `bigint_compare` as given in Fig. 5.3 where we do not pay attention to

Table 5.1 Steps to add two big integers paying attention to their signs

Sign of a	Sign of b	Steps to find $a + b$
+	+	Add $\|a\|$ and $\|b\|$.
+	−	$\|a\| > \|b\|$, subtract $\|b\|$ from $\|a\|$.
		$\|a\| < \|b\|$, subtract $\|a\|$ from $\|b\|$. Negate answer.
−	+	$\|a\| > \|b\|$, subtract $\|b\|$ from $\|a\|$. Negate answer.
		$\|a\| < \|b\|$, subtract $\|a\|$ from $\|b\|$.
−	−	Add $\|a\|$ and $\|b\|$. Negate answer.

Table 5.2 Steps to subtract two big integers paying attention to their signs

Sign of a	Sign of b	Steps to find $a - b$
+	+	$\|a\| \geq \|b\|$, subtract $\|b\|$ from $\|a\|$.
		$\|a\| < \|b\|$, subtract $\|a\|$ from $\|b\|$. Negate answer.
+	−	Add $\|a\| + \|b\|$.
−	+	Add $\|a\| + \|b\|$. Negate answer.
−	−	$\|a\| \leq \|b\|$, subtract $\|a\|$ from $\|b\|$.
		$\|a\| > \|b\|$, subtract $\|b\|$ from $\|a\|$. Negate answer.

```
signed int *bigint_negate(signed int *a) {
    a[0] = -a[0];
    return a;
}

int bigint_ucompare(signed int *a, signed int *b) {
    signed int i;

    if (abs(a[0]) < abs(b[0])) return -1;
    if (abs(a[0]) > abs(b[0])) return 1;

    for(i=1; i <= abs(a[0]); i++) {
        if (a[i] > b[i]) return 1;
        if (b[i] > a[i]) return -1;
    }

    return 0;
}
```

Fig. 5.4 Big integer helper functions to compare the absolute value of two big integers, `bigint_ucompare`, and to negate a big integer, `bigint_negate`

the sign. It returns -1 if $|a| < |b|$, 0 if $|a| = |b|$, and +1 if $|a| > |b|$. With these in hand, we can implement addition and subtraction using low-level functions that ignore the sign of their arguments. After that, we will use a higher level version of addition and subtraction that does look at the sign of the arguments to decide how to proceed.

```
1 signed int *bigint_uadd(signed int *a, signed int *b) {
2     signed int *c, *x, *y, *t;
3     int i, m, n, yy, carry=0;
4
5     if (abs(a[0]) > abs(b[0])) {
6         n = abs(a[0]);  m = abs(b[0]);
7         x = a;          y = b;
8     } else {
9         n = abs(b[0]);  m = abs(a[0]);
10        x = b;          y = a;
11    }
12
13    c = (signed int *)malloc((n+1)*sizeof(signed int));
14    c[0] = n;
15
16    for(i=n; i>0; i--) {
17        yy = (i>(n-m)) ? y[i-(n-m)] : 0;
18        c[i] = x[i] + yy + carry;
19        if (c[i] > 9999) {
20            carry = 1;
21            c[i] -= 10000;
22        } else
23            carry = 0;
24    }
25
26    if (carry) {
27        t = (signed int *)malloc((n+2)*sizeof(signed int));
28        for(i=1; i<=n; i++)  t[i+1] = c[i];
29        t[0] = n+1;
30        t[1] = 1;
31        free(c);
32        c = t;
33    }
34    return c;
35 }
```

Fig. 5.5 Low-level big integer addition using the school method

Our low-level addition routine will pay no attention to the sign of the arguments. It will assume they are positive. This routine implements addition using the simple method we learned in school. We add digits, from right to left, carrying on every 10,000. We will add a new first digit to the output, if necessary, based on any carry in the top digit position. This is directly analogous to addition in any other base. We call the routine `bigint_uadd` and it is given in Fig. 5.5.

Recall that we will be using a high-level driver routine to decide when to call the low-level add and subtract routines. Because of this, `bigint_uadd` will ignore the sign of its input and treat all arguments as positive. Lines 5 through 11 set up the addition. We use extra pointers, `x`, and `y`, and assign them the arguments `a` or `b` so that x will always point to the larger number (line 5). We also use n and m to hold the number of digits in these numbers. We do this to align the digits properly when we move from right to left across the number adding as we go. Since y will always have as many or fewer digits as x we can use this fact in the addition to substitute zero for digits that do not exist in y. Line 13 allocates memory for the addition while line 14 sets the length of the output. We know that when adding two big integers it is possible that we overflow in the most significant digit. If that happens, we adjust for it later in the routine and initially assume that the answer will fit in n digits which is the number of digits in the larger of a or b. Lines 16 through 24 perform the actual addition, digit by digit, from the least significant digit to the most significant. We use yy to hold the digit value of the lesser of a and b (in terms of number of digits). Line 17 either returns the actual digit of y or zero if we are looking at digits of x that have no match in y. Line 18 sets the output digit in c. We add the current digit of x and y and any carry that came over from the previous digit addition. Note that we set carry to zero in line 3 when it was declared.

After the addition of the digits we need to check if we caused a carry. We are using base 10,000 so we know that any two digits added together will still fit in a single signed int. We simply need to see if the result is larger than 9999. If so, we subtract the base and set the carry for the next digit. Otherwise, we ensure that the carry is clear (line 23).

The addition loop continues until all the digits of x have been examined. At the end of this we have the answer in c but we need to do one more check. We need to check for overflow of the most significant digit of c. This is line 26. If the carry is set from the last digit addition we have overflowed the n digits we allotted to c and need to add one more most significant digit. Lines 27 through 33 do this by creating a new output array in t with n+1 digits. Line 28 copies the digits from c to t leaving room for the new most significant digit. Lines 29 and 30 set the number of digits in t and set the most significant digit to 1 since that is the carry. Lines 31 and 32 release the old memory used by c and point c to t. Line 34 returns the pointer to the new big integer which holds the sum of a and b. This addition routine can, temporarily, use double the memory because of the copy when c overflows. For our present purposes this inefficiency is tolerable and will seldom happen in practice (how often? we don't really know, but we believe it will be infrequent).

In a similar fashion, we implement the low-level subtraction routine, bigint_usub, in Fig. 5.6. This routine, like bigint_uadd, ignores the sign of its inputs. It also assumes that a is always larger, in terms of magnitude, than b. This means that we will never underflow and end up with a negative answer. Our high-level driver routine will arrange things so that this assumption is always true. Lines 5 through 8 get the number of digits of our arguments in n and m and allocate room for our answer in c which will always need n+1 digits. We set c[0] to the number of digits.

Lines 10 through 19 perform the subtraction, from least significant digit to most significant using the same trick we used in bigint_uadd to get the digit of the smaller number in y, which may be zero if we are looking at digits b does not have. Line 12 does the subtraction for the current digit and puts it in c. Note that we

```
 1 signed int *bigint_usub(signed int *a, signed int *b) {
 2     signed int *c, *t;
 3     int i, m, n, y, borrow=0, zero=0;
 4 
 5     n = abs(a[0]);
 6     m = abs(b[0]);
 7     c = (signed int *)malloc((n+1)*sizeof(signed int));
 8     c[0] = n;
 9 
10     for(i=n; i>0; i--) {
11         y = ((i-(n-m)) > 0) ? b[i-(n-m)] : 0;
12         c[i] = a[i] - y - borrow;
13         if (c[i] < 0) {
14             borrow = 1;
15             c[i] += 10000;
16         } else
17             borrow = 0;
18         zero += c[i];
19     }
20 
21     if ((c[1] == 0) && (zero != 0)) {
22         t = (signed int *)malloc(n*sizeof(signed int));
23         for(i=2; i<=n; i++)  t[i-1] = c[i];
24         t[0] = n-1;
25         free(c);
26         c = t;
27     }
28 
29     if (zero == 0) {
30         free(c);
31         c = (signed int *)malloc(2*sizeof(signed int));
32         c[0] = 1;
33         c[1] = 0;
34     }
35 
36     return c;
37 }
```

Fig. 5.6 Low-level big integer subtraction using the school method

subtract any borrow that happened in the previous digit subtraction. We set `borrow` to zero when it is defined in line 3. How do we know if we borrowed? If the digit answer is negative, we need to borrow a base value (10,000) from the next higher digit. The check for this is in line 13. If true, `borrow` is set to 1 (we only ever need to borrow 10,000) and we add the borrow into the current digit value to bring it positive. If no borrow is needed, we set `borrow` to zero and continue to the next most significant digit. Line 18 appears to be curious. It adds the current digit value to a single variable we are calling `zero`. This is set to zero when defined in line 3. We use this variable to decide if the result of the subtraction is exactly zero. This

```
1 signed int *bigint_add(signed int *a, signed int *b) {
2
3     if ((a[0]>0) && (b[0]>0))
4         return bigint_uadd(a,b);
5
6     if ((a[0]<0) && (b[0]<0))
7         return bigint_negate(bigint_uadd(a,b));
8
9     if ((a[0]>0) && (b[0]<0)) {
10        if (bigint_ucompare(a,b) == -1)
11            return bigint_negate(bigint_usub(b,a));
12        else
13            return bigint_usub(a,b);
14    }
15
16    if ((a[0]<0) && (b[0]>0)) {
17        if (bigint_ucompare(a,b) == 1)
18            return bigint_negate(bigint_usub(a,b));
19        else
20            return bigint_usub(b,a);
21    }
22 }
```

Fig. 5.7 High-level big integer addition

allows us to simplify the return value, `c`, in lines 29 through 34, if possible. If we are subtracting two large integers and they happen to be exactly the same value the result, without this check, would use as much memory as the arguments themselves with every digit set to zero. This check lets us save that memory and return a very simple result that is only one digit long. The new result releases the memory used by the initial result and sets up a single digit big integer that is exactly zero instead. We also check to see if the subtraction has left us with a leading zero. This is purpose of lines 21 through 27. If the leading digit is zero, but the entire value is not zero, the leading digit is removed by allocating a new value in line 22 and copying the digits from `c` to this value ignoring the initial zero. Once the loop in lines 10 through 19 ends the answer is in `c` and we return it in line 36.

Our final `bigint_add` and `bigint_sub` routines look at the signs of their arguments and proceed to call the low-level versions of add and subtract as appropriate. Figure 5.7 shows `bigint_add` while Fig. 5.8 shows `bigint_sub`.

Our high-level addition routine examines the signs and magnitudes of the arguments to determine how to get the proper result. The checks in the code are directly analogous to those of Tables 5.1 and 5.2. For addition, if the arguments are both positive we simply add (line 3). If the arguments are both negative, we add and make the answer negative (line 6). If the signs are mixed we need to consider the relationship between the magnitudes of the numbers. If `a` is positive and `b` is negative (line 9) we see two cases: $|a| < |b|$ which is handled in line 11 and $|a| \geq |b|$ which is handled in line 13. Similarly, for `a` negative and `b` positive (line 16) we consider cases when $|a| > |b|$ in line 17 and when $|a| \leq |b|$ in line 20.

```
1  signed int *bigint_sub(signed int *a, signed int *b) {
2
3      if ((a[0]>0) && (b[0]>0)) {
4          if (bigint_ucompare(a,b) == -1)
5              return bigint_negate(bigint_usub(b,a));
6          else
7              return bigint_usub(a,b);
8      }
9
10     if ((a[0]<0) && (b[0]<0)) {
11         if (bigint_ucompare(a,b) == 1)
12             return bigint_negate(bigint_usub(a,b));
13         else
14             return bigint_usub(b,a);
15     }
16
17     if ((a[0]>0) && (b[0]<0))
18         return bigint_uadd(a,b);
19
20     if ((a[0]<0) && (b[0]>0))
21         return bigint_negate(bigint_uadd(a,b));
22 }
```

Fig. 5.8 High-level big integer subtraction

For high-level subtraction in Fig. 5.8 we follow Table 5.2. If both `a` and `b` are positive we look at $|a| < |b|$ in line 4 or otherwise in line 7. Similarly, if both are negative we look at $|a| > |b|$ in line 11 and otherwise in line 14. The simpler cases here are if `a` is positive and `b` is negative, line 17, or `a` is negative and `b` is positive, line 20. The signs of the arguments will fit one of these cases for both addition and subtraction so we know that all possibilities have been considered. Note that for subtraction we always ensure that the first argument to `bigint_usub` is the same size or larger than the second.

Multiplication Big integer multiplication can be implemented using the school method without much difficulty. We do that in Fig. 5.9 where we pay no attention to the signs of the arguments. Figure 5.10 is our high-level routine that manages the signs. Let's look at Fig. 5.10 first. The rules of multiplication say that when the signs of the multiplicand and multiplier are the same, either both positive or negative, that the answer is positive. This check is made in line 3. If the signs are the same, we simply call `bigint_umult` as that will always give a positive result. If the signs are different, we drop to line 7 and negate the result since the result should be negative.

Figure 5.9 is really three separate functions all of which are used to multiply the numbers stored in `a` and `b`. Lines 1 through 7 define a simple helper function that duplicates an existing big integer. It allocates space for the new integer and copies all the bytes of `a` to `b` by calling the C library function, `memcpy`. It then returns the reference to this new big integer.

```
signed int *bigint_copy(signed int *a) {
    signed int *b;

    b = (signed int *)malloc((abs(a[0])+1)*sizeof(signed int));
    memcpy((void *)b, (void *)a, (abs(a[0])+1)*sizeof(signed int));
    return b;
}

signed int *bigint_umultd(signed int *a, signed int d, unsigned int s) {
    signed int *p, carry=0, n;
    int i;

    n = abs(a[0]) + s + 2;
    p = (signed int *)malloc(n*sizeof(signed int));
    memset((void *)p, 0, n*sizeof(signed int));
    p[0] = n-1;

    for(i=abs(a[0]); i>=1; i--) {
        p[i+1] = a[i] * d + carry;
        if (p[i+1] > 9999) {
            carry = p[i+1] / 10000;
            p[i+1] = p[i+1] % 10000;
        } else
            carry = 0;
    }

    if (carry) p[1] = carry;
    return p;
}

signed int *bigint_umult(signed int *a, signed int *b) {
    int i, n;
    signed int *m, *x, *y;
    signed int *p=(signed int *)NULL, *q=(signed int *)NULL;

    if (abs(a[0]) > abs(b[0])) {
        x = a;  y = b;
    } else {
        x = b;  y = a;
    }

    m = (signed int *)malloc(2*sizeof(signed int));
    m[0] = 1;  m[1] = 0;

    for(i=abs(y[0]); i>=1; i--) {
        if (p != NULL) free(p);
        if (q != NULL) free(q);
        p = bigint_umultd(x, y[i], abs(y[0])-i);
        q = bigint_uadd(m, p);
        if (m != NULL) free(m);
        m = bigint_copy(q);
    }

    if (p != NULL) free(p);
    if (q != NULL) free(q);
    return m;
}
```

Fig. 5.9 Low-level big integer multiplication using the school method

```
signed int *bigint_mult(signed int *a, signed int *b) {

    if (((a[0]>0) && (b[0]>0)) ||
        ((a[0]<0) && (b[0]<0)))
        return bigint_umult(a,b);

    return bigint_negate(bigint_umult(a,b));
}
```

Fig. 5.10 High-level big integer multiplication

Lines 9 through 29 define `bigint_umultd` which multiplies a big integer in `a` by a single digit, `d`. When we multiply two numbers using the school method we accumulate partial products, shifting them for the power of the base we are currently multiplying by. We then add these together to get the final result. For example,

$$
\begin{array}{rrl}
 & 1776 & \leftarrow \textit{multiplicand} \\
\times & 1492 & \leftarrow \textit{multiplier} \\
\hline
 & 3552 & \leftarrow \textit{1st partial product} \\
 & 159{,}840 & \\
 & 710{,}400 & \\
+ & 1776{,}000 & \leftarrow \textit{4th partial product} \\
\hline
 & 2649{,}792 &
\end{array}
$$

We can get the same result by starting with a product of zero and adding each digit times the multiplicand after shifting to the left as many places as the location of the digit in the multiplier. This is exactly what we are doing when we write and then sum the partial products. This same approach works regardless of base. In `bigint_umultd` we do the same thing. The multiplicand is in `a`, the digit to multiply by is in `d` and the number of places to shift the result to the left is in `s`. In Line 13 of Fig. 5.9 `n` is set to the number of output digits. We need to have room for all the digits in `a`, which we get with `abs(a[0])`. To this we add `s`, these are extra digits at the end of the number that we will initialize to zero. This takes the place of the zeros added to the end of the partial products in the example above. We need to add one more array element to hold the sign and length of the number and to this we add still one more digit to handle any overflow (carry in the last digit multiplied). This is the origin of the + 2 in line 13. Line 14 allocates the product and points `p` to it. Line 15 erases the entire number and sets it to zero while line 16 sets the length of the number. This initializes the partial product.

Lines 18 through 25 do the multiplication by the digit, `d`. We look at all the digits of the multiplicand `a` from least significant to most significant. This defines the loop in line 18. Line 19 does the multiplication of the digit of `a` by `d`. There are two things to note in this line. First, we add in `carry`, just as we did for addition, but in this case the carry may be more than one. Second, we assign this result to `p[i+1]` and not p[i]. This is done to leave a leading zero in p to handle any final overflow. Recall that array elements further from p[0] are less significant. Lines 20 and 24 handle any carry. If the resulting multiplication is greater than 9999 we have a carry. Unlike addition, the carry is not simply one but is the value of p[i+1] divided by the base. The actual value that should be in p[i+1] is the remainder when dividing by the base. If the carry has not happened, line 24 clears carry.

When the loop exits it is possible that carry is not zero. Line 27 checks for this and sets the leading zero already allocated to the carry. Line 28 returns the partial product. A full implementation would do something about the leading zero in p, but we ignore it here.

The function that does the actual unsigned multiplication is bigint_umult. The result of the multiplication will be in m which we initialize to zero in lines 42 and 43 of Fig. 5.9. Before this, lines 36 through 40 set pointers x and y so that x always points to the input with the largest number of digits, which may be the same as the other input. We do this because we need fewer passes through the loop in line 45 if we make the multiplier the value with the least number of digits.

The loop in lines 45 through 52 accumulates the partial products of the digits of y (the shorter of a and b) with x. This code is inefficient because it requires copying and free-ing memory more than is necessary but this makes the algorithm easier to follow. We use p and q to point to temporary results. In this case, p points to the partial product returned by the call to bigint_umultd. This call multiplies x by the current digit of y and shifts the result abs(y[0])-i places to the left. As we move from right to left over the digits of y we need to shift each partial product the current digit position to the left and this is exactly abs(y[0])-i. When this call finishes the partial product is in p and it must be added to our accumulated product. We reuse bigint_uadd to do this and store the new running product value in q. We really want this result in m so we release memory used by m and call bigint_copy to copy q to m. This is necessary so that the next pass through the loop can update q. This is a consequence of using plain C and not a language with automatic garbage collection. At the end of the loop we have the final product in m. To avoid leaking memory we must free p and q at the end of the loop. Finally, we return the product in line 56.

Clearly this implementation is pedagogical as there are more efficient ways to implement the multiplication routine. The school approach is well known to be of order n^2 which is denoted $O(n^2)$. Below we will discuss alternate big integer multiplication routines that do better than this. If a routines runs in $O(n^2)$ time it means that as n, here the number of the digits, increases the running time increases as the square of the number of digits (with an unstated multiplier that is generally ignored). An $O(n^2)$ routine may be perfectly acceptable for small inputs (relatively few digits) but quickly becomes unusable when the number of digits becomes large.

Division Division of big integers is rather difficult to implement quickly. The school algorithm for division is typically implemented but for the purposes of this book even that is too long. So, in order to provide a minimalist but complete package for big integer arithmetic we will instead implement the simplest approach possible. Fear not, however, we will discuss more comprehensive algorithms for division, along with multiplication, in the next section.

So, how to implement division with big integers? Thinking back to the simplest possible ways of working with numbers we remember that division of two integers is really counting the number of times the divisor can be subtracted from the dividend before we hit zero or go below zero if the division would leave a remainder. In this case, we ignore the remainder and leave implementing it as an exercise.

As with multiplication, we implement division as a high-level function that handles the signs of the numbers and a low-level function that deals with the magnitude only. Figure 5.11 shows our "division by repeated subtraction" approach. We divide `a` by `b`, ignoring signs, by counting the number of times we can subtract `b` from `a`, updating `a` in the process, before we hit zero or go negative. This is the quotient which we store in `q`. Lines 4 through 6 initialize the quotient and set it to zero. Likewise, lines 8 through 10 initialize a constant value of `1` which is added to `q` each time through the loop.

Line 12 copies `a` so we can subtract from it without altering it in memory. The loop of lines 14 through 26 do the work. We need to manage memory carefully to avoid leaks so we use auxiliary pointers `x` and `y` to hold intermediate results. Line 17 subtracts the divisor in `b` from `t` which is the portion of `a` remaining. Line 18 bumps the quotient counter by adding the constant `1` in `c`. Lines 21 and 22 copy the partial results of the loop to `t` and `q`. We do this so we can free memory before we lose the pointer address. Languages with automatic garbage collection can avoid this sort of inelegance. Line 24 checks to see if we have reached a point where the remainder in `t` is less than `b`, the divisor. If so, we are done and break out of the loop. After cleaning up memory the quotient is returned.

It is well-known that this approach to division, while correct, is terribly inefficient, especially for fixed length integers. Be that as it may, the algorithm is straightforward and simple to understand. All we need do now is account for the signs of the dividend and divisor. This is done in Fig. 5.12 through the `bigint_div` routine.

In order to handle the signs of `a` and `b` it is necessary to check several conditions. Along the way we check for situations that result in invalid outputs for the case of integer division or division by zero. Lines 5 through 11 do this. If the divisor is zero (`bigint_iszero(b) == 0`) or the dividend is less than the divisor, we return a constant value of zero. For the second case this is correct but division by zero should instead signal an error. For clarity, we ignore the error and return the incorrect value of zero instead. This would be corrected if implementing a full big integer package.

We extract the signs of the arguments in lines 13 and 14. We then ensure that they are positive so we can pass the values to `bigint_udiv`. We will reset the signs of the arguments when done, see lines 25 and 26. We know from school that if the signs of `a` and `b` are the same, either both positive or both negative, that the quotient must be positive. So, if this is the case, we execute line 20 to perform the division. If the signs of the arguments are opposite, either `a` positive and `b` negative or vice versa, we must make the answer negative as well. In this case, line 22 is what gets executed. After restoring the proper signs of the arguments we return the quotient in line 28.

```
signed int *bigint_udiv(signed int *a, signed int *b) {
    signed int *q, *t, *c, *x=NULL, *y=NULL;

    q = (signed int *)malloc(2*sizeof(signed int));
    q[0] = 1;
    q[1] = 0;

    c = (signed int *)malloc(2*sizeof(signed int));
    c[0] = 1;
    c[1] = 1;

    t = bigint_copy(a);

    while (1) {
        if (x != NULL) free(x);
        if (y != NULL) free(y);
        x = bigint_usub(t,b);
        y = bigint_uadd(q,c);
        free(t);
        free(q);
        t = bigint_copy(x);
        q = bigint_copy(y);

        if (bigint_ucompare(t,b) < 1)
            break;
    }

    free(x);
    free(y);
    free(t);
    free(c);
    return q;
}
```

Fig. 5.11 Low-level big integer division

Division by repeated subtraction is unsatisfying in the long run and can be incredibly inefficient. However, it completes our implementation of basic arithmetic for big integers and therefore is worth looking at even if it is not how things would be done if we were developing a big integer package for wide-spread use.

The algorithms we implemented above are adequate for simple purposes but, especially for multiplication and division, people have certainly done far better. In the next section we look at some of these improved algorithms without implementing them. This will give an appreciation for what goes into efficient algorithm implementation and design and will give us a background we will use when discussing particular big integer libraries that you may someday wish to use in your own programs.

```
signed int *bigint_divide(signed int *a, signed int *b) {
    int sa, sb;
    signed int *c;

    if ((bigint_iszero(b)) ||
        (bigint_ucompare(a,b) == -1)) {
        c = (signed int *)malloc(2*sizeof(signed int));
        c[0] = 1;
        c[1] = 0;
        return c;
    }

    sa = (a[0] < 0) ? -1 : 1;
    sb = (b[0] < 0) ? -1 : 1;
    a[0] = abs(a[0]);
    b[0] = abs(b[0]);

    if (((sa<0) && (sb<0)) ||
        ((sa>0) && (sb>0))) {
        c = bigint_udiv(a,b);
    } else {
        c = bigint_negate(bigint_udiv(a,b));
    }

    a[0] = sa*a[0];
    b[0] = sa*b[0];

    return c;
}
```

Fig. 5.12 High-level big integer division

5.4 Alternative Multiplication and Division Routines

In this section we take a look at some alternate algorithms for multiplication and division. A true big integer package would implement some if not all of these alternatives. By necessity we are cherry-picking algorithms. A particularly nice treatment of big integer algorithms is found in Chap. 9 of Crandall and Pomerance [1] and readers interested in a more thorough mathematical treatment are directed there.

Multiplication In the previous section we implemented the school method for multiplication of big integers. As we noted, this method runs $O(n^2)$ in the size of the inputs. While this is adequate for numbers with perhaps a thousand digits, it becomes intractable if we are working with very large numbers with tens of thousands, hundreds of thousands or even millions of digits. To work with truly large big integers we need to improve our multiplication algorithm. Fortunately, this problem has been investigated and a number of alternate multiplication algorithms exist. Here we will look at three of them: Comba multiplication, Karatsuba multiplication and FFT multiplication. Each has an advantage over the simple school method and with respect to each other.

Our first possible improvement to multiplication is the Comba algorithm [2]. Comba noted that the school method of multiplication is geared towards humans and not computers. After each single digit multiplication of a column we carry to the next column. In Comba multiplication we instead do not carry but store the entire product for each column position. We then repeat for the next digit to multiply adding in any existing value in that column (from a previous digit multiplication). When all single digit multiplications have been performed the resulting vector of column values is adjusted from left to right by adding any carry to the next digit to the left and leaving the remainder in the column. When this is done the vector will contain the output value. This process is described in Fig. 5.13 where we work in base 10 for clarity. If we were implementing this algorithm for our big integer representation we would be using base 10,000 instead.

In (a) we work out the multiplication using the school method but instead of carrying after each multiplier digit multiplication we keep the value in the proper column. When each digit of the multiplier has been used we sum the results to get an intermediate answer that still needs to be adjusted for the carries we did not perform in the previous steps. The last step is to move from right to left replacing each column value, n, with $n\%10$, and adding to the next digit the result of $n/10$

			2	3	7	
×			1	4	8	
			16	24	56	← *237×8*
		8	12	28		← *237×4*
+	2	3	7			← *237×1*
	2	11	35	52	56	← *column sums*
	3	5	0	7	6	← *carries, right to left*

			2	3	7	
×			1	4	8	
			16	24	56	← *column 1*
		8	28	52	56	← *column 2*
	2	11	35	52	56	← *column 3*
	3	5	0	7	6	← *carries, right to left*

Fig. 5.13 Comba column-wise multiplication of 237 × 148 in base 10. **(a)** Writing out the multiplication by hand keeping each column without carries to the next. After all digits have been multiplied we add the columns. Finally, we perform the carries from right to left. **(b)** The same but as would be implemented in a computer. In this case, a single output number stores the columns adding the new value for that column to the previous. Once all the digits of the multiplier have been visited the output is adjusted by carrying from right to left to achieve the same answer as in **(a)**

using integer division. This handles all the carries we did not do in the lines above. In (b) we do the same as in (a) but each time we multiply a digit we immediately add the result to the proper column of the single output value. Then, when done, we handle the carries.

Comba multiplication is generally believed to be about 30% more efficient than the school method. However, there are some things to keep in mind before using it. Depending upon the way digits of the big integer are stored it is possible to overflow a particular digit value because a column may need to sum the single digit multiplication for n digits where n is the minimum number of digits in the multiplicand and multiplier. For that reason, Comba multiplication is limited in the number of digits in the product and is not suitable for multiplying truly large numbers. Additionally, while a 30% improvement is helpful, the algorithm is still essentially the school method and therefore is still $O(n^2)$. We can do better than this.

Karatsuba multiplication [3] improves on the school method by simplifying the multiplication. The essential insight involves simplifying the multiplication of two large integers, a and b, by replacing the multiplication with three multiplications of numbers with about half as many digits as a and b. If we choose m, $m < n$, where n is the number of digits, we can rewrite a and b as,

$$\begin{aligned} a &= a_1 B^m + a_0 \\ b &= b_1 B^m + b_0 \end{aligned}$$

where a_0, a_1, b_0, and b_1 are all less than B^m where B is the base of our numbers, for example, $B = 10{,}000$. If we then calculate ab we get,

$$ab = (a_1 B^m + a_0)(b_1 B^m + b_0) = a_1 b_1 B^{2m} + (a_1 b_0 + a_0 b_1) B^m + a_0 b_0$$

which we can rewrite as,

$$\alpha B^{2m} + \beta B^m + \gamma$$

with,

$$\begin{aligned} \alpha &= a_1 b_1 \\ \beta &= a_1 b_0 + a_0 b_1 \\ \gamma &= a_0 b_0 \end{aligned}$$

This is seemingly no improvement. Instead of the single multiplication ab we now have *four* multiplications to get α, β and γ, but we press on anyway and will soon see the benefit. We can remove one of the four multiplications by rewriting our expressions. First we look at $(a_1 + a_0)(b_1 + b_0)$,

$$\begin{aligned} (a_1 + a_0)(b_1 + b_0) &= a_1 b_1 + a_1 b_0 + a_0 b_1 + a_0 b_0 \\ &= (a_1 b_0 + a_0 b_1) + a_1 b_1 + a_0 b_0 \end{aligned}$$

where the first term on the right is β. This means we can determine β with one less multiplication,

$$\begin{aligned} a_1 b_0 + a_0 b_1 &= (a_1 + a_0)(b_1 + b_0) - a_1 b_1 - a_0 b_0 \\ \beta &= (a_1 + a_0)(b_1 + b_0) - \alpha - \gamma \end{aligned}$$

One thing to note is that $(a_1 + a_0)(b_1 + b_0)$ can itself be calculated using Karatsuba multiplication meaning this algorithm is recursive. This is one of its strong suits.

So, where is the benefit? It comes from the fact that the three multiplications we have replaced *ab* with are numbers with roughly half the number of digits as *a* or *b*. While a formal complexity analysis is beyond the scope of this book we see that the additions and subtractions are linear, $O(n)$, and as n grows they become less important. The recursive nature makes the analysis a bit tricky but the final result is that the algorithm as a whole is $O(n^{lg(3)})$ where $lg(3)$ is the log base 2 of 3. Compare this to the $O(n^2)$ of the school method and as Fig. 5.14 shows, this difference quickly becomes important for large n.

If n is too small, say less than four digits, it is more efficient to use the school method to multiply *ab*. This means we can use the school method as the terminating condition of the recursive application of Karatsuba multiplication.

Our final multiplication algorithm is that of Schönhage and Strassen [4]. This algorithm is based deeply in abstract algebra and Fourier transform theory but

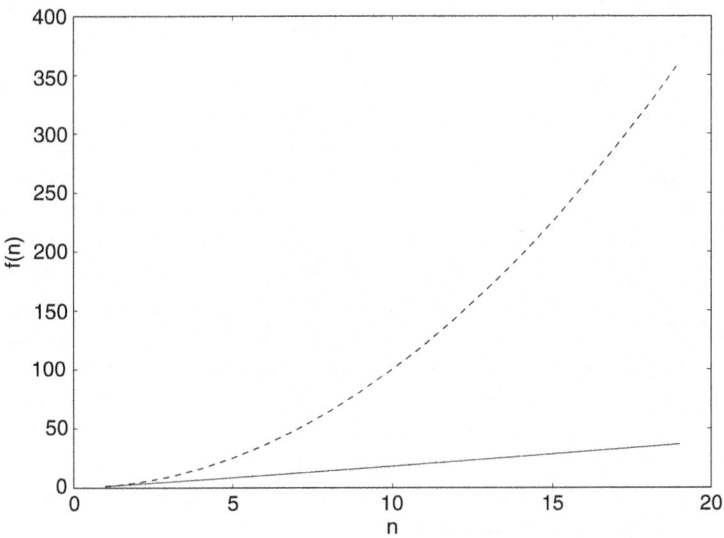

Fig. 5.14 $O(n^{lg(3)})$ (*solid blue line*) versus $O(n^2)$ (*dashed green line*) growth as n increases. This plot shows that the Karatsuba method is a significant improvement over the school method for suitable input lengths (n)

we will describe it at a high level. Here we discuss the variant presented as Algorithm 9.5.23 in [1].

Before diving into the algorithm itself we need to address a few concepts related to the operation of the algorithm which are its connection to abstract algebra and a bit about convolution, multiplication, and Fourier transforms.

The algorithm actually calculates multiplication modulo a specific integer, 2^n+1. This means that the result, which is an integer, is really a number from $\mathbb{Z}_{2^n+1}$. If you are not familiar with the notation, $\mathbb{Z}$ refers to the set of integers and $\mathbb{Z}_m$ refers to the set of integers modulo m. For example, $\mathbb{Z}_4 = 0, 1, 2, 3$ and operations that go above or below "wrap around" as needed so that in $\mathbb{Z}_4$ we have $2 + 3 = 1$. This is not a problem, really, since we select n so that $2^n + 1$ is larger than the product of our arguments. Call the arguments x and y so we select n such that $2^n + 1 > xy$. In this case, we never "wrap around" and we get the desired product.

In reality, the algorithm operates on a *ring* [5] defined in $\mathbb{Z}_{2^n+1}$ and uses only integer operations. A ring is an enhancement of a group which is itself a generalization of a binary operation acting on a set. In a ring, along with the first binary operation of the group, which is often denoted as $+$ even though it need not be addition, we add a second operation, $\bullet$, which need not really be multiplication. This is the abstract algebra connection. In our case, addition is addition, multiplication is multiplication and the set is $\mathbb{Z}_{2^n+1}$, which is a (possibly very large) set of integers. If we select n so that $n \geq \lceil \lg x \rceil + \lceil \lg y \rceil$ then $2^n + 1 > xy$ which is what we want so that the algorithm produces the actual product.

In the Comba algorithm for multiplication we saw that multiplying the multiplicand by a digit of the multiplier, especially when we collected the resulting value without carrying, was in essence doing convolution. A convolution is passing one value over another, so here we "pass" the single multiplier digit "over" the multiplicand digits, and collect the product. This means that multiplication of two numbers is related to convolution. This is where the Fourier connection comes from. We need not know anything about Fourier transforms at present beyond the fact that they map a sequence of values from one space to another space. We denote a Fourier transform of x, where x is a vector of values, by $\mathscr{F}(x)$ and recognize that this forward transform has an inverse transform, $\mathscr{F}^{-1}$, so that $\mathscr{F}^{-1}(\mathscr{F}(x)) = x$. In truth, the Fourier transform is such a powerful mathematical tool that all readers should be familiar with it. Please take a look at Bracewell [6] or a similar text to familiarize yourself with this elegant addition to the mathematical toolbox.

One of the many properties of the Fourier transform is that multiplication in one space is equivalent to convolution in another. This is a key hint in understanding what the Schönhage and Strassen algorithm is doing. It will calculate the convolution of x and y through the Fourier transform and multiplication when in Fourier space. In essence, the algorithm does,

$$\begin{array}{ll} X' = \mathscr{F}(X), Y' = \mathscr{F}(Y) & \leftarrow \text{map to Fourier space} \\ Z' = X'Y' & \leftarrow \text{calculate a product} \\ Z = \mathscr{F}^{-1}(Z') & \leftarrow \text{return to } X \text{ and } Y \text{ space} \end{array}$$

1. [Initialize]
   Choose an FFT size, $D = 2^k$ dividing $\boldsymbol{n}$.
   Set a recursion length, $n' \geq 2M + k$, $n = DM$ such that D divides n', $n' = DM'$.
2. [Decomposition]
   Split x and y into arrays A and B of length D such that $A = A_0, A_1, \ldots, A_{D-1}$ and $B = B_0, B_1, \ldots, B_{D-1}$ as residues modulo $2^{n'} + 1$.
3. [Prepare FFT]
   Weight A and B by $(2^{jM'} A_j) \bmod (2^{n'} + 1)$ and $(2^{jM'} B_j) \bmod (2^{n'} + 1)$ respectively for $j = 0, 1, \ldots, D-1$.
4. [Perform FFT]
   $A = \mathscr{F}(A)$ and $B = \mathscr{F}(B)$ done in-place.
5. [Dyadic stage]
   $A_j = A_j B_j \bmod (2^{n'} + 1)$ for $j = 0, 1, \ldots, D-1$. Storing product in Fourier space in A.
6. [Inverse FFT]
   $A = \mathscr{F}^{-1}(A)$. A now contains the convolution of the decomposition of x and y. This is the product.
7. [Normalization and Sign Adjustment]
   Scale A assigning to C to account for scale factor in Fourier transform.
8. [Composition]
   Perform carry operations to C to give final sum: $xy \bmod (2^n + 1) = \Sigma_{j=0}^{D-1} C_j 2^{jM} \bmod (2^n + 1)$ which is the desired product.

Fig. 5.15 The Schönhage and Strassen big integer multiplication algorithm as presented in [1] with paraphrasing and explanatory text added

which at first seems backwards but isn't because the result in z is the convolution of X and Y which is what we want. It is important to note that X and Y are not x and y, the numbers to be multiplied, but decompositions of them as performed in the first part of the algorithm. We are intentionally glossing over the fact that the Fourier transform referred to by $\mathscr{F}$ is not the one familiar to scientists and engineers but a generalization of it that operates via integer operations only. See Definition 9.5.3 of [1], particularly the integer-ring DFT variant. We are now ready to look at the Schönhage and Strassen algorithm which we present in Fig. 5.15 with apologies to Crandall and Pomerance for the paraphrasing.

In step 1 we select a size for our Fourier transforms and n'. The selection of n' is not arbitrary. It is selected so that the algorithm will work on smaller sets in step 5 should the multiplication there, which is also of the form $xy \bmod N$ where $N = 2^{n'} + 1$, be done by a recursive call to the algorithm itself. In step 2 we then split the input values, x and y, into vectors A and B. It is these vector representations that will be mapped to Fourier space to perform the convolution. Recall that the convolution in this space will mimic the school method of multiplication.

In Step 3 we weight the vectors A and B to prepare for the FFT, remembering that the FFT here implements integer-only operations. The multiplication can be performed using only shifts in a binary representation of the numbers. Also, the modulo operation can likewise be implemented efficiently as Crandall and Pomerance also show in [1].

Step 4 performs the actual FFT in-place to save memory as these numbers may be very large. Step 5 multiplies the components in Fourier space, again this is convolution in the original space, and it is this step that requires something other than shifts and basic integer operations. Notice that the form of this multiplication matches what this very algorithm does so it is possible to make a recursive call here which is the motivation for selecting n' properly. Other multiplication methods could be used here as well including the basic school method. Notice that the result is stored in A, again to save memory. Step 6 performs the inverse FFT to return to the original space and Step 7 normalizes the result by applying a scale factor that is typical of the FFT, either split between the two transforms or applied in one direction. During this process the vector C is created. Lastly, step 8 accounts for the carries (ala Comba above) and returns the result which is the desired product. A complexity analysis of this algorithm shows that it runs in $O(n \log n \log \log n)$ which is significantly better than the $O(n^2)$ of the school method.

Division For division we look at two algorithms. The first is that found in Knuth [7] known as Algorithm D which is an implementation of long division suitable for application to big integers. The second is the divide-and-conquer division algorithm of Burnikel and Ziegler [8].

In [7] Knuth gives a thorough description his long division algorithm which follows the school method except for one twist. The twist has to do with trial divisions based on an observation, which he proves, that assuming the quotient of an $n + 1$ digit number x divided by an n digit number y to be the quotient of the top two digits of x by the leading digit of y means that your assumption will never be more than 2 away from the true value. A few examples of this fact,

$$\begin{array}{r|r|r}
24{,}356/5439 = 4 & 24/5 = 4 & 0 \\
95{,}433/6796 = 14 & 95/6 = 15 & 1 \\
12{,}345{,}678/9876{,}543 = 1 & 12/9 = 1 & 0 \\
8888{,}888{,}888/777{,}777{,}777 = 11 & 88/7 = 12 & 1
\end{array}$$

where the right-most column is the difference between the exact quotient and the trial quotient. It is important to note, however, that this fact is only true if the divisor is greater than or equal to the one half the base, here base is 10 so the divisors are all 5 or larger. The normalization step of the algorithm takes case of this case by multiplying both dividend and divisor by a value, which can be done with a simple shift, since the quotient will be the same and the remainder can be unnormalized at the end.

The essence of the Knuth algorithm, paraphrased from [7], is given below,

1. [Normalize]
   Scale inputs u and v by a single digit, d so that the leading digit of v, called v_1, is $v_1 \geq \lfloor b/2 \rfloor$ where b is the base. Note that u and v are big-endian vectors representing the digits of the numbers in base b.
2. [Initialize]
   Loop m times where m is the difference between the number of digits in u and v. Naturally, if m is negative the quotient is 0 and the remainder is v since $v > u$. Each pass through the loop is division of the n digits of u starting at j, the loop index, by v.

3. [Calculate Trial Quotient]
   Set $\hat{q} = \lfloor (u_j b + u_{j+1})/v_1 \rfloor$, the trial quotient. If $u_j = v_1$ set $\hat{q} = b - 1$. Knuth adds a test here that will ensure that the trial quotient is never more than one greater than it should be.
4. [Multiply and Subtract]
   Subtract $\hat{q}$ times v from the n digits of u, starting at position j, replacing those digits. This is the subtraction normally done in long division by hand. Since it is possible $\hat{q}$ is too large this result may be negative. If so, it will be handled in step 6.
5. [Test Remainder]
   Set the j-th digit of the quotient to the trial value in $\hat{q}$. If the result of step 4 was negative, do to Step 6, otherwise, go to Step 7.
6. [Add Back]
   Decrease the j-th digit of the quotient by one and add v back to u starting with the j-th digit. This accounts for the few times the trial quotient is too large.
7. [Loop]
   Go back to Step 2 while $j \leq m$.
8. [Unnormalize]
   The correct quotient is in q and the remainder is u_{m+1} through u_{m+n} divided by d, the scale factor used in the first step.

The run time of this algorithm is $O(mn)$ which is adequate for smaller input sizes but becomes cumbersome if the inputs are very large in much the same way the school multiplication algorithm does.

Burnikel and Ziegler [8] implement two algorithms which work together to perform division of big integers with specific numbers of digits. They then describe how to apply these algorithms to arbitrary division problems. The first algorithm, which they name $D_{2n/1n}$, divides a number of $2n$ digits by a number of n digits. They do this by breaking the number up into super digits. If we wish to find A/B and use this algorithm we must break A into four digits, (A_1, A_2, A_3, A_4) and break B into two digits (B_1, B_2) so that each digit has a length of $n/2$. In other words,

$$A = A_1\beta^{3n/2} + A_2\beta^n + A_3\beta^{n/2} + A_4$$
$$B = B_1\beta^{n/2} + B_2$$

With this separation of A and B into super digits we can examine the $D_{2n/1n}$ algorithm,

This algorithm is deceptively simple. The recursive nature of it becomes clear when we look at algorithm $D_{3n/2n}$ which in turn calls $D_{2n/1n}$ again. First we present $D_{3n/2n}$ and then we discuss them together,

The pair of algorithms, $D_{2n/1n}$ and $D_{3n/2n}$, work together on successively smaller inputs to return the final solution to the division problem. We know that $D_{2n/1n}$ requires A to have twice as many digits as B and when it is called in the first [Case 1] of $D_{3n/2n}$ we are also passing it arguments that meet the $2n$ to n digit requirement so the recursion will end.

Algorithm $D_{2n/1n}$

1. [Given]
   $A < \beta^n B$, $\beta/2 \le B < \beta^n$, n even, β is the base. Find $Q = \lfloor A/B \rfloor$ and remainder $R = A - QB$.
2. [High Part]
   Compute Q_1 as $Q_1 = \lfloor [A_1, A_2, A_3]/[B_2, B_1] \rfloor$ with remainder $R_1 = [R_{1,1}, R_{1,2}]$ using algorithm $D_{3n/2n}$.
3. [Low Part]
   Compute Q_2 as $Q_2 = \lfloor [R_{1,1}, R_{1,2}, A_4]/[B_2, B_1] \rfloor$ with remainder R using algorithm $D_{3n/2n}$.
4. [Return]
   Return $Q = [Q_1, Q_2]$ as the quotient and R as the remainder.

Algorithm $D_{3n/2n}$

1. [Givens]
   $A = [A_1, A_2, A_3]$ and $B = [B_1, B_2]$ from $D_{2n/1n}$. Find $Q = \lfloor A/B \rfloor$ and remainder $R = A - QB$.
2. [Case 1]
   If $A_1 < B_1$ compute $\hat{Q} = \lfloor [A_1, A_2]/B_1 \rfloor$ with remainder R_1 using $D_{2n/1n}$.
3. [Case 2]
   If $A_1 \ge B_1$ set $\hat{Q} = \beta^n - 1$ and set $R_1 = [A_1, A_2] - [B_1, 0] + [0, B_1] = [A_1, A_2] - \hat{Q}B_1$.
4. [Multiply]
   Compute $D = \hat{Q}B_2$ using Karatsuba multiplication.
5. [Remainder]
   Compute $\hat{R} = R_1\beta^n + A_4 - D$.
6. [Correct]
   As long as $\hat{R} < 0$ repeat:

   a. $\hat{R} = \hat{R} + B$
   b. $\hat{Q} = \hat{Q} - 1$.

7. [Return]
   Return $\hat{Q}$ and $\hat{R}$.

It is interesting to compare the steps in $D_{3n/2n}$ with those of the Knuth algorithm above. We get a "trial" quotient, $\hat{Q}$, from one of the cases and then determine the remainder in step 5. Then in step 6 we correct the quotient in case we selected too large of an answer. This is exactly what is done in the Knuth algorithm in step 6. Here we are structuring the problem in such a way that the digits always match the desired pattern and using recursion to divide up the work. In the Knuth algorithm we move through the divisor digit by digit.

The school division method of Knuth is $O(mn)$ which would be $O(n^2)$ for the case we are considering here since the dividend has twice as many digits as the divisor. What about the Burnikel and Ziegler algorithm? As they discuss in [8] the performance of this algorithm depends upon the type of multiplication used in step 4 of $D_{3n/2n}$. If using Karatsuba multiplication they determine the performance to be $2K(n) + O(n \log n)$ with $K(n)$ the performance of Karatsuba multiplication itself. As they note, if the school method of multiplication is used instead we get $n^2 + O(n \log n)$ which is actually worse than long division itself. So, the choice of multiplication algorithm is crucial in determining the performance.

We have examined, superficially, several alternatives to the basic arithmetic algorithms we developed for our big integer implementation. We have seen that they out perform the naive algorithms and that they extend our abilities in working with big integers to those that are much larger than what we would be able to work with otherwise. Computer memory is inexpensive so the limiting factor is now algorithm performance. Every small improvement counts. There are many other algorithms we could have examined but cannot for space considerations. There is ongoing research in this area and tweaks and small improvements are happening all the time. But, for now, we leave the algorithms and move on to more practical considerations, namely, implementations. It is fun to implement some of these basic algorithms ourselves, but the complex and highly performant ones are often tricky and it is wise to use proven libraries when speed and accuracy count.

5.5 Implementations

Most programming languages do not define big integer operations as intrinsic to the language. In this section we will look at some that do, specifically Python and Scheme. We will also look at some extensions to existing languages such as C/C++, Java and JavaScript which give big integer functionality to developers using these languages. We start first with the extensions.

Libraries When discussing software libraries time is naturally a factor. Projects start and finish and are often abandoned as people move on to other work making an orphan of the library which becomes less and less relevant as time moves on. Here we will look at one library that has considerable popular support and is a member of the Gnu Software suite, namely, the *GNU Multiple Precision Arithmetic Library* or GMP for short [9]. This library is open source software and is distributed via the GNU LPGL and GNU GPL licenses which allows the library to be used in its existing form in commercial code while enabling additions to the library to be passed on to other users. It has been in constant development since 1991 which is why we include it here as it is likely to be supported for some time to come.

GMP is a C and C++ library for a wide-range of mathematical operations involving multi-precision numbers. At its core it consists of several main packages: signed integer arithmetic functions (*mpz*), rational arithmetic functions (*mpq*), floating-point functions (*mpf*), and a low-level implementation (*mpn*). Additionally, a C++ class wrapper is available for the *mpz*, *mpq* and *mpf* packages. Here we will consider only the integer operations of the *mpz* package. The GMP library goes to great lengths to ensure efficiency and optimal performance.

The *mpz* package uses much the same format for numbers as we have used above. It defines a structure in `gmp.h` called `__mpz_struct` which is `typedef`-ed to `mpz_t`. The struct is,

```
typedef struct
{
    int _mp_alloc;
    int _mp_size;
    mp_limb_t *_mp_d;
} __mpz_struct;
```

where mp_limb_t points to a "limb" or digit. Its actual size depends on the system the code is compiled for but is typically an unsigned long int which is 4 bytes on the 32-bit system we assume in this book. The _mp_alloc field counts the number of limbs allocated while _mp_size is the number of limbs used. As we did above, the sign of the number is stored in the sign of _mp_size. If _mp_size is zero the number is zero. This saves allocating a limb and setting it to zero.

Input of big integers is via strings as was done in our code above. Output is to file streams or strings depending upon the function used. Integers must be created and initialized before being assigned but may be freely assigned after that. When done, they must be cleared. This code initializes a big integer, assigns it a value and then prints the value to stdout in two different number bases,

```
1  #include <stdio.h>
2  #include <gmp.h>
3
4  int main() {
5      mpz_t A;
6
7      mpz_init(A);
8
9      mpz_set_str(A, "1234567890123456789", 0);
10     mpz_out_str(NULL, 10, A);
11     printf("\n");
12     mpz_out_str(NULL, 2, A);
13     printf("\n");
14
15     mpz_clear(A);
16 }
```

The GMP library is included in line 2 (linking with -lgmp is also required). A new big integer, A, is defined in line 5 and initialized in line 7. Note that A is not passed as an address. The definition of mpz_t is a one element array of __mp_struct which passes it as an address already. In line 9 we assign A from a string, the 0 means use the string itself to determine the base, in this case base 10. Lines 10 and 12 output A first as a base 10 number and then in binary. Finally, we release A in line 15.

Basic big integer functions in *mpz* include the expected addition (mpz_add), subtraction (mpz_sub) and multiply (mpz_mul). Also included are negation (mpz_neg) and absolute value (mpz_abs). For efficiency there are also special functions that combine operations as it is sometimes better to combine them than perform them individually. For example, there are mpz_addmul which adds the product of two numbers to an existing number as well as mpz_submul which subtracts the product.

For division *mpz* gives us many options. These options reflect various rounding methods and implement different algorithms which may be much faster than general algorithms if certain conditions are met. If we select only the `tdiv` functions, which round towards zero, like C does, we have three functions,

`mpz_tdiv_q`	-	divide n by d returning only the quotient, q.
`mpz_tdiv_r`	-	divide n by d returning only the remainder, r.
`mpz_tdiv_qr`	-	divide n by d returning both quotient, q, and remainder, r.

There is also a general `mod` function which always returns a positive result (`mpz_mod`). If it is known in advance that the numerator n is exactly divisible by the denominator d one may use `mpz_divexact` which is much faster than the generic division routines. Filling out the division options are `mpz_divisible` to check if n is exactly divisible by d and `mpz_congruent_p` to check if n is congruent to c modulo d meaning there is an integer q such that $n = c + qd$.

What algorithms are implemented by these top-level functions? For addition and subtraction, which are at the lowest level for Intel architecture implemented in assembly, we can expect the normal school method. What about multiplication? As we are now well aware, there are many options.

The GMP library uses one of seven multiplication algorithms for N digit by N digit multiplication (as of GMP version 6.0.0). Which is used depends upon the size of N and whether or not it exceeds a preset threshold. The algorithms are, in order of use (i.e., larger N uses algorithms further down the list),

1. School method. This is the algorithm we implemented above.
2. Karatsuba. The Karatsuba method discussed above.
3. Toom-3. A three-way Toom extension to Karatsuba. See [7].
4. Toom-4. A four-way Toom algorithm.
5. Toom-6.5. A six-way Toom-like algorithm.
6. Toom-8.5. An eight-way Toom-like algorithm.
7. FFT. The Schönhage and Strassen method as discussed above.

All of these algorithms are optimized by the authors of GMP and at a lowest level are somewhat different than might be found in the original papers. If the operands to the multiplication are of very different sizes, N digits by M digits, then below one threshold the school method is used while FFT is used for very large numbers. In between, the Toom algorithms are used based on heuristics given the sizes N and M.

For division, four main algorithms are in use. For numbers below a threshold (50 digits) the long division method of Knuth is used. Above this threshold the divide-and-conquer method of Burnikel and Ziegler (modified) is used. Division by very large numbers uses an algorithm based on Barrett reduction [10]. Exact division uses Jebelean's method [11]. Switching between methods is automatic, the user need only call the top-level functions.

If GMP is for C and C++, what about other languages that lack intrinsic big integer functionality? Java uses the `BigInteger` class. This class comes with its own implementations, in Java, for low-level operations which exactly match those found in GMP's *mpn* package. This was done to make it possible to easily replace the Java code with high performance C code if desired. So, we see that C, C++ and Java all make use of the GMP library.

For JavaScript a common choice to add support for big integers is the `BigInteger.js` library which is distributed via the MIT License and implements all functionality in pure JavaScript so it can run within a web browser. This implementation is very generic and uses a base of 10,000,000 to take advantage of the simplicity this provides input and output just as we did with our implementation above. Addition and subtraction use the school method as does multiplication. There is a comment in version 0.9 to add Karatsuba in the future. Division uses the Knuth algorithm.

Languages We know that Python has intrinsic support for big integers. Let's take a look at what it is doing under the hood. The Python interpreter is written in plain C and big integers are found in the source code file (Python 2.7) `longobject.c`. This file contains implementations for all the operations supported by Python long integers, it does not use the GMP or other libraries. For addition and subtraction the expected school method is used. For multiplication only two algorithms are available, the school method and Karatsuba if the number of digits exceeds 70. Division is long division *ala* Knuth. The numbers themselves use a base of 2^{15}. While the native implementations are adequate, for serious work in Python with big integers one would do well to wrap the GMP library. Surprisingly, given the current popularity of Python, this does not appear to have been done with any level of seriousness.

Another language which requires in its very definition the support of big integers is Scheme [12]. This is a small dialect of Lisp and has been used widely for instruction in computer programming. Indeed, it is used in the very popular *Structure and Interpretation of Computer Programs* by Abelson and Sussman [13] which has trained a generation of programmers from MIT though it has fallen out of style in recent years. Regardless, the text comes highly recommended.

The version of Scheme we will consider is *Scheme48* available online from *s48.org*. This version of Scheme conforms to the R5RS version of the language [14] and includes big integer support for exact numeric computation. Section 6.2.3 of the R5RS report strongly encourages implementations to support "exact integers and exact rationals of practically unlimited size and precision" which *Scheme48* does.

The Scheme interpreter itself is written in C and places big integer operations in the file `bignum.c`. Looking at this file we see that the implementations are based on earlier work from MIT. As expected, the addition and subtraction routines use the school method. The multiplication routine is limited to the school method as well, regardless of the size of the input. This limits the usefulness of Scheme for high-performance work with big integers unless using an external library like GMP. Similarly, we see that division is the Knuth algorithm, again regardless of the size of the operands.

5.6 Rational Arithmetic with Big Integers

Once we have the ability to represent any integer p with an arbitrary number of digits it logically follows that one might consider how to represent fractions, which are the ratio of two integers, using big integers. In this section we look at rational arithmetic with big integer components. If we group two big integers together, p and q, we can treat them as the numerator and denominator of a fraction, p/q, and perform arithmetic with them. For this section we will take advantage of the big integers provided by Python and extend them to implement basic rational arithmetic of arbitrary precision. It is proper to consider rational numbers in computers in a chapter on big integers because rational arithmetic involves fractions that grow quickly to have large numerators and denominators which implies that they will quickly exceed the capacity of fixed-length integers.

Elementary mathematics tell us how to implement the four basic arithmetic operations using fractions. They are,

$$\text{Addition} \qquad \frac{a}{b} + \frac{c}{d} = \frac{ad+bc}{bd}$$

$$\text{Subtraction} \qquad \frac{a}{b} - \frac{c}{d} = \frac{ad-bc}{bd}$$

$$\text{Multiplication} \qquad \frac{a}{b} \times \frac{c}{d} = \frac{ac}{bd}$$

$$\text{Division} \qquad \frac{a}{b} \div \frac{c}{d} = \frac{ad}{bc}$$

where a,b,c, and d are themselves big integers. However, there is one twist we must bear in mind. If we were to implement these operations directly we would quickly find our numerators and denominators are becoming very large indeed. This is because after every operation we need to reduce the resulting fraction to lowest terms, as no doubt our grade school teachers mentioned to us many times! In order to do this we need to find, for each numerator and denominator, the greatest common divisor or GCD. Thankfully, this is straightforward to find as we will soon see.

Storing Rationals in Python Python is a modern object-oriented language. To store a rational number we need to store two integers, the numerator and the denominators. Since we are developing a rational arithmetic class, we will store the numerator and denominator as member variables `n` and `d`. Additionally, we will overload the basic math operations so that we can work with rationals in a manner similar to other numbers. Input and output will be straightforward as we simply display the number by printing the numerator, the fraction symbol (/), and the denominator. We will adopt the convention that the sign of the rational is stored in the numerator.

A Basic Rational Arithmetic Class The first part of our `Rational` class defines the class and the constructor, `__init__`. Since Python uses automatic garbage collection we do not need in this case to define a specific destructor. So far, then, our class looks like this,

```
1  from types import *
2
3  class Rational:
4
5      def __init__(self, n,d=None):
6          if (d == None):
7              t = float(n).as_integer_ratio()
8              n = t[0]
9              d = t[1]
10         g = self.__gcd(n,d)
11         self.n = n/g
12         self.d = d/g
13         if (self.d < 0):
14             self.n *= -1
15             self.d = abs(self.d)
```

where we consistently use four spaces for indenting as is the Python standard. In line 1 we include the `types` module which is part of the standard Python library. We will use this to determine whether or not arguments are `Rational` objects or integers. Our constructor is in `__init__` and takes `self` as the first argument as all Python methods do. This is a reference to the object itself and this is how we will access member data and call other class methods. The constructor takes two arguments, `n` and `d`, which are the numerator and denominator of the new rational object. Lines 6 through 9 check to see if a single argument was given. If so, it is assumed to be a floating-point number and converted to a rational by calling the `as_integer_ratio` method of the `float` class. In line 10 we call the `__gcd` method to calculate the GCD of the given numbers. We will explore this method in more detail shortly. Note that we are using the standard Python naming convention of two leading underscores (__) which marks the method as being private to our class. The important point here is that each method in our class representing an operation will return a new `Rational` object. This means that the only time we need consider reducing the fraction is when the new object is created as it is not changed once created. In lines 11 and 12 we set the local values for the numerator (`self.n`) and denominator (`self.d`) dividing each by the GCD so that the fraction is stored in lowest terms. For simplicity, we ignore error states such as a denominator of zero. Lines 13, 14 and 15 ensure that the new fraction places its sign in the numerator as is our convention. As this point, we have a new `Rational` object and it is initialized with the given numerator and denominator in lowest terms. Now lets look at the `__gcd` method itself.

Reducing Fractions Using the GCD In school we learned that fractions are not unique. For example, p/q is a fraction with a particular value which we can say, when expressed as a decimal number, is c since $c = p/q$. But, if we multiply our fraction by a new fraction which is equal to 1, $\alpha/\alpha = 1$, we get $\alpha p/\alpha q$ which is, of course, still c. Therefore, when working with fractions we were constantly encouraged to express our answers in lowest terms (factor out the α). When working with rational numbers on a computer this is especially important to prevent our numerators and denominators from growing too large too quickly. To reduce the fraction we need to find α which is the greatest common divisor of both p and q and divide p and q by it.

The iterative form of the algorithm of Euclid [15] is a good way to calculate the GCD even for large numbers. Our implementation of it is,

```
def __gcd(self, a,b):
    while b:
        a,b = b, a % b
    return abs(a)
```

Arithmetic and Other Methods We are now ready to implement basic arithmetic operations. Python defines a set of methods that users can override in their own classes to capture situations which syntactically look like more ordinary operations. For example, the seemingly simple statement,

```
a + b
```

is really equivalent to,

```
a.__add__(b)
```

which means that a Python class which overrides the `__add__` method will be able to participate in mathematical expressions. This is exactly what we require in order to make using the `Rational` class as natural as possible.

Python has a large set of methods that we could override in our class. We will do some here and leave others to the exercises. In particular, we will override the following methods,

`__add__`	Addition (+)
`__sub__`	Subtraction (−)
`__mul__`	Multiplication (×)
`__div__`	Division (/)
`__str__`	String output for `str()` and `print`
`__repr__`	String output for `repr()`
`__float__`	Floating-point value for `float()`
`__int__`	Integer value for `int()`
`__long__`	Integer value for `long()`

In addition, we will override the "reverse" forms for the arithmetic operators. These are called if instead of `a+2` we write `2+a`. For operators that commute, addition and multiplication, we really only need call the non-reversed form. For division and subtraction we have to be more careful.

So, how do we implement addition? Like so,

```
1  def __add__(self, b):
2      t = type(b)
3      if (t == InstanceType):
4          n = self.n*b.d + self.d*b.n
5          d = self.d*b.d
6      elif (t == IntType) or (t == LongType):
7          n = self.n + self.d*b
8          d = self.d
9      return Rational(n,d)
10
11 def __radd__(self, b):
12     return self.__add__(b)
```

where we first look at the type of the argument, `b`, in line 2. We not only want to work with other rational numbers but also with integers and long integers. If the argument is another rational number we move to line 4 and add to get the numerator followed by the denominator. Otherwise, if the argument is an integer, we move to line 7 to add the integer to the fraction. Line 9 returns the new rational answer. Recall that the constructor is what applies the GCD so it is not explicitly called here. Line 11 defines the reverse case method. Since addition is commutative, $a + b = b + a$, we simply fall back to the main addition method.

Subtraction is virtually identical except for the reversed call,

```
1  def __sub__(self, b):
2      t = type(b)
3      if (t == InstanceType):
4          n = self.n*b.d - self.d*b.n
5          d = self.d*b.d
6      elif (t == IntType) or (t == LongType):
7          n = self.n - self.d*b
8          d = self.d
9      return Rational(n,d)
10
11 def __rsub__(self, b):
12     t = type(b)
13     if (t == IntType) or (t == LongType):
14         n = b*self.d - self.n
15         d = self.d
16         return Rational(n,d)
```

where `__sub__` is identical to `__add__` except for the change from $+$ to $-$. However, as subtraction is not commutative, we must implement it explicitly for the reverse case. This is done in lines 13 through 16. Why is there no check for `InstanceType`? Because Python is smart enough to know that if both arguments are objects the main method will be called instead.

Multiplication follows addition,

```
def __mul__(self, b):
    t = type(b)
    if (t == InstanceType):
        n = self.n * b.n
        d = self.d * b.d
    elif (t == IntType) or (t == LongType):
        n = self.n * b
        d = self.d
    return Rational(n,d)

def __rmul__(self, b):
    return self.__mul__(b)
```

while division follows subtraction in implementing the reverse call explicitly,

```
def __div__(self, b):
    t = type(b)
    if (t == InstanceType):
        n = self.n * b.d
        d = self.d * b.n
    elif (t == IntType) or (t == LongType):
        n = self.n
        d = self.d * b
    return Rational(n,d)

def __rdiv__(self, b):
    if (type(b) == IntType):
        n = b * self.d
        d = self.n
        return Rational(n,d)
```

and in both cases the proper steps for multiplication and division of fractions are performed.

The remaining methods are convenience methods,

```
def __str__(self):
    if (self.n == 0):
        return "0"
    elif (self.d == 1):
        return "%d" % self.n
    else:
        return "%d/%d" % (self.n, self.d)

def __repr__(self):
    return self.__str__()

def __float__(self):
    return float(self.n) / float(self.d)

def __int__(self):
    return int(self.__float__())

def __long__(self):
    return long(self.__float__())
```

which allow us to interact with Python in a more natural way. Use `__str__` to return a string representation of the fraction as the output of `str()` or `print`. In this method we check for special cases like zero or an integer (denominator of one). The `__repr__` method responds to the `repr()` function by returning exactly the same string `print` would receive. The `__float__`, `__int__` and `__long__` methods perform the necessary conversion to return the rational as a float, integer, or long integer, respectively. These are used by the `float()`, `int()`, and `long()` functions. These are the minimal set of methods for our `Rational` class.

Using the Rational Class Because of our design choices in overriding key default Python methods we can use the `Rational` class in a quite natural way. The following examples are from an interactive session and illustrate how to use the class,

```
1 >>> from Rational import *
2 >>> a = Rational(2,3)
3 >>> b = Rational(355,113)
4 >>> c = a + b
5 >>> c
6 1291/339
7 >>> (a+b) * c
8 1666681/114921
9 >>> d = (a + b) * c
10 >>> d = 2*d
11 >>> d
12 3333362/114921
13 >>> float(b)
14 3.1415929203539825
15 >>> d / d
16 1
17 >>> 1/d
18 114921/3333362
19 >>>
```

where `>>>` is the Python interactive shell prompt and replies from Python are lines without the prompt.

In line 1 we import the `Rational` class. Lines 2 and 3 define `a` and `b` to be $2/3$ and $355/113$, respectively. Line 4 adds `a` and `b` and assigns the result to `c` which is then displayed in line 6. Line 9 uses all three rationals in an expression and displays it. Note already how the numerator and denominator are growing. We define `d` in line 9 and multiply it by an integer in line 10. Line 13 shows us that $355/113$ is a good approximation of π by requesting the floating point version of it. We use line 15 as a sanity check to make sure that a rational number acts rationally. Finally, line 17 checks that the inverse of a rational is, in fact, what you get when you flip the numerator and the denominator.

This concludes our simple Python `Rational` class. As promised, the exercises will offer you the opportunity to extend the class even further.

5.7 When to Use Big Integers and Rational Arithmetic

The sections above demonstrate how to implement big integers, how to operate on them efficiently, and where to find performant big integer libraries, however, none of the above actually talks about *when* and *why* one might want to use big integers. We remedy that in this section.

In modern computing undoubtedly the largest use of big integers is in cryptography. Cryptographic systems depend upon the fact that certain operations are easy to do in one direction but next to impossible to do, in a reasonable period of time, in the other direction. Most of these systems depend upon the following two facts,

1. The *fundamental theorem of arithmetic* states that all numbers are either prime or can be factored, uniquely, into a specific set of primes.
2. Factoring large integers is a tedious process even though multiplying them is straightforward.

This means that while it is easy to find $n = pq$ with p and q both integers it is hard to factor n to see that it is in fact what you get if you multiply p and q. This is the second fact above. The first says that if p and q are prime then indeed the *only* way to factor n is to find the specific pair, p and q. If p and q are too small, trial and error will eventually find them, but if they are very large, on the order of several hundred bits or about 100 decimal digits, then finding the factors of their product will take a very long time. This is the heart of modern cryptography. Let's see briefly how multiplying large integers plays into it.

First we consider classic Diffie-Hellman key exchange [16]. This system allows two parties, traditionally known as Alice and Bob, to exchange a secret number to be used as a key to encrypt a message shared between them. We are not concerned with the encryption but rather how they exchange the secret key so that an outsider (traditionally known as Carol or Eve) cannot learn what it is. Let's look at the steps and then we will see where big integers play into it,

The heart of the exchange relies on the choice of p and g. We must choose p to be prime and g to be a primitive root modulo p. The exact definition of a "primitive root modulo p" does not concern us here but it a number theory concept. What does concern us is that p and g are public and known by Alice, Bob and Carol. Additionally, Alice and Bob each select a secret number only they know. We'll call Alice's number a and Bob's number b. These numbers are known only to the respective person. For our example, we will select $p = 37$ and $g = 13$, which is a primitive root of 37. With this setup, the steps necessary to exchange a secret key are,

1. Alice chooses $a = 4$. Only Alice knows this number.
2. Bob chooses $b = 11$. Only Bob knows this number.
3. Alice sends Bob $A = 13^4 \bmod 37 = 34$. Carol sees A.
4. Bob sends Alice $B = 13^{11} \bmod 37 = 15$. Carol sees B.
5. Alice computes the secret key, $s = 15^4 \bmod 37 = 9$. Carol cannot compute s.
6. Bob computes the secret key, $s = 34^{11} \bmod 37 = 9$.
7. Alice and Bob both use s as the key for their encryption system.

Recall that a and b are secret but $p = 37$, $g = 13$, $A = 34$ and $B = 15$ are public. When the exchange is complete, Alice and Bob have both calculated $s = (g^a)^b \bmod 37 = (g^b)^a \bmod 37 = 9$ which they will use as the secret key for their encryption system without Carol knowing what s is.

Big integers come into this system with the choice of p, a and b. If these are very large, hundreds of digits long instead of small as in our example above, it becomes virtually impossible to determine s from the values that are known publicly. Even though A is known and it is known that $A = g^a \bmod p$, the reverse operation, known as the *discrete logarithm*, is very hard to do so a remains unknown publicly. It is this difficulty that makes the key exchange system secure and big integers are essential because without them the numbers involved would be too small to offer security against brute force attacks.

Another cryptographic system that relies on big integers is RSA public key encryption [17]. This system was published by Rivest, Shamir, and Adelman (hence RSA) of MIT in 1977. In this system users generate two keys, a public key and a private key. They then publish their public key allowing anyone to encrypt a message using it. The holder of the corresponding private key, however, is the only one who can decrypt it. For simplicity we ignore the extra twist of signing the message using the sender's private key. The essence of the algorithm depends upon two primes, p and q, their product, $n = pq$, a public key exponent, e, and its multiplicative inverse, d. The person generating the keys keeps d as the private key and releases both n and e together to be the public key. With this public key anyone can encrypt a message but only the holder of the private key as decrypt it. The length of n is the key length which is related to the strength of the algorithm.

If Bob has released his public key, the pair n and e, then Alice can send Bob a secure message by turning her message into an integer, m, the exact method for this is irrelevant but must be agreed upon by both Alice and Bob, and computing $c = m^e \bmod n$. She then sends c to Bob. Bob, in turn, recovers Alice's message by computing $m = c^d \bmod n$ using the private key, d. Since only Bob has d, he is the only one who can recover the message.

As in Diffie-Hellman, the security of the encryption depends upon the choice of large primes for p and q. When these are very large numbers it becomes virtually impossible to recover them from knowledge of just n. The exponent, e, and private key, d are calculated from p and q so knowledge of p and q would lead to a breakdown of the entire algorithm. Fortunately, big integers come to the rescue and let us work with the very large primes that make the system secure. Practical implementations of encryption algorithms which rely on big integers are found in *OpenSSL* [18] which implements the secure socket layer for internet data transfer and *libgcrypt* [19] which is an open source package implementing numerous algorithms including RSA.

Big integers are of practical importance to cryptography but they are also used for research purposes by mathematicians. For example, the open source package *Pari* [20] has been in existence for some time and it supports arbitrary precision calculations for mathematicians interested in number theory and other disciplines. Additionally, sometimes it is helpful to have the ability to restrict the precision of calculations in order to explore the effect of mathematical operations. In the paper *Chaos, Number Theory and Computers* [21] Adler, Kneusel and Younger explored the effect of the precision used when calculating values for the logistic map, $x_{n+1} = Ax_n(1-x_n)$, $A \leq 4$, $0 < x < 1$, which can lead to anomalous behavior. The software for this paper made use of the intrinsic big integer arithmetic found in the Scheme language to calculate logistic map values with a set precision.

What about rational arithmetic? As we've seen, adding rational arithmetic once big integers are available is straightforward. Many languages and libraries do support rational arithmetic. For example, the `mpq` part of GMP implements rational functions and while Python supports big integers natively, Scheme supports rationals natively including syntactically so that `(+ 2/3 3/4)` is a perfectly valid function call yielding `17/12` as its answer. Another use for rational arithmetic would be as a way to increase the precision of calculations beyond that of native floating point.

5.8 Chapter Summary

In this chapter we introduced the concept of a big integer by which we mean an integer larger than any native type supported by the computer. We implemented basic big integer operations in C and discussed improved algorithms for these basic operations which are of more practical utility. Next we looked at important implementations of big integer libraries as well as programming languages which support big integers intrinsically. We then examined rational numbers built from big integers and implemented a simple rational number class in Python. Lastly, we discussed some prominent uses of big integers to give examples of when big integers might be the tool of choice.

Exercises

5.1 Big integer libraries typically represent integers in *sign-magnitude* format as we did in this chapter. Another possibility would be to use a complement notation, call it B's-complement where B is the base, which is the analog of 2's-complement used almost exclusively for fixed-length integers. However, very few big integer packages actually do this. Why?

5.2 In Sect. 5.2 we gave a C function for `bigint_equal` to test if two big integers are equal. This function made use of `bigint_compare`. Write the remaining comparison functions for $<$, $>$, $\leq$, $\geq$ and $\neq$. *

5.3 Figure 5.11 implements a function to perform big integer division. Extend this function to return the remainder as well. *

5.4 Add a new function to our library, `bigint_pow(a,b)`, which returns a^b. Be sure to handle signs and special cases. *

5.5 Add another new function to our library, `bigint_fact(a)`, which returns $a!$. Recall that the argument must be ≥ 0. *

5.6 Using the big integer functions defined in this chapter create a simple arbitrary precision desk calculator in C which takes as input expressions of the form "*A op B*", as a string, where *A* and *B* are big integers and *op* is one of the four basic arithmetic operations and returns the result. If you wish, you may use postfix notation and a stack. In postfix notation the operands come first followed by the operation, "*A B op*". You can use Python to check your answers. **

5.7 Extend the basic `Rational` Python class outlined above to compare rational numbers by overriding the `__lt__`, `__le__`, `__gt__`, `__ge__`, `__eq__`, and `__ne__` methods to implement comparison operators $<$, $\leq$, $>$, $\geq$, $==$, and $\neq$. *

5.8 Extend the basic `Rational` Python class outlined above by adding `__neg__`, `__abs__` methods for negation and absolute value. *

5.9 Extend the basic `Rational` Python class outlined above by adding methods to handle shortcut operations. N.B., these, unlike all the others, destructively update the existing rational object. They do not return a new object. The method names are: `__iadd__`, `__isub__`, `__imul__`, and `__idiv__` corresponding to operators `+=`, `-=`, `*=`, and `/=`. Be careful when handling integer arguments and be aware that the updated value may no longer be in lowest terms. *

5.10 Use the `Rational` class to create a simple interactive rational desk calculator. This should take as input expressions of the form "*A op B*", as a string, where *A* and *B* are rationals (with a "/" character) or integers and *op* is one of the four basic arithmetic operations and return the result. If you wish, you may use postfix notation and a stack. **

References

1. Crandall R., Pomerance C., Prime Numbers: a Computational Perspective, Springer (2005).
2. Comba P., Exponentiation Cryptosystems on the IBM PC. IBM Systems Journal 29, 4:526–536 (1990).
3. Karatsuba, A., Ofman, Y., Multiplication of Many-Digital Numbers by Automatic Computers, Proceedings of the USSR Academy of Sciences 145: 293–294 (1962). Translation in the academic journal Physics-Doklady, 7, pp. 595–596 (1963).
4. Schönhage, D., Strassen, V., Schnelle multiplikation grosser zahlen, Computing, 7(3–4), 281–292 (1971).
5. Saracino D., Abstract Algebra: A First Course, Addison-Wesley (1980).
6. Bracewell, R., The Fourier transform and its applications, New York: McGraw-Hill (1986).
7. Knuth, D., The Art of Computer Programming, Volume 2: Seminumerical Algorithms, Addison-Wesley Professional (2014).
8. Burnikel, C., Ziegler, J., Im Stadtwald, D., Fast recursive division (1998).
9. Granlund, T., gmp-GNU multiprecision library, Version 6.0.0 (2014).
10. Barrett, P., Implementing the Rivest Shamir and Adleman Public Key Encryption Algorithm on a Standard Digital Signal Processor, Advances in Cryptology - CRYPTO' 86. Lecture Notes in Computer Science 263. pp. 311–323 (2006).
11. Jebelean T., An algorithm for exact division, Journal of Symbolic Computation, volume 15, 169–180 (1993).
12. Sussman, G. Steele, Jr, G., The First Report on Scheme Revisited, Higher-Order and Symbolic Computation 11 (4): 399–404 (1998).
13. Abelson, H., Sussman, G., Structure and interpretation of computer programs, 2nd Ed, MIT Press (1996).
14. Kelsey, R., Clinger, W., Rees, J. (eds.), Revised5 Report on the Algorithmic Language Scheme, Higher-Order and Symbolic Computation, Vol. 11, No. 1, (1998).
15. Heath T., The Thirteen Books of Euclid's Elements, 2nd ed. (Facsimile. Original publication: Cambridge University Press, 1925), Dover Publications (1956).

16. Diffie, W., Hellman, M., New directions in cryptography, Information Theory, IEEE Transactions on, 22(6), 644–654 (1976).
17. Rivest, R., Shamir, A., Adleman, L., A method for obtaining digital signatures and public-key cryptosystems, Communications of the ACM, 21(2), 120–126 (1978).
18. http://www.openssl.org/.
19. http://www.gnu.org/software/libgcrypt/.
20. The PARI Group, PARI/GP version `2.7.0`, Bordeaux, 2014, http://pari.math.u-bordeaux.fr/.
21. Adler C., Kneusel R., Younger, B., Chaos, Number Theory, and Computers, Journal of Computational Physics, 166, 165–172 (2001).

Chapter 6
Fixed-Point Numbers

Abstract Fixed-point numbers implement floating-point operations using ordinary integers. This is accomplished through clever use of scaling and allows systems without explicit floating-point hardware to work with floating-point values effectively. In this chapter we explore how to define and store fixed-point numbers, how to perform signed arithmetic with fixed-point numbers, how to implement common trigonometric and transcendental functions using fixed-point numbers, and, lastly, discuss when one might wish to use fixed-point numbers in place of floating-point numbers including an example of an emerging use case involving modern neural networks.

6.1 Representation (Q Notation)

A fixed-point number is a scaled floating-point number. The scaling can be in decimal or binary. We will focus on binary here because this makes the scaling a power of two which means scaling up or down is simply a matter of shifting bits up or down. One reason to use fixed-point arithmetic is speed versus floating-point in software, so this rapid shifting is an advantage. Let's take a look at what scaled arithmetic actually entails.

If we are working with numbers in a known range, say between -3 and $+3$, we normally use a `float` or `double` to represent the numbers and go on our merry way. However, what if we are in a situation where floating-point math is expensive in terms of speed or memory? One option is to use scaled arithmetic, which is fixed-point arithmetic. Before we can implement any arithmetic functions, however, we need to understand the way scaling is used so let's look at an example. If we have 16-bit integers, C `short int`, we can represent values between $-65{,}536$ and 65,535, including the end points, if we use two's-complement notation. This means that we need to scale our -3 to $+3$ range numbers so that they fit in the range $-65{,}536$ to 65,535. Additionally, we want to make the scale factor a power of two and make it as large as possible while still keeping our scaled integer in the 16-bit range. In this case, we need two bits to store the integer part and one additional bit to store the sign for a total of three bits. This leaves $16-(2+1) = 13$ bits for representing the fractional part. This means that the scale factor is $2^{13} = 8192$ and the resolution of our fixed-point numbers will be 2^{-13} or about 0.000122. Here "resolution" means that the difference between any two consecutive 16-bit integers,

R.T. Kneusel, *Numbers and Computers*, DOI 10.1007/978-3-319-50508-4_6

in this interpretation, is a fixed difference of 0.000122. More formally, to implement scaled arithmetic, we are really using integers to represent rational numbers with an implied denominator which is the scale factor. So, for our example, we are using 16-bit integers to represent fractions where the integer is the numerator and the denominator is fixed at $2^{13} = 8192$. We will look at how to convert floating-point numbers to and from fixed-point notation and then things should be clearer but before doing that, let's introduce some notation that will make discussing fixed-point numbers much easier, this is Q notation.

The Q notation was introduced by Texas Instruments (for example, see [1]) as a shorthand for specifying the size of the integer used to represent a fixed-point number as well as the number of bits reserved for the integer and fractional part. Formally, the notation is $Qm.n$ where m is the number of bits used in the integer part while n is the number of bits in the fractional part. Signs are stored using two's-complement notation meaning the entire integer is a two's-complement number. As this is the expected way fixed-length integers are stored in modern computers this is a good thing. This also means that there must be one additional bit for the sign of the fixed-point number and the entire length of the fixed-point number is $m + n + 1$ bits. Our example above used two bits for the integer part and 13 for the fractional part. In Q notation this is $Q2.13$ which uses $2 + 13 + 1 = 16$ bits, exactly the number of bits in a C `signed short` integer. If the number of integer bits is zero it is dropped from the Q notation so that $Q31$ means a 32-bit integer with no integer part and 31 bits reserved for the fractional part.

Armed with Q notation we can see that our example above is a good choice for representing numbers in the range -3 to $+3$ using 16 bits. The range of a fixed-point number $Qm.n$ is $[-2^m, 2^m - 2^{-n}]$ which for $Q2.13$ is about $[-4, 3.999878]$, beyond the $[-3, 3]$ range we required. By picking the smallest number of bits for the integer part we maximize the number of bits for the fractional part thereby making the fixed-point numbers as accurate as possible over the desired range for the given total number of bits.

What if instead of selecting $Q2.13$ we select $Q7.8$? In this case the interval between values is 2^{-8} or about 0.0040 which is 32× larger than $Q2.13$ meaning our numbers are much less accurate. To see this, consider a number line such as that of Fig. 6.1 which shows graphically the reduced accuracy we would have if we did not keep the integer part of the fixed-point number as close to the working range as possible.

We now understand that a fixed-point number is a fraction with a denominator that is a power of two. We also know how to indicate, via the Q notation, how we are interpreting this fraction so that the entire range of the fixed-width integer we are using to represent the fraction is used. With this in mind, we can begin to piece together our own library of fixed-point functions in C. In this case C is a very natural choice as one of the primary uses of fixed-point is in embedded development where often one is working with microcontrollers without floating-point hardware. While it is possible, and often done in the past, to emulate floating-point hardware in software, in general, fixed-point math is much faster. And in the microcontroller world, C is still the language of choice.

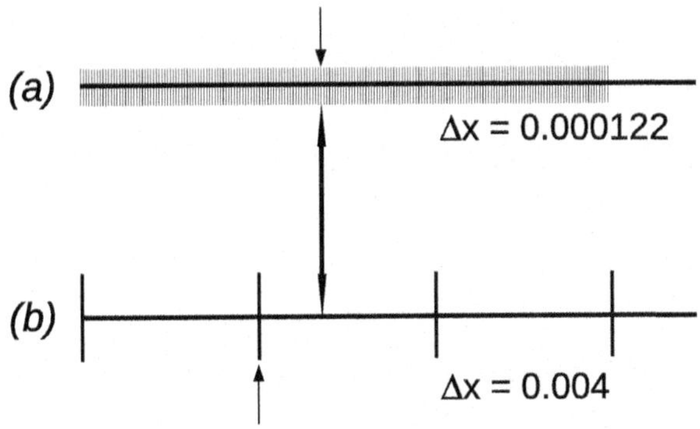

Fig. 6.1 The difference between numbers representable in **(a)** $Q2.13$ and **(b)** $Q7.8$. For $Q2.13$ each hash mark is 0.000122 apart while for $Q7.8$ the hash marks are 0.0040 apart. If we want to represent a floating point value, indicated by the *double arrow*, we must select the closest hash mark in each subfigure (*small arrows*). This is why if the range is known we should select the smallest integer width encompassing that range

Existing fixed-point libraries are highly optimized to take full advantage of the performance found in small systems. To this end, they often implement one or two types of fixed-point numbers. For example, the open source *libfixmath* project [2] implements $Q15.16$ only using signed 32-bit integers. This is a reasonable choice for a general-purpose library, actually, as it covers the range $[-32{,}768, 32{,}767.99998]$ which likely includes many of the values one would encounter in daily practice, especially in the embedded development world. The resolution is 2^{-16} or about 0.0000153 which is also quite reasonable. For our library, however, we will opt to implement a variable set of fixed-point numbers. This means that when the library is initialized we will pass in our desired $Qm.n$ by passing m explicitly. This will determine how we will interpret values internally and allow us to tailor the library to maximize the floating-point accuracy of our numbers. The library will always use a signed 32-bit integer to represent the number. This means that we need only specify m, the number of bits for the integer part, and we know that n will be $32-(m+1)$ bits so that the total number of bits used is always 32 (recall one extra bit for the sign). Additionally, when the library is initialized for a particular configuration we will recalculate certain constants which will be used when we implement trigonometric and transcendental functions. Naturally, this recalculation of floating-point constants is a cheat if we are using hardware that lacks floating-point but one could imagine calculating these ahead of time or a situation where slow software-based floating-point is used by the compiler and we wish to use faster fixed-point routines. In the end, our library is meant to be pedagogical, after all.

We start the library with code to set up for a particular $Qm.n$ and to calculate some constants in that representation for future use. Note that we create a new type, `fix_t`, to distinguish between ordinary integers and integers we mean to interpret as fixed-point numbers as well as `fixx_t` for a double precision 64-bit version. So, our initialization routine becomes,

```
1  #define MASK_SIGN 0x80000000
2  #define DEG2RAD 0.0174532925
3  #define PI 3.141592653589793
4  #define EXP_ITER_MAX 50
5  #define EXP_TOLERANCE 0.000001
6
7  typedef int32_t fix_t;
8  typedef int64_t fixx_t;
9  int32_t M=0,N=31;
10 fix_t PIh, PId, d2r, r2d;
11 fix_t f90,f180,f270,f360;
12 fix_t st1,st2,st3,st4;
13 fix_t p0,p1,p2,p3,p4,p5;
14 fix_t sin_tbl[91];  // sine lookup table
15
16 void Q_init(int32_t m) {
17     int i;
18
19     M = ((m >= 0) && (m < 32)) ? m : 0;
20     N = 32 - (m+1);
21
22     for(i=0; i<91; i++)
23         sin_tbl[i] = Q_to_fixed(sin(i*DEG2RAD));
24
25     PId = Q_to_fixed(2*PI);
26     PIh = Q_to_fixed(PI/2);
27     d2r = Q_to_fixed(DEG2RAD);
28     r2d = Q_to_fixed(1.0/DEG2RAD);
29     f90 = Q_to_fixed(90*DEG2RAD);
30     f180= Q_to_fixed(180*DEG2RAD);
31     f270= Q_to_fixed(270*DEG2RAD);
32     f360= Q_to_fixed(360*DEG2RAD);
33     st1 = Q_to_fixed(1.0/6);
34     st2 = Q_to_fixed(1.0/120);
35     st3 = Q_to_fixed(1.0/5040);
36     st4 = Q_to_fixed(1.0/362880);
37     p0  = Q_to_fixed(1.17708643e-05);
38     p1  = Q_to_fixed(9.99655930e-01);
39     p2  = Q_to_fixed(2.19822760e-03);
40     p3  = Q_to_fixed(-1.72068031e-01);
41     p4  = Q_to_fixed(5.87493727e-03);
42     p5  = Q_to_fixed(5.79939713e-03);
43 }
```

where `Q_init` takes the number of integer bits in `m`. We check that `m` is reasonable in line 19 and set our global `M` (line 9) to it. In line 20 we set `N` so that we always use 32 bits for the entire number. Note that we are including two standard libraries (not listed). The *math.h* library is necessary for our conversion routines between floating-point and fixed-point and to be explicit about integer types we include *stdint.h*.

This latter library defines the integer types in a consistent way so that `int32_t` is a `signed int` of 32-bits while `uint32_t` is an `unsigned int` of 32-bits.

We introduce several `#define` values. We use `MASK_SIGN` to mask values to get the sign of the number. This will be handy when we check for overflow. `DEG2RAD` is the constant that converts an angle in degrees to radians. This is used in line 23 to build the sine look up table. We define `PI` to be π for convenience. We will use `EXP_ITER_MAX` and `EXP_TOLERANCE` when we implement the exponential function.

Besides `M` and `N` there are several other globals which in a fixed library would themselves be constants. Here they are variables so that we can adapt to different Q representations. In line 10 we define variables to hold fixed-point representations of 2π, $\pi/2$, `DEG2RAD`, and its reciprocal, respectively. The specific assignments for a new Q notation are in lines 25 through 28.

The variables `f90`, `f180`, `f270`, and `f360` are set to $\pi/2$, π, $3\pi/2$ and 2π radians. These values are needed when implementing the sine function. Variables `st1`, `st2`, `st3`, and `st4` are set to coefficients of the Taylor series expansion of $\sin x$. Lastly, `p0` through `p5` are the coefficients of the 5-th degree polynomial fit to $\sin x$. These values are determined empirically by curve fitting.

We will add two convenience functions to the library which one would likely not include if working in a constrained environment. These functions let us easily map a floating-point value (C `double`) to a fixed-point value and back. To find the $Qm.n$ representation of a floating-point value simply scale it by the denominator, 2^n, and round to the nearest integer,

$$z_{fixed} = \lfloor 2^n z_{float} + 0.5 \rfloor$$

Similarly, to change fixed-point back to floating-point simply multiply by the reciprocal of the denominator,

$$z_{float} = 2^{-n} z_{fixed}$$

These functions are easily implemented in C as,

```
fix_t Q_to_fixed(double z) {
    return (fix_t)floor(pow(2.0,N)*z + 0.5);
}

double Q_to_double(fix_t z) {
    return (double)z * pow(2.0,-N);
}
```

What about comparisons of fixed-point numbers? In this case, since we are scaling by a fixed power of two (ignoring the case of comparing different Q notations), normal integer comparisons will still work as expected so we do not need to implement anything specific for fixed-point numbers. An example will illustrate. If we use an eight bit integer and $Q3.4$ notation to store 2.6875 we interpret each bit as follows,

bit value $\rightarrow$	0	0	1	0	1	0	1	1
place value $\rightarrow$	*sign*	2^2	2^1	2^0	2^{-1}	2^{-2}	2^{-3}	2^{-4}

where the integer we store has the value $2b_{16} = 43$. If we then look at storing 2.3125 in the same format we get an integer with the value of 37. Clearly, since $2.3125 < 2.6875$ we also have $37 < 43$ following directly from the fact that if the Q notation is the same then the place value for each bit is the same and we can compare bit by bit as is done when comparing two signed integers. Negative values are stored in two's-complement format so we again see that the comparison of two signed integers arrives at the same relationship as the two floating-point values those integers represent when interpreted as fixed-point numbers. With this understood, we are now ready to look at implementing basic arithmetic.

6.2 Arithmetic with Fixed-Point Numbers

Arithmetic with fixed-point numbers is arithmetic using fractions with an implied denominator. Another way to consider fixed-point numbers is that they are scaled by the fractional part. So, if x is a fixed-point representation of X, a floating-point number, then $x = Xd$ where $d = 2^n$ is the scale factor, with rounding to the nearest integer. Similarly, for fixed-point number $y = Yd$, where Y is the actual floating-point number, we have,

$$x + y = Xd + Yd = (X + Y)d$$

and,

$$x - y = Xd - Yd = (X - Y)d$$

which means that addition and subtraction of fixed-point numbers naturally results in a properly scaled answer. If the magnitude of $X + Y$ or $X - Y$ is too large to fit in m bits then the resulting number will overflow or underflow the representation and a loss of precision, possibly significant, will occur.

For multiplication we have,

$$xy = (Xd)(Yd) = XYd^2$$

where the answer is now scaled by d^2 instead of d. To get back to the same scaling we divide by d to get a final answer, XYd, which is what we expect for multiplication of two $Qm.n$ numbers returning a $Qm.n$ result.

There are several things to keep in mind with multiplication of fixed-point numbers. First, multiplication may lead to an intermediate result that requires double the precision to hold since it is scaled by d^2 instead of d. Second, a loss of fractional precision may well happen when we divide by d to get the final answer. This loss will be in the lowest bits of the fractional part. Once the division has been made the resulting value can be put back into a single precision integer. This means keeping the lowest b bits of the double precision $2b$-bit integer. Note that the division is simply a shift to the right by n bits for a $Qm.n$ number. Finally, if the integer part of the multiplication has overflowed meaning the result has an integer value that cannot fit in m bits, there will be a catastrophic loss of precision in the answer. Therefore, it is important to check for this overflow when working with the double precision version.

There is one other detail we did not mention yet. This is the issue of what happens when we divide the double precision product by the denominator. Recall that this division is really a shift to the right a specified number of bits. If we do this shift without concern for any rounding issues we will introduce bias in our results. To avoid this, we take the extra step of adding a value that is equal to 2^{n-1} which is the highest bit position that will be discarded by the right shift of n bits. This is tantamount to adding 0.5 to a floating point value before applying the *floor* operation so that the resulting integer is the integer the number is closest to instead of the closest integer less than the number. This will increase the precision of our answers in the long run and cause errors to average out over a series of calculations without constantly losing precision by applying a *floor* operation. For multiplication this extra addition is done before the rescaling by 2^{-n} so that we use the bits that will be discarded.

The algorithm for multiplication of two $Qm.n$ fixed-point numbers, x and y, is,

1. Cast x to x' which is a sign-extended $2b$-bit integer. Cast y to y', also a sign-extended $2b$-bit integer.
2. $t' = x' \times y'$. This is the result scaled by $(2^n)(2^n) = 2^{2n}$ and stored in a $2b$-bit integer, t'.
3. $t' = t' + 2^{n-1}$. This is the rounding to avoid truncation bias. We will divide by 2^n so we add 2^{n-1} which is the largest bit value we are about to discard.
4. $t' = t' \times 2^{-n}$. This is the result scaled by 2^n. Note that this is a right shift by n bits.
5. Cast t' to t by keeping the lowest $m + n + 1$ bits. This is the product in $Qm.n$ format. If any bits of t' above bits $m + n + 1$ are set overflow has happened and our result, t, will be incorrect.

Lastly, let's look at fixed-point division. We want to divide two $Qm.n$ numbers, x by y, and return a quotient that is also a $Qm.n$ number. This means we want an integer that represents X/Y scaled by 2^n. This is what we get if we implement the following algorithm,

1. Cast x to $x' = x \times 2^n$ where x' is a $2b$-bit integer. This means that x' is x scaled by 2^{2n}, hence using a $2b$-bit integer to store the result.
2. Correct for truncation bias by adding $y/2$ to x', $x' = x' + y/2$. Note that $y/2$ is also a $2b$-bit integer. This addition is adding $\frac{Y \times 2^n}{2} = Y \times 2^{n-1}$ to the value x' which is scaled by 2^{2n}. We will see below why this is helpful.

3. Calculate a quotient, $q' = x'/y$ where y has been properly sign-extended to $2b$-bits. Because of the initial scaling of x this quotient is already properly scaled so we keep the lowest b bits to return the $Qm.n$ answer.

Let's look more closely at this algorithm. In the end, we want a number that is $(X/Y)\times 2^n$ which is the scaled $Qm.n$ representation of the quotient of $x/y = X/Y$. By scaling x in step 1 we have $x' = x\times 2^n = X\times 2^{2n}$. The bias correction in step 2 seems particularly mysterious but we see that we are adding $y/2 = (Y\times 2^n)/2 = Y\times 2^{n-1}$. This means that the division in step 3 is,

$$x'/y = \frac{X\times 2^{2n} + Y\times 2^{n-1}}{Y\times 2^n} = (X/Y)\times 2^n + \frac{1}{2}$$

which is exactly the quotient we want, properly scaled to $Qm.n$ plus the bias correction of 1/2.

Now that we know how to do basic arithmetic with fixed-point numbers, let's implement these operations for our library. We will implement two versions of addition, subtraction, and multiplication, one that checks for overflow and one that does not. First we present versions that do not check for overflow conditions as they are the simplest,

```
1 fix_t Q_add(fix_t x, fix_t y) { return x+y; }
2 fix_t Q_sub(fix_t x, fix_t y) { return x-y; }
3
4 fix_t Q_mul(fix_t x, fix_t y) {
5     fixx_t ans = (fixx_t)x * (fixx_t)y;
6
7     ans += 1<<(N-1);
8     return (fix_t)(ans >> N);
9 }
```

Addition and subtraction are as simple as can be if we do not check for overflow. Adding (subtracting) two two's-complement integers is mathematically identical to adding two "floating-point" values stored in two's-complement format which is what a fixed-point number actually is. There is an implied radix point between bits n and $n-1$, counting bits from zero, but the process of addition (subtraction) of the fractional part is the same as for the integer part so we simply add or subtract the two values and will arrive at the proper answer.

Multiplication is much the same but if we think about the school method of multiplication we see that we will end with, since both numbers are in $Qm.n$ format, a number that is scaled by $2n$ bits. Similarly, it may also need $2m$ bits for the integer part. So, we cannot use a single precision integer of 32-bits to hold the result, we must use a double precision integer with 64-bits. This is the `fixx_t` data type. In line 5 we cast our inputs to 64-bits then multiply them. In line 8 we shift down `N` bits to get the answer to fit into a 32-bit integer. We'll skip line 7 for the moment.

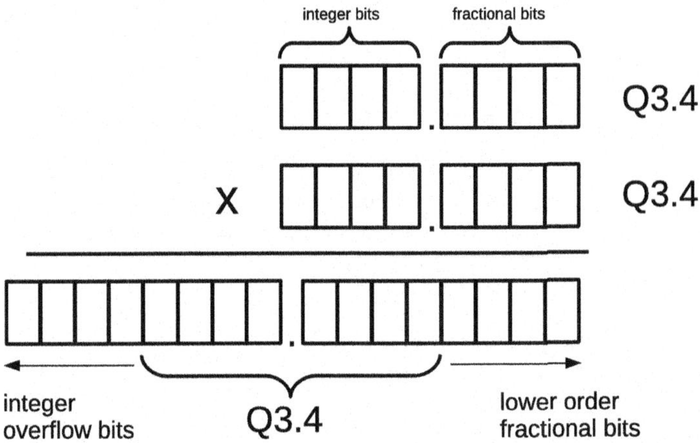

Fig. 6.2 Multiplication of two fixed-point numbers. In this case two 8-bit fixed-point $Q3.4$ numbers are multiplied. The radix point is shown to separate the fractional bits from the integer bits (including the sign). The product requires 16-bits of which the middle eight are the desired answer as indicated. Fractional bits below this range are of lower order (smaller) and are discarded when the product is shifted down $n = 4$ bits. If any bits above the indicated range are set, and not the extended sign of the answer, then overflow has happened

Pictorially, the multiplication looks like Fig. 6.2 where we see the result of multiplying two 8-bit $Q3.4$ values using a 16-bit intermediate. The effect is the same when two 32-bit values are multiplied using a 64-bit intermediate. This figure shows clearly why the product must be shifted to the right to remove the extraneous lower-order bits. In the general case this shift is by `N` bits as in line 8 above. Finally, line 9 casts back to a 32-bit integer type, losing any overflow in the integer part of the fixed-point product, and returns.

We skipped line 7 above. This is the correction to remove truncation bias that was mentioned in Step 2 of the multiplication algorithm. We add a value which is equal to the highest bit position we will be discarding. This is much the same as adding 0.5 to a decimal number before keeping only the integer part so that the integer will round to the nearest value, not just the lowest value,

$$2068.34 \rightarrow \lfloor 2068.34 + 0.5 \rfloor = \lfloor 2068.85 \rfloor = 2068$$

$$2068.64 \rightarrow \lfloor 2068.65 + 0.5 \rfloor = \lfloor 2069.15 \rfloor = 2069$$

which is what we would expect from a round-to-the-nearest-integer operation.

What if we do check for overflow? Starting with addition and subtraction we have,

```
1  fix_t Q_add_over(fix_t x, fix_t y) {
2      uint32_t ux = x;
3      uint32_t uy = y;
4      uint32_t s = ux + uy;
5
6      if ((ux & MASK_SIGN) == (uy & MASK_SIGN)) {
7          if ((s & MASK_SIGN) != (ux & MASK_SIGN)) {
8              printf("Q_add_over: overflow!\n");
9              return 0;
10         }
11     }
12     return x+y;
13 }
14
15 fix_t Q_sub_over(fix_t x, fix_t y) {
16     uint32_t ux = x;
17     uint32_t uy = y;
18     uint32_t s = ux - uy;
19
20     if ((ux & MASK_SIGN) != (uy & MASK_SIGN)) {
21         if ((s & MASK_SIGN) != (ux & MASK_SIGN)) {
22             printf("Q_sub_over: overflow!\n");
23             return 0;
24         }
25     }
26     return x-y;
27 }
28
```

where in each case we return the same value as above, see lines 12 and 26, but before that we use unsigned 32-bit integers to examine the sign of the result. We use unsigned integers because overflow is undefined in the C standard for signed integers. This tip we borrow from the *libfixmath* project.

For addition we know overflow happened if the signs of the arguments are the same but the sign of the answer is not the same as the arguments. In line 6 we mask the unsigned representation, same bit pattern as the signed representation, keeping only the highest bit. This is the sign bit for the signed interpretation of the number. If these are the same for both arguments, we then check whether the sign of the result differs from the sign of the first argument (line 7). If they do, we know overflow happened and we report it in line 8 and return zero in line 9. Naturally, a complete implementation would set an error indicator and that would be used to know that the return value is not valid.

The situation is similar for subtraction except instead of checking for the same sign in the arguments line 20 checks to see if the signs are different. If so, we check to see if the sign of the answer is the same as the sign of the first argument. If not, we know overflow has happened and we report it in lines 22 and 23. Otherwise, we return the difference in line 26 knowing that the two's-complement representation will handle the signs properly.

Multiplication with overflow detection is,

```
1  fix_t Q_mul_over(fix_t x, fix_t y) {
2      fixx_t xx,yy,tt;
3      uint32_t ux = x;
4      uint32_t uy = y;
5
6      xx = (fixx_t)x;
7      yy = (fixx_t)y;
8      tt = xx * yy;
9      tt += 1<<(N-1);
10     tt >>= N;
11
12     if ((ux & MASK_SIGN) == (uy & MASK_SIGN)) {
13         if (((fix_t)tt & MASK_SIGN) != (ux & MASK_SIGN)) {
14             printf("Q_mul_over: overflow!\n");
15             return 0;
16         }
17     }
18     return (fix_t)tt;
19 }
```

where we multiply as before but check our result starting at line 12. If the signs of the arguments are the same (line 12) but the signs of the result and first argument are different (line 13) we know overflow has happened and we report it in lines 14 and 15. If not, we return the product in line 18.

Lastly, we present division,

```
1  fix_t Q_div(fix_t x, fix_t y) {
2      fixx_t xx,yy;
3
4      xx = (fixx_t)x << N;
5      xx += (fixx_t)y / 2;
6      return xx / (fixx_t)y;
7  }
```

which implements the algorithm for division outlined above. We cast the dividend to a 64-bit integer and then scale it up by `N` bits in line 4. This sets `xx` to $x \times 2^n$ using 64-bits to hold the result. Line 5 is the truncation bias correction term which adds a 64-bit version of `y` divided by 2. The actual division is performed in line 6 which as we saw above returns an already properly scaled quotient. Since the return type for the function is `fix_t` the result of the division in line 6 is cast to the proper 32-bit size.

Before we finish this section let's take a quick look at the effect the truncation bias correction term has on repeated multiplication. To do this we will look at powers of $p = \pi - 3$ using $Q2.29$ which is capable of representing this value accurately to nine decimal places. First, we calculate $(\pi - 3)^{10}$, then we calculate p^{10} with truncation bias correction and without to show the effect of a series of ten multiplications in a row,

$$(\pi - 3)^{10} = 0.00000000323897 \text{ floating-point value}$$
$$(\pi - 3)^{10} = 0.00000000372529 \;\; Q2.29, \text{truncation bias correction}$$
$$(\pi - 3)^{10} = 0.00000000186265 \;\; Q2.29, \text{no bias correction}$$

Notice how the truncation bias correction term keeps the final value reasonably close to the actual value while without it the error accumulates and a serious loss of precision occurs. Clearly, truncation bias correction is an essential component of fixed-point arithmetic.

6.3 Trigonometric and Other Functions

Now that we have implemented basic arithmetic operations in fixed-point, what about trigonometric and other functions like log and exp? To implement these functions we have several options. We can use a look-up table, which is very fast, but potentially either not accurate enough because there are not enough entries in the table, or it uses too much memory because the table is big enough to be accurate. We can approximate the value by using a polynomial. Lastly, we can approximate the value by summing a Taylor series expansion. Let's look at each of these options for the sine function. We can calculate the cosine from the sine and once we have both sine and cosine we can calculate the tangent. We will ignore the inverse trigonometric functions. After this, we will show implementations for square root, logarithm and exponential functions.

Trigonometric Functions For our fixed-point library we will implement the three basic trigonometric functions: sine, cosine, and tangent. All of these will be based on the sine which we will implement in three different ways so that we can compare the performance and accuracy of each method. The cosine is, of course, easy to derive from the sine since,

$$\cos(\theta) = \sin(\frac{\pi}{2} - \theta)$$

and, by definition,

$$\tan\theta = \frac{\sin\theta}{\cos\theta}$$

meaning we get everything once we have the sine implemented. We will implement the sine using a table, a Taylor series expansion, and a polynomial fit. First the table,

```
1  fix_t Q_sin_table(fix_t x) {
2      int32_t sign;
3      fix_t angle;
4
5      if (x < 0) x += PId;
6      sin_val_sign(&x, &sign);
7
8      angle = Q_mul(x,r2d);
9      angle += 1 << (N-1);
10     return sign*sin_tbl[angle >> N];
11 }
```

which uses the globally defined `sin_tbl` array defined above with `Q_init`. This function also uses a helper function, `sin_val_sign`, which we describe below. This function will be used by our other sine implementations, too. It takes an angle, in radians, and updates it to be in the range $[0, \pi/2]$ along with a sign. This is equivalent to paying attention to which quadrant the angle puts us in and then using the appropriate angle and sign for that quadrant to give us the right sine value. Recall that `Q_init` initializes `sin_tbl` with the sine for each degree from $[0, 90]$. This means that we only need consider input angles in that range.

For `Q_sin_table` we first see if the argument is negative. If so, we add 2π to make it positive. So, the input to the function must be in the range $[-2\pi, 2\pi]$. This is line 5. Note that we use normal addition thereby making an implicit assumption that we will not overflow (nor to we check for it). In line 6 we call `sin_val_sign` to adjust our input angle to be in the range $[0, \pi/2]$ and set `sign` accordingly. In line 8 we convert our adjusted input angle, assumed to be in radians, to degrees. This is because we built the table by degrees and want to index it by degree. Line 9 adds 0.5 to the angle, this is the bias correction, as we are interested in mapping `angle` to the nearest degree so we can use the look-up table. In line 10 we truncate `angle`, dropping the fractional part by shifting it down `N` bits. This means that `angle » N` is an integer in the range $[0, 90]$ degrees. With this integer we index our sine table and, after multiplying by `sign`, return the final answer. This function is fast but has the overhead of building the table in the first place for each call to `Q_init` and the memory used by the table. Additionally, while most accurate for inputs which have no fractional component, it is not as accurate as it could be for arbitrary inputs. The exercises will challenge the reader to increase the accuracy by linear interpolation.

Let's look at `sin_val_sign` which we used above and will use again,

```
void sin_val_sign(int32_t *angle, int32_t *sign) {

    if ((*angle >=0) && (*angle <= f90)) {
        // quadrant I
        *sign = 1;
    } else {
        if ((*angle > f90) && (*angle <= f180)) {
            // quadrant II
            *sign = 1;
            *angle = f180-*angle;
        } else {
            if ((*angle > f180) && (*angle <= f270)) {
                // quadrant III
                *sign = -1;
                *angle = *angle-f180;
            } else {
                // quadrant IV
                *sign = -1;
                *angle = f360-*angle;
            }
        }
    }
}
```

This function accepts as input via pointers the current angle and a sign value which it will set. By comparing with the constants `f90`, `f180`, `f270`, and `f360`, each of which is set in `Q_init`, the function determines which quadrant `angle` falls in. It then sets `sign` appropriately for that quadrant and updates the value of `angle` so that the sine of the new angle, times `sign`, will be the proper sine for the original angle. This means we only ever work with angles in the range $[0, \pi/2]$.

We get `Q_cos_table` and `Q_tan_table` for free via `Q_sin_table`,

```
fix_t Q_cos_table(fix_t x) {
    return Q_sin_table(Q_sub(PIh,x));
}

fix_t Q_tan_table(fix_t x) {
    return Q_div(Q_sin_table(x), Q_cos_table(x));
}
```

where we have made use of the trigonometric identities given above. Note the use of the constant `PIh` for $\pi/2$ which is set in `Q_init` for the given *Qm.n* fixed-point representation.

Now we consider implementing sine using a Taylor series expansion. Please see any calculus book for a description of a Taylor series. Here we simply use the well-known series expansion for $\sin x$,

$$\sin x = x - \frac{x^3}{3!} + \frac{x^5}{5!} - \frac{x^7}{7!} + \frac{x^9}{9!} + \cdots$$

after mapping the input angle to $[0, \pi/2]$ as before. This leads to,

```
1  fix_t Q_sin_taylor(fix_t x) {
2      fix_t x3,x5,x7,x9;
3      fix_t ans;
4      int32_t sign;
5
6      if (x < 0) x += PId;
7      sin_val_sign(&x, &sign);
8
9      x3 = Q_mul(Q_mul(x,x),x);
10     x5 = Q_mul(Q_mul(x3,x),x);
11     x7 = Q_mul(Q_mul(x5,x),x);
12     x9 = Q_mul(Q_mul(x7,x),x);
13
14     ans = x;
15     if (x3 != 0) ans -= Q_mul(st1,x3);
16     if (x5 != 0) ans += Q_mul(st2,x5);
17     if (x7 != 0) ans += Q_mul(st3,x7);
18     if (x9 != 0) ans += Q_mul(st4,x9);
19
20     return sign*ans;
21 }
```

where lines 6 and 7 check for a negative input, adding 2π if it is, and we map the argument to $[0, \pi/2]$ while setting `sign`. Lines 9 through 12 calculate powers of the input, x. We are implementing the series directly to make it as easy as possible to see the translation from the equation to the code so we calculate all the powers we need in advance. A little algebraic manipulation will show that we only really need x and x^2 plus some well-placed parentheses. See the exercises for another challenge to the reader. Line 14 sets `ans` to the first term in the series. Lines 15 through 18 add subsequent terms if the power of x for that term is not zero. This may happen if $x < 1$ and we do not have enough precision in the n fractional bits of the $Qm.n$ number to represent the power. The magic constants `st1`, `st2`, `st3`, and `st4` are the reciprocals of the factorials in the series representation. These are calculated in advance for the given Q representation when `Q_init` is called. Finally, line 20 multiplies by the sign and returns the final estimate for $\sin x$. As before, we get $\cos x$ and $\tan x$ for free,

```
fix_t Q_cos_taylor(fix_t x) {
    return Q_sin_taylor(Q_sub(PIh,x));
}

fix_t Q_tan_taylor(fix_t x) {
    return Q_div(Q_sin_taylor(x), Q_cos_taylor(x));
}
```

where the form of the functions matches the table versions above but is based on `Q_sin_taylor` instead.

Our final $\sin x$ implementation comes from a set of points calculated from a floating-point version of sine that is then curve fit to a 5-th degree polynomial. We calculated $\sin x$ for 100 equally spaced points from $[0, \pi/2]$ using the C library version of `sin`. We then fit a 5-th degree polynomial to these points to find the coefficients p_0, p_1, p_2, p_3, p_4, and p_5 which minimize the squared difference between $\sin x$ and $f(x)$ where,

$$f(x) = p_0 + p_1x + p_2x^2 + p_3x^3 + p_4x^4 + p_5x^5$$

for each of the 100 data points x_i, $i = 0, 1, \ldots, 99$. This is standard least-squares curve fitting. We used the Python *numpy* package [3] to calculate the coefficients (see *polyfit*). These coefficients are mapped to the current Q notation in `Q_init` and set to the variables `p0`, `p1`, `p2`, `p3`, `p4`, and `p5`, respectively. With these parameters we are able to approximate the sine in the range $[0, \pi/2]$ with,

```
1  fix_t Q_sin_poly(fix_t x) {
2      fix_t x2,x3,x4,x5;
3      fix_t ans;
4      int32_t sign;
5  
6      if (x < 0) x += PId;
7      sin_val_sign(&x, &sign);
8  
9      x2 = Q_mul(x,x);
10     x3 = Q_mul(x2,x);
11     x4 = Q_mul(x3,x);
12     x5 = Q_mul(x4,x);
13 
14     ans = p0 + Q_mul(p1,x) + Q_mul(p2,x2) + Q_mul(p3,x3) +
15           Q_mul(p4,x4) + Q_mul(p5,x5);
16     return sign*ans;
17 }
```

where as before, lines 6 and 7 adjust for negative arguments and map x to $[0, \pi/2]$ while setting `sign`. Lines 9 through 12 calculate the powers of x for the 5-th degree polynomial. Lines 14 and 15 directly implement the polynomial using `Q_mul` to multiply with a proper intermediate number of bits and ordinary addition thereby ignoring any potential overflow. Lastly, line 16 multiplies by `sign` and returns the answer. As before, we can write $\cos x$ and $\tan x$ using the polynomial sine,

```
fix_t Q_cos_poly(fix_t x) {
    return Q_sin_poly(Q_sub(PIh,x));
}

fix_t Q_tan_poly(fix_t x) {
    return Q_div(Q_sin_poly(x), Q_cos_poly(x));
}
```

Comparison of Trigonometric Functions All three sets of trigonometric functions implemented above have a version of sine at their heart. Therefore, we can compare them by looking at the accuracy of the sine functions. If we generate the sine, using the table, series expansion, or polynomial fit, for 100 randomly selected angles in $[0, 2\pi)$ we can compare the resulting values to the actual floating-point sine. Any deviation can be used to assess the quality of the sine approximation.

We generate the table with code like this,

```
1  Q_init(7);
2  srand(time(NULL));
3
4  for(i=0; i<100; i++) {
5      f = 2*3.14159265*((double)rand()/RAND_MAX);
6      a = Q_to_fixed(f);
7      printf("%0.8f %0.8f %0.8f %0.8f %0.8f\n", f, sin(f),
8          Q_to_double(Q_sin_table(a)),
8          Q_to_double(Q_sin_taylor(a)),
9          Q_to_double(Q_sin_poly(a)));
10 }
```

where we include the `stdlib.h` C library and use `srand` to seed the random number generator which we call in line 5 to set `f` to a random value in the range $[0, 2\pi)$. We note that it is well known that `rand` is a very poor pseudo-random number generator and that we have a very (very) small chance of selecting 2π as `rand` returns a number from $[0, \mathit{RAND\_MAX}]$ and we are dividing by `RAND_MAX`. However, for our purposes we accept the inefficiency and take the risk. The reader is encouraged, however, to develop an understanding of pseudo-random number generation, which is fascinating, by referring to one of the many good texts on the subject. For example, see Gentle's *Random Number Generation and Monte Carlo Methods* [4].

Now that we have our randomly selected sine values we can calculate the delta between the floating-point value and the three approximations for each angle and then report the absolute maximum deviation ($|\Delta|$), the mean ($\bar{x}$), and the standard deviation (σ) of the deltas. These will give us a feel for how the functions perform over the their full range. The results of this analysis for *Q*7.24 numbers are in Table 6.1.

Table 6.1 Comparison of the difference between the floating-point sine and each of the fixed-point sine functions using summary statistics for 100 randomly selected angles over the entire range $[0, 2\pi)$

Function	$\|\Delta\|$	$\bar{x}$	σ
`Q_sin_table`	8.4×10^{-3}	-6.5×10^{-5}	3.8×10^{-3}
`Q_sin_taylor`	8.2×10^{-3}	-7.8×10^{-6}	8.2×10^{-3}
`Q_sin_poly`	7.7×10^{-6}	-1.1×10^{-7}	3.9×10^{-6}

We see that while all three approximations are adequate with relatively small errors, on average, it is clear that `Q_sin_table` is the least accurate, as expected because it forces the argument to the nearest degree, and that `Q_sin_poly` is to be preferred over `Q_sin_taylor` in terms of accuracy. If we count the number of multiplications we see that `Q_sin_poly` is also more efficient than `Q_sin_taylor` with nine multiplications to as many as twelve.

Square Root and Transcendental Functions In addition to basic trigonometric functions we want to add square root, natural logarithm and exponential functions to our library. Let's look at how we can implement these functions starting first with square root.

Perhaps the fastest way to implement a fixed-point square root function is to use Newton's method. This is an iterative method to find the roots of a function by solving $f(x) = 0$. The method selects an initial guess for the root, x_0, and then iterates,

$$x_{i+1} = x_i - \frac{f(x_i)}{f'(x_i)}$$

until convergence where $f'(x)$ is the first derivative of $f(x)$. In this case, we want $f(x) = x^2 - n = 0$ which will find the value of x which when squared equals a specific number n. This means that x will be equal to the square root of n which is exactly what we want. If we work through the steps we have,

$$\begin{aligned}
f(x) &= x^2 - n \\
f'(x) &= 2x \\
x_{i+1} &= x_i - \frac{f(x_i)}{f'(x_i)} \\
&= x_i - \frac{x_i^2 - n}{2x_i} \\
&= x_i - \frac{1}{2}(x_i - \frac{n}{x_i}) \\
&= \frac{1}{2}(x_i + \frac{n}{x_i})
\end{aligned}$$

which implies that we need only an initial guess followed by some number of iterations of the final equation above to generate a good approximation to the square root of n. It is known that this method converges quickly so we only need a handful of iterations.

Translating this into code gives,

```
1  fix_t Q_sqrt(fix_t n) {
2      fix_t a;
3      int32_t k;
4
5      if (n <= 0) return -(1<<N);
6      if (n == (1<<N)) return (1<<N);
7
8      a = n >> 1;
9
10     for(k=0; k < 10; k++)
11         a = Q_mul(1 << (N-1),Q_add(Q_div(n,a),a));
12
13     return a;
14 }
```

where we first check for negative arguments and return -1 if we have one in line 5. Recall that our fixed-point numbers are binary fractions scaled by `N` bits so shifting `1` up `N` bits gives us a value equal to one in our current Q notation. Line 6 is a quick check to see if our argument is one. If so we immediately return the square root (1). Line 8 is our initial guess which we store in `a`. We are looking for a square root so our initial guess is $n/2$ which is exactly what we get by shifting the argument one bit position to the right. The loop in lines 10 and 11 performs the actual iteration. We fix the number of iterations at ten though a more sophisticated approach would perhaps be sensitive to the number of bits in the fractional part of our numbers and adjust the iteration limit accordingly. Line 11 is a direct translation of our final equation above using our previously defined arithmetic functions. Note that for simplicity we are not checking for any overflow conditions. Also note the use of $1 << (N-1)$ which is exactly 0.5 for `N` bits in the fractional part. Finally, line 13 returns `a` which is the approximate square root of `n`.

Repeating a random analysis similar to the one performed above for the sine functions we can compare `Q_sqrt` to the floating-point `sqrt` function for 100 randomly selected values from $[1, 1000)$ using $Q10.21$ format numbers. If we do this we get the following results,

$\lvert\Delta\rvert$	$\bar{x}$	σ
3.6×10^{-7}	-1.3×10^{-7}	1.4×10^{-7}

which indicates that this approach gives quite accurate answers in general.

We consider the exponential function, e^x, next as we will need this function in order to efficiently implement the natural logarithm. The Taylor series expansion of e^x is particularly well suited to implementation on a computer. The series itself is,

$$e^x = 1 + x + \frac{x^2}{2!} + \frac{x^3}{3!} + \frac{x^4}{4!} + \cdots = \sum_{i=0}^{\infty} \frac{x^i}{i!}$$

Looking at the series shows that if the current term in the series is t_i then the next term in the series is $t_{i+1} = t_i(\frac{x}{i})$ which means that we need not calculate the factorials completely but will get them by multiplying the current term by $\frac{x}{i}$ and adding it to a running total. If we accumulate terms until we get a term that is below a given threshold, say too small to add anything meaningful to the total given the number of bits used for the fractional part of the fixed-point number, we have a particularly nice implementation. Writing this in C using `EXP_TOLERANCE` to be the cut-off value for our terms, 0.000001 or the like, and adding a maximum number of iterations (`EXP_ITER_MAX` = 50) gives the following,

```
1  fix_t Q_exp(fix_t x) {
2      int32_t i;
3      fix_t e, c, t, tol;
4
5      tol = Q_to_fixed(EXP_TOLERANCE);
6      i = 0;
7
8      c = 1 << N;
9      e = 1 << N;
10     t = x;
11
12     do {
13         e = Q_add(e,t);
14         c = Q_add(c, 1<<N);
15         t = Q_mul(t, Q_div(x,c));
16         i++;
17     } while ((abs(t) > tol) && (i < EXP_ITER_MAX));
18
19     return e;
20 }
```

where line 5 sets the tolerance and line 6 initializes the iteration counter. Line 8 sets `c` to one. This is the term counter and is set to one because the first term in the series (1) is used to initialize the accumulator, `e` in line 9. The current term is then the second term in the series, which is $x = x^1/1!$, and we initialize `t` with it. The loop in lines 12 through 17 adds new terms to the accumulator (line 13) while bumping the term counter (line 14). Line 15 calculates the next term in the series from the previous term and sets `t` accordingly. Line 16 increments the iteration counter while line 17 checks to see if we can exit the loop by either calculating a term below the tolerance or by reaching the limit on iterations. At this point `e` contains the sum of the series, which is e^x, and it is returned in line 19.

Repeating the random value test for 100 random arguments in the range $[-4, 4)$ for $Q10.21$ format numbers gives,

$\|\Delta\|$	$\bar{x}$	σ
2.0×10^{-5}	-9.2×10^{-7}	4.5×10^{-6}

showing that `Q_exp` does give good results.

With e^x implemented we can now add our final library function, the natural log ($\log x$) which is implemented using Newton's method just as square root was. In this case, our function is $f(x) = e^x - c = 0$ which is zero when x is the log of c. The first derivative of this function is $f'(x) = e^x$ which means our iteration function is,

$$x_{i+1} = x_i - \frac{f(x_i)}{f'(x_i)} = x_i - \frac{e^{x_i} - c}{e^{x_i}} = x_i - 1 + ce^{-x_i}$$

where we again set our initial guess to $x_0 = c/2$. Translating to C code gives,

```
1  fix_t Q_log(fix_t x) {
2      fix_t a;
3      int32_t k;
4  
5      if (x <= 0) return -(1<<N);
6      if (x == (1<<N)) return 0;
7  
8      a = x >> 1;
9  
10     for(k=0; k < 7; k++) {
11         a = Q_add(a, Q_mul(x,Q_exp(-a)));
12         a = Q_sub(a,1<<N);
13     }
14 
15     return a;
16 }
```

where a sanity check is made in line 5 to disallow negative arguments (return -1) and a quick check for a known value ($\log 1 = 0$) is in line 6. Line 8 sets a to our initial guess of $c/2$. Lines 10 through 13 are the iteration loop. This is why we needed to implement e^x before we could implement the logarithm. Line 11 calls `Q_exp` as part of the update to a. Line 12 subtracts one from the sum in line 11. The number of iterations is hard-coded to seven which gives good results. Again, a more sophisticated approach would be aware of the number of bits in the fractional part of our fixed-point numbers and set the loop limit accordingly. Line 15 returns a as our approximation of the natural log.

A final repeat of the random value test with 100 random arguments from [0.5, 10) gives,

$\|\Delta\|$	$\bar{x}$	σ
1.3×10^{-5}	4.0×10^{-7}	4.4×10^{-6}

again showing good agreement with the floating-point C library function.

At this point our C library of fixed-point functions is complete.

6.4 An Emerging Use Case

In this section we detail an emerging, and important, use case for fixed-point numbers: machine learning. As of this writing, Fall 2016, there is currently a revolution underway in terms of machine learning. Machine learning is a branch of artificial intelligence that uses statistically-learned models to enable machines to make predictions mimicking many feats that were previously restricted to humans. For example, object detection in images, speech recognition, automatic language translation, medical record interpretation, advanced game playing, etc. are all areas where machine learning has demonstrated its usefulness. Personal assistants for our smartphones, like Siri, Google Now and Cortana, are prime examples of machine learning in action.

At its core, the machine learning revolution is powered by advanced neural networks. It is beyond the scope of this book to describe all that this entails but we will illustrate the relevant points so that the link between neural networks and fixed-point numbers is clear.

What is a Neural Network? For our purposes a neural network can be thought of as a feed-forward, directed, acyclic graph. On one end are a set of inputs, the "feature vector", which represents the thing we want to make a prediction on and at the other end is one or more outputs which are the predictions. A classic example is a network that accepts an image of a hand-written digit and outputs which digit the image represents, $0 \ldots 9$.

A neural network itself consists of layers, two of which are mentioned above. The power of the network comes from the addition of one or more hidden layers between the input and output. Each layer accepts input from the layer below and passes its output to the layer above. For each layer there are a set of nodes ("neurons") which accept multiple inputs, sum them after multiplying each input by a weight value, and applying a nonlinear function to this total in order to output a value for the node (the activation). The final layer output, often characterized as a probability, is typically used, for classification, to select the largest output value as the output for the given input. For the digit image example, the output is a set of ten numbers, one for each possible digit, and the input is assigned the label of the largest of these outputs.

A traditional neural network is illustrated in Fig. 6.3. The networks in current use are significantly more complex and for many tasks include convolutional, pooling and other layers. We will ignore the various architectures since in the end what we are concerned with here are the weights and biases assigned to the connections between the layers and nodes, regardless of the way those connections are architected.

Neural networks are usually trained using labeled training data. These are sets of input feature vectors and the known label that goes with them. This training process modifies the weights and biases (the connections between nodes) to so that the model represented by the network maximizes the likelihood of the training data given the network model. How this is done, while crucially important practically, is of no interest to us here. We will assume that we have a neural network that has

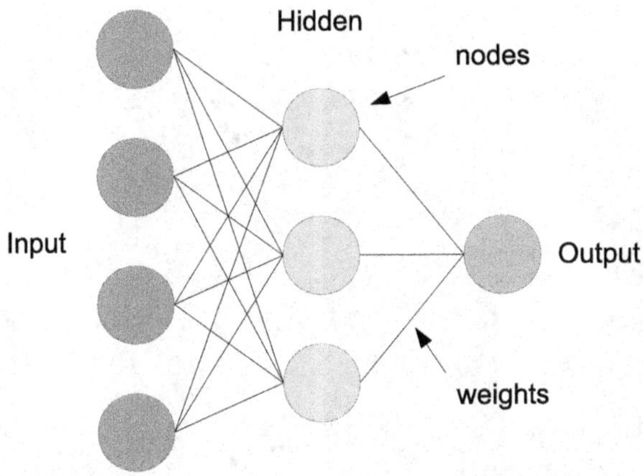

Fig. 6.3 A traditional neural network. Input values (*left*) are passed to each node of the hidden layer (*middle*) and the output of the hidden layer is passed to the output (*right*). Each connection between layers has an associated weight value (not shown) which is multiplied by the input value. Additionally, each node has an associated bias value which is added to the sum of the weights and inputs (also not shown) before passing the total through the activation function for the node. When put together, application of a traditional neural network boils down to the multiplication of input vectors by a weight matrix, the addition of a bias vector, and passing each element of the resulting vector through the nonlinear function to become the input of the layer above

been trained successfully. What we are concerned with are the weights and biases that go with the final set of connections between the layers of the network. These are the magic values that turn the generic network into one that is capable of making meaningful predictions.

Using Fixed-Point Numbers in Neural Networks Let's consider a simple neural network example which illustrates using fixed-point numbers and their effect on the predictions made by a trained network. All of the files and data referred to in this section are available on the website associated with this book or by contacting the author at `numbersandcomputers@gmail.com`. The network we will train and then evaluate using fixed-point numbers is a simple one. It accepts as input a small 40×40 pixel image of one of two types of keys imaged end on using a laser. Full resolution examples for the two classes are shown in Fig. 6.4. For training we make the images grayscale, threshold them to fill in some of the variation in intensity, reduce them to 40×40 pixels from 400×400, and finally scale them [0, 1] by dividing the pixel intensities by 255. We then treat the 40×40 images as 1600 element vectors ($= 40 \times 40$). So, in the end, we are training the network with feature vectors of 1600 elements each with a label of 0 or 1. The output of the network can be considered to be the probability of membership in class 1. If the output is ≥ 0.5 we assign the image to class 1, otherwise we assign it to class 0.

The network we will train is a traditional neural network with a single hidden layer of 500 nodes. We will use a rectified linear activation function. This function is currently very popular and is trivial to implement: if the input to the function is

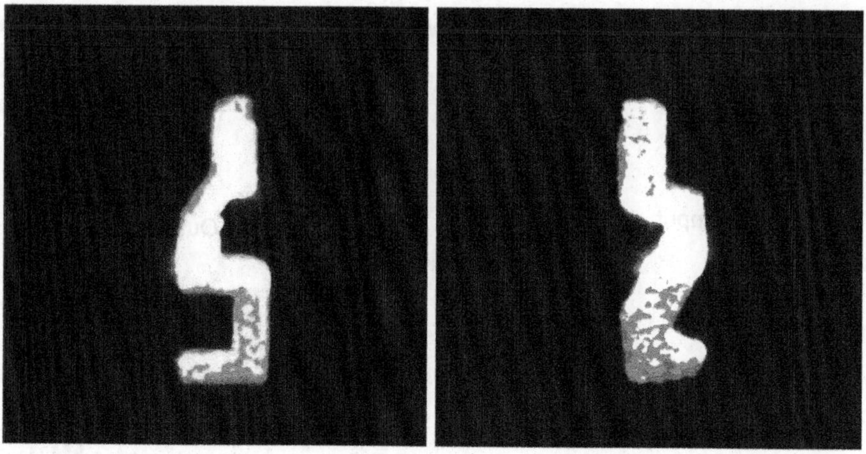

Fig. 6.4 Examples of the two key types in the keys data set. The images are end-on views of the keys created by combining multiple laser scans. The images used for training are 40×40 pixel grayscale versions of these images. Class 0 is on the *left*, class 1 is on the *right*

positive, return it, otherwise return zero. We train the network using the Scikit Learn library, specifically the `MLPClassifier` class,

```
1  import numpy as np
2  from sklearn.neural_network import MLPClassifier
3  import cPickle
4
5  trn  = np.load("train.npy")
6  trnl= np.load("train_labels.npy")
7
8  mlp = MLPClassifier(hidden_layer_sizes=(500,), max_iter=400,
9            alpha=1e-4, algorithm='sgd', verbose=10, tol=1e-4,
10           random_state=1, activation='relu')
11
12 mlp.fit(trn, trnl)
13 cPickle.dump(mlp, open("classifier.pkl","w"))
14
15 mlp.coefs_[0].tofile(open("weight0.raw","w"))
16 mlp.coefs_[1].tofile(open("weight1.raw","w"))
17 mlp.intercepts_[0].tofile(open("bias0.raw","w"))
18 mlp.intercepts_[1].tofile(open("bias1.raw","w"))
```

where lines 5 and 6 load the previously calculated training data (line 5) and associated labels (line 6). Line 8 defines the neural network and training parameters. The important part here is that there is one hidden layer of 500 nodes and that the rectified linear activation function is used. Actual training is done with the single call in line 12. Line 13 stores the object representing the neural network for possible future use. Lines 15 through 18 store the weights and biases for the input to hidden layer and the hidden layer to output. These are the numbers that tailor the network to the task of separating the two key type images. We will use these files below in a C program to apply the network to held-out test images. Note that in this case the trained classifier achieves 100% accuracy on the test data when using the full double precision floating-point weights and biases. We will examine how switching to fixed-point calculations with varying fractional precision affect these results. Achieving 100% accuracy on held-out test data is not typical in real-world machine learning problems but is a reflection of the simplicity of the task we are asking the network to learn in this case.

In order to use the trained network with our C library for fixed-point arithmetic we need a function capable of applying an input feature vector in order to generate an output. For the true double precision weights and biases we can use the following,

```
1  double weight0[1600][500];
2  double weight1[500];
3  double bias0[500];
4  double bias1;
5  double test[263][1600];
6  int    labels[263];
7
8  double ApplyClassifier(double *vec) {
9      double d, z[500];
10     int i,j;
11
12     for(i=0; i < 500; i++) {
13         d = 0.0;
14         for(j=0; j < 1600; j++) {
15             d += vec[j]*weight0[j][i];
16         }
17         d += bias0[i];
18         z[i] = (d < 0) ? 0.0 : d;
19     }
20
21     d = 0.0;
22     for(i=0; i < 500; i++)
23         d += z[i]*weight1[i];
24     d += bias1;
25
26     return 1.0 / (1.0 + exp(-d));
27 }
```

where lines 1 through 6 define arrays and variables holding the weights, biases, labels and training data. For example, since there are 1600 inputs (the 40×40 image) and 500 hidden nodes the first weight matrix (`weight0`, line 1) is 1600×500 so that there is a weight for each input to each hidden layer node. Similarly, as there are 500 hidden layer nodes there are also 500 bias values, one for each node (`bias0`, line 3). To go from the hidden layer to the single output value we only need 500 weights, one for each hidden node (`weight1`, line 2). Finally, as we have a single output value we need only a single bias value in `bias1`.

Lines 5 and 6 hold the test data. These are images that were not used during training so that we can test the network to see how well it is performing. We have 263 test images so that means we have 263 feature vectors (line 5) and 263 corresponding labels (line 6). We will apply each of these 263 feature vectors to the network below.

Lines 8 through 26 apply a new feature vector (`vec`) to the network and return a prediction. This network uses a sigmoid, $1/(1 + e^{-x})$, to convert the final output value in `d` to a probability estimate of belonging to class 1 (line 26). The sigmoid is restricted to the range [0, 1] which explains the "probability" interpretation.

The function `ApplyClassifier` first applies the input vector to the hidden layer nodes (lines 12 through 19) and then applies those outputs (`z`) to the output (lines 21 through 24). Line 12 loops over each of the 500 hidden nodes multiplying each of the 1600 input feature vector values by the corresponding weight (line 15) and stores the running total in `d`. Line 17 adds in the bias value for that hidden node. Line 18 applies the rectified linear activation function setting the output in `z` to zero or `d` depending on the sign of `d`.

At this point `z`, a 500 element array, contains the 500 activations of the hidden layer. To calculate the output value these need to be multiplied by each of the 500 weights of the second weight vector (line 23) with the final single bias value added in line 24 so that `d` now contains a single value representing the sum of the hidden layer outputs. Line 26 applies the sigmoid to return the prediction itself.

The output of `ApplyClassifier` can be compared against a threshold, typically 0.5, and if equal to or greater we can assign the input image to class 1. If we call `ApplyClassifier` for each of the 263 test images we will have a record of predictions against images with known labels. Once we have these we can calculate standard metrics typically used by the machine learning community to interpret the performance of a classifier. In this case, we will calculate the accuracy and the area under the *receiver operating characteristics* (ROC) curve. The accuracy is a natural measure and in this case helpful because the test data set is balanced (each class shows up equally often, statistically) and because we will be comparing the same network architecture repeatedly. The higher the accuracy the better because it means that the classifier correctly assigned class labels to the test data. We report accuracy as a fraction [0, 1].

The area under the ROC curve is a more meaningful measure which is appropriate when comparing different classifiers. Entire books have been written about what it really is and how to best calculate and interpret it. For our purposes we can consider the area under the ROC curve, denoted AUC, as a measure between 0.5 and 1.0. A perfect classifier will have an AUC of 1.0 while a classifier that simply guesses the class label will have an AUC of 0.5. So, like accuracy, higher AUC is better.

The code that loads the stored weights, biases, labels and test feature vectors is straightforward, a series of calls to `fopen`, `fread`, and `fclose` that we need not waste space listing here. The code to apply the network to each test image and to generate a prediction is also straightforward,

```
1 for(i=0; i < 263; i++) {
2     pred = ApplyClassifier(&test[i][0]);
3     printf("%d %0.8f %0.8f \n", labels[i], 1.0-pred, pred);
4 }
```

where the output prediction [0, 1] is in `pred` for test image number `i`. Line 3 is a bit unusual. It outputs the known class label in `labels` and then the probability of class 0 followed by the probability of class 1. Since this is a binary classifier and we are predicting likelihood of class 1 membership the likelihood of class 0 is simply `1-pred`. This output format allows us to use existing tools to simply calculate the accuracy and AUC for the test data. The tools are included in the archive for this section but are too extensive to list here in the text.

The above is a lengthy description of how to apply the trained network to the test data. As already mentioned, when using the double precision floating-point weights and biases the test set is perfectly classified meaning the accuracy is 1.0 as is the AUC.

In order to use our fixed-point library with different precisions we need to replace `ApplyClassifier` with a version that uses only the library. A slavish translation to fixed-point arithmetic gives,

```
double ApplyClassifierFixed(double *vec) {
    fix_t d, z[500];
    int i,j;

    for(i=0; i < 500; i++) {
        d = 0;
        for(j=0; j < 1600; j++) {
            d = Q_add(d, Q_mul(Q_to_fixed(vec[j]),
                               Q_to_fixed(weight0[j][i])));
        }
        d = Q_add(d, Q_to_fixed(bias0[i]));
        z[i] = (d < 0) ? 0.0 : d;
    }

    d = 0;
    for(i=0; i < 500; i++)
        d = Q_add(d, Q_mul(z[i], Q_to_fixed(weight1[i])));
    d = Q_add(d, Q_to_fixed(bias1));

    return Q_to_double(Q_div(Q_to_fixed(1.0),
                 Q_add(Q_to_fixed(1.0),
                 Q_exp(Q_mul(Q_to_fixed(-1.0),d)))));
}
```

where we have replaced every calculation with a call to our fixed-point library. For clarity we convert constants each time through to preserve the symmetry with `ApplyClassifier` above. The calculations are the same.

We can now explore the effect of varying precisions by setting the fixed-point precision using `Q_init` to set the integer portion and then call `ApplyClassifierFixed` for each test feature vector. For each precision we can calculate the accuracy and AUC as well as the average deviation of the output predictions from the gold standard values found using double precision weights and biases. If we do this we get,

N	Accuracy	AUC	$\|\Delta\|$
21	1.0000	1.0000	0.0001
11	1.0000	1.0000	0.0124
10	1.0000	1.0000	0.0252
9	1.0000	1.0000	0.0438
8	1.0000	1.0000	0.0956
7	0.9316	0.9635	0.2728
6	0.8707	0.9416	0.3774
4	0.5361	0.5863	0.5858

where N is the number of bits in the fractional part of the fixed-point representation, accuracy and AUC as above and $|\Delta|$ is the average absolute difference between the gold standard predictions and the fixed-point predictions.

How should we interpret these results? The gold standard uses double precision floating-point numbers with 53 binary bits of precision in the significand. The table shows that with only 21 bits of precision in the fractional part we have virtually identical output to the double precision values. This illustrates a general effect seen with many neural networks: that they are largely insensitive to the precision used for the weights and biases. Indeed, most software packages for neural networks use 32-bit floating point numbers to save memory and disk space. However, we also see above that we still make accurate predictions even when, for this test set, we use as few as 8 bits of precision in the fractional part. It is only when we get below eight bits that the output starts to degrade as illustrated by accuracies and AUCs of less than 1.0. Indeed, with only 4 fractional bits our classifier degrades to the point where it is in effect only guessing the class label.

If we look at the $|\Delta|$ column we get an idea of why the predictions are so robust. Even with 8 bits of precision we are averaging less than 0.1 deviation from the output of the full precision network. Neural network predictions are typically very one-sided meaning that outputs cluster around zero or one with few hovering around 0.5. If the average deviation by dropping to 8 bits of fractional precision is < 0.1 it is unlikely that this will force significant misclassifications.

However, dropping to 7 bits of precision causes the deviation to jump to nearly 0.3 which can, in some cases, cause a change in the predicted class. We see this with a drop in accuracy to 93%. The case for 6 bits is even worse with a drop in accuracy to 87% because of the larger average deviation of about 0.4. The 4 bit case is simply too little precision and as expected the classifier is basically guessing due to the 0.6 deviation from the gold standard.

This experiment illustrates that extreme precision is not necessary to implement accurate and effective neural networks. The effect is seen in more advanced network architectures as well, such as convolutional neural networks, which are the standard for object detection in images. For example, see [7]. This effect has encouraged research into specialized hardware implementing fixed-point arithmetic for low power applications [8].

6.5 When to Use Fixed-Point Numbers

Perhaps the most obvious time to use fixed-point numbers is when working on a system that does not have floating-point capability in hardware. This includes many small systems, like microcontrollers and digital signal processors (DSP), but also includes older computers that lack floating-point hardware. Using fixed-point numbers on these systems makes sense and is much faster than using software to emulate floating-point hardware. For example, older Macintosh computers based on the Motorola 68,000 and later processors often used software to emulate floating-point hardware. Called SANE (Standard Apple Numerics Environment), it implemented IEEE 754 floating point arithmetic in software for early Macintosh computers that lacked the floating-point hardware found in the Motorola 68,040 or did not use the 68,881 floating-point coprocessor. The SANE library provided maximum numeric compatibility ensuring that floating-point operations on one computer matched those of another that used hardware instead of software for the calculations. However, the simplicity of fixed-point arithmetic would have greatly improved performance on these early systems.

In 1993, Id Software released DOOM [5], one of the earliest first-person shooter (FPS) video games. FPS games give the user the illusion of being in a three-dimensional environment and as such require a lot of geometric graphics processing at high speed to enable realistic (limited for DOOM, granted) rendering of the environment. However, in 1993 only Intel 80,486DX microprocessors had built-in hardware floating point abilities. As processor speeds were very low compared to current standards, only tens of megahertz, it was crucial that calculations be as fast as possible while still using only software. Id's solution was to use fixed-point arithmetic for all graphics calculations. Specifically, they used $Q15.16$ format numbers which fit in a 32-bit signed integer.

Another reason to consider fixed-point math is power consumption. The floating-point unit on processors uses more power than the integer unit hence fixed-point arithmetic will use less power than the same calculations performed with hardware floating-point. This is particularly important in mobile applications.

Lastly, many popular audio and video codecs are designed to be implemented using fixed-point arithmetic so that simple devices that lack floating-point hardware can be used. For example, the open source Shine project [6] is a fast fixed-point MP3 encoder which can be used on simple computers.

6.6 Chapter Summary

In this chapter we introduced Q notation to specify fixed-point numbers. These numbers interpret an integer as a floating point number with a fixed number of fractional bits. We learned how to convert floating-point numbers to fixed-point and back. We also implemented arithmetic operations in C for fixed-point numbers using signed 32-bit integers. Additionally, we implemented trigonometric functions in three different ways and compared them for accuracy. To round out the library we then added square root, natural logarithm, and exponential functions. Lastly, we talked about when to use fixed-point numbers and gave some examples of places where fixed-point numbers have been used including an important emerging use case involving machine learning.

Exercises

6.1 The function `Q_sin_table` rounds its argument to the nearest degree and then returns the sine stored in the precomputed table for that degree. This can be made more accurate by linearly interpolating between the degrees that bound the input angle. For example, if the input value, as a fixed-point number in some $Qm.n$ representation, is equivalent to 45.4° then `Q_sin_table` will round down to 45° and return 0.70710678 as the sine. The actual sine of 45.4° is 0.71202605 so the returned value underestimates the true value.

In linear interpolation the values that bound the input, in this case $45° < 45.4° < 46°$, are used to determine the line between the values, $y = mx + b$, with $m = (y_2 - y_1)/(x_2 - x_1)$ and $b = y_1 - mx_1$ where $x_1 = 45$, $x_2 = 46$, $y_1 = \sin(x_1)$ and $y_2 = \sin(x_2)$. Then, $\sin(45.4)$ is equal to $m(45.4) + b$.

In this case, then, the improved answer is $\sin(45.4) = m(45.4) + b = 0.71199999$ with $m = 0.01223302$ and $b = 0.15662088$ for a difference on the order of 2.6×10^{-5} when compared to the true result. This is better than the difference of 4.9×10^{-3} by simply rounding to the nearest degree. Add linear interpolation to `Q_sin_table`. **

6.2 Add a function, `Q_ipow`, to the library that takes two arguments. The first is a fixed-point number, x, and the second is an integer, n, which may be negative. The return value of the function is x^n. *

6.3 Add a function, `Q_pow`, to the library that takes two arguments. The first is a fixed-point number, x, and the second is also a fixed-point number, y. The return value of the function is x^y. (Hint: $x^y = e^{y \log(x)}$.) *

6.4 Add a function, `Q_asin`, to the library that takes a sine value, positive and negative, and returns the angle for that sign, in radians. Use a Taylor series expansion,

$$\sin^{-1} x = x + \left(\frac{1}{2}\right)\left(\frac{x^3}{3}\right) + \left(\frac{1}{2}\right)\left(\frac{3}{4}\right)\left(\frac{x^5}{5}\right) + \left(\frac{1}{2}\right)\left(\frac{3}{4}\right)\left(\frac{5}{6}\right)\left(\frac{x^7}{7}\right) + \cdots$$

where you may need terms to x^9 or higher. (Hint: the coefficient of term t_{i+1} can be created from t_i.) **

6.5 Create a Python class, `Fixed`, that duplicates the arithmetic functionality of the C fixed-point library. Be sure to overload the operators for $+$, $-$, $/$, $*$ so that the fixed-point class can be used in expressions. **

6.6 The function `Q_sin_taylor` uses terms out to x^9 and calculates them explicitly for input x. Show via algebraic manipulation that the same truncated series sum can be found using only x and x^2. Then, update `Q_sin_taylor` to use this form. Note that you will need to introduce new constants in place of the values used for `st1` through `st4`. Update `Q_init` as well to set these new values. **

References

1. TMS320C64x DSP Library Programmer's Reference Literature Number: SPRU565B, Texas Instruments, October 2003.
2. https://code.google.com/p/libfixmath/.
3. http://www.numpy.org/.
4. Gentle, J. E. Random number generation and Monte Carlo methods. Springer, 2003.
5. Doom. Id Software. 1993. Video game.
6. https://github.com/savonet/shine.
7. Courbariaux, Matthieu, Jean-Pierre David, and Yoshua Bengio. "Training deep neural networks with low precision multiplications." arXiv preprint arXiv:1412.7024 (2014).
8. Gupta, Suyog, et al. "Deep learning with limited numerical precision." CoRR, abs/1502.02551 392 (2015).

Chapter 7
Decimal Floating Point

Abstract Decimal floating-point is an emerging standard which uses base 10 instead of base 2 to represent floating-point numbers. In this chapter we will take a look at how decimal floating-point numbers are stored using the IEEE 754-2008 standard as our reference. We will then look at a C language software implementation of decimal floating-point which conforms to the IEEE standard and give some examples of its use. We follow this with a look at the Python `decimal` library. We end with some thoughts on decimal floating-point and its use.

7.1 What is Decimal Floating-Point?

Modern computers make extensive use of floating-point arithmetic. Virtually all of this is accomplished with binary floating-point numbers. Binary floating-point numbers use base 2 and generally represent their numbers as a significand (mantissa), also binary, multiplied by 2 raised to some exponent. The standard for this type of floating-point number is given in the IEEE 754-2008 document [1] which has been universally adopted and includes hardware implementations for maximum performance. However, it is known that there are often accuracy issues with binary floating-point because of rounding errors or lack of precision when converting decimal numbers to binary. For example, decimal 0.1 is an infinitely repeating binary fraction which therefore must be truncated when stored in a finite number of bits.

Decimal floating-point (DFP) was introduced to address these issues. In DFP, the base is 10, not 2, so that decimal numbers may be represented exactly (to within the number of digits in the significand). The IEEE standard that defines binary floating-point also defines decimal floating-point formats and operations. This is the standard we will refer to in this chapter. Even though there is limited hardware support for DFP, notably from IBM, we will focus exclusively on software implementations here.

A binary floating-point number is represented as,

$$\pm d_0.d_1d_2d_3\ldots d_{p-1} \times 2^e, \;\; 0 \le d_i < 2$$

R.T. Kneusel, *Numbers and Computers*, DOI 10.1007/978-3-319-50508-4_7

where $d_0.d_1d_2d_3 \ldots d_{p-1}$ is the p digit *significand* (or *mantissa*), 2 is the base and e is the integer exponent. Similarly, a decimal floating-point number is represented as,

$$\pm d_0.d_1d_2d_3 \ldots d_{p-1} \times 10^e, \quad 0 \le d_i < 10$$

where the digits of the significand are decimal digits. This implies, as we will be storing our DFP numbers on a computer, that there is some sort of encoding for the significand so the digits $0 \ldots 9$ can be stored efficiently. This is not necessary in a binary floating-point number. For a description of the difference between a floating-point number, regardless of base, and a real number see the introduction to Chap. 3.

7.2 The IEEE 754-2008 Decimal Floating-Point Format

In this section we introduce decimal floating-point using the IEEE 754 format. We will describe the way in which DFP numbers are stored, using the fixed *decimal32*, *decimal64* and *decimal128* storage formats, and cover special values such as zero, infinity, and Not-a-Number (NaN). These values have direct analogues in binary floating-point.

Storage Formats The IEEE 754 standard specifies DFP numbers as a sign, coefficient and exponent so that the actual number is,

$$n = (-1)^{sign} \times \mathit{coefficient} \times 10^{exponent}$$

with the sign a single bit, 0 if positive and 1 for negative, and the exponent in excess binary notation. Excess notation means that the actual exponent is found by subtracting the bias from the stored value. There are two equivalent storage formats for the coefficient field, one as a binary integer and the other as a packed decimal format. In this chapter we will be working with a software library which uses the packed decimal format so that is the format we will concentrate on. The few existing hardware implementations of DFP use the packed format most often as well.

While the actual DFP value is specified by the sign, coefficient and exponent the actual way in which DFP numbers are stored in memory is fairly complicated. The *decimal64* format is made up of four different fields,

where S is the sign (1 bit), CF is the combination field (5 bits), $BXCF$ is the biased exponent continuation field (8 bits), and CCF is the coefficient continuation field

(50 bits). For a *decimal32* number the *BXCF* is 6 bits and the *CCF* is 20 bits. For *decimal128* these fields are 12 bits and 110 bits respectively.

Let's start by decoding an example of a *decimal64* number, in this case 0.1. While 0.1 is infinitely repeating when expressed in binary in decimal it is simply a single digit. If we encode 0.1 as a DFP number it will use eight bytes of memory (64 bits). We can then dump each of these bytes as a sequence of hexadecimal numbers,

$$0.1 \rightarrow \texttt{22 34 00 00 00 00 00 01}$$

from which we see that the fields are, in binary,

$$\begin{aligned} S &= \texttt{0} \\ CF &= \texttt{01000} \\ BXCF &= \texttt{10001101} \\ CCF &= \texttt{0000000000 0000000000 0000000000 0000000000 0000000001} \end{aligned}$$

where we have written the *CCF* as five groups of ten binary digits. We'll see why momentarily. For now, let's consider the other fields. The sign is straightforward enough. It is zero so the number is positive since $(-1)^0 = 1$. The *CF* field requires some explanation. It is a five bit field which encodes the two leftmost binary digits of the bias exponent along with the leftmost digit of the coefficient. There are ten possible coefficient digits and three different values of the leftmost two bits of the exponent for a total of 30 possible five bit combinations. As $2^5 = 32$ this means that 30 of the possible bit patterns encode the values we want and we still have two left over for other uses as we will see below. Table 7.1 shows all the bit patterns for the *CF* field and how to interpret them to get the exponent and coefficient digit values.

Table 7.1 Decoding the *CF* field

Digit	00	01	10
0	00000	01000	10000
1	00001	01001	10001
2	00010	01010	10010
3	00011	01011	10011
4	00100	01100	10100
5	00101	01101	10101
6	00110	01110	10110
7	00111	01111	10111
8	11000	11010	11100
9	11001	11011	11101

Locate the *CF* bit pattern to read off the leftmost two bits of the biased exponent (column) and the leftmost digit of the coefficient (row). After [2]

For our example we have `01000` which corresponds to exponent bits `01` and a leftmost coefficient digit of 0.

The remainder of the ten bit exponent is in the *BXCF* field. Combining this with the decoded *CF* value gives $0110001101_2 = 18D_{16} = 397$ for the biased exponent. The bias value for *decimal64* is 398 so the unbiased exponent is $397 - 398 = -1$. The bias value for *decimal32* is 101 while the bias for *decimal128* is 6176. The minimum and maximum exponents for each of the three encodings is therefore,

	Minimum	Maximum
decimal32	−101	90
decimal64	−398	398
decimal128	−6176	6111

The last part we need to decode is the coefficient field itself. We already know the leftmost digit of the coefficient field from the *CF* field. It is zero. The remainder of the coefficient is encoded in the *CCF* field. This field encodes the decimal digits of the coefficient in a packed format called *densely packed decimal* or DPD. This format uses ten bits to encode three decimal digits. This is more efficient than classical binary-coded decimal which would require 12 bits to encode three decimal digits See Chap. 2 for an explanation of binary-coded decimal.

The DPD encoding takes ten bits labeled *pqr stu v wxy* and maps them to three decimal digits represented in binary as *abcd efgh ijkm* where each letter is a binary digit (*l* is skipped to avoid confusion with 1). This group of ten bits is called *declet* and can be decoded by mapping through Table 7.2.

For our example we have five declets which corresponds to 15 decimal digits, three digits per declet, plus one additional leading (leftmost) digit of zero. This means that a *decimal64* number has 16 digits of precision. For *decimal32* the precision is 7 digits while for *decimal128* it is 34 digits. Four of the five declets in our example are all zeros mapping to three zero digits. The rightmost declet is `0000000001` which means that in Table 7.2 *v* is zero corresponding to the first row of the table. Therefore, the three decimal digits of this last declet are,

0pqr	0stu	0wxy
0000	0000	0001

Table 7.2 Unpacking a declet (*pqr stu v wxy*) by translating it into three groups of binary digits representing three decimal digits (*abcd efgh ijkm*)

vxwst	*abcd*	*efgh*	*ijkm*
`0---`	0pqr	0stu	0wxy
`100--`	0pqr	0stu	100y
`101--`	0pqr	100u	0sty
`110--`	100r	0stu	0pqy
`11100`	100r	0pqu	100y
`11110`	0pqr	100u	100y
`11111`	100r	100u	100y

After [2]

corresponding to 001 in decimal. So, finally, we have decoded the entire DPF number: sign, coefficient, and exponent. The final value is, as we expect,

$$n = (-1)^0 \times 0000000000000001 \times 10^{-1} = 1 \times 10^{-1} = 0.1$$

It should be noted that, unlike IEEE binary floating-point, DFP does not have a single exponent for each number. This means that there are possibly multiple representations of the same number since $1 \times 10^{-1} = 10 \times 10^{-2}$, etc. The standard defines a preferred exponent, based on the arguments to an operation, in order to produce consistent output. The set of representations mapping to the same number are called its *cohort*.

Packing numbers into a DPD declet is straightforward. For the example above we produced three BCD digits (four bits each, the value in binary is the value of the digit) and called the bits of each *abcd*, *efgh*, and *ijkm*. Using these labels for the bits we can take any set of three BCD digits and encode them into a DPD declet using Table 7.3. Recall that the ten bits of the declet are labeled *pqr stu v wxy* which can be read from the table by finding the row matching the *aei* bit values from the input BCD numbers. We will not work an example here but save using Table 7.3 for the exercises.

Special Values Just as IEEE 754 defines special values for binary floating-point, so it does for decimal floating-point. These are zero, which may be signed or unsigned but interpreted as zero regardless, infinities, and Not-a-Number (NaN). Let's consider each of these.

For *decimal64* the eight bytes representing +0 are `22 38 00 00 00 00 00 00`. Parsing this gives *S* of 0, *CF* of 01000, and *BXCF* of 10001110. As before, the *CF* implies that the first two bits of the exponent field are 01 while the first digit of the coefficient field is zero. Glancing at the remaining digits of the bytes representing zero makes it clear that the entire coefficient field is zero. What about the exponent?

Table 7.3 Packing three BCD numbers (*abcd efgh ijkm*) into a declet (*pqr stu v wxy*)

aei	*pqr*	*stu*	*v*	*wxy*
`000`	bcd	fgh	0	jkm
`001`	bcd	fgh	1	00m
`010`	bcd	jkh	1	01m
`011`	bcd	10h	1	11m
`100`	jkd	fgh	1	10m
`101`	fgd	01h	1	11m
`110`	jkd	00h	1	11m
`111`	00d	11h	1	11m

Locate the row by matching the first bit of each BCD number to the value under the *aei* column, then, the bits of the declet are read from the remaining columns of that row. After [2]

Combining the 01 from the *CF* field with the bits of the *BXCF* field gives,

$$e = 0110001110_2 = 18E_{16} = 398$$

which unbiased is $e = 398 - 398 = 0$. So, zero is represented in DFP as $(-1)^0 \times 0 \times 10^0$ which is as we would expect. What happens if we consider -0? The only value that changes is the sign giving $(-1)^1 \times 0 \times 10^0$, which is still zero.

If a number is too large, positive or negative, to be represented, a special value of infinity is used. To signal infinity the *CF* field is set to one of the two 5-bit values that are not used to encode the exponent. In this case, the *CF* field will be `11110`. The sign bit determines whether the value is $+\infty$ or $-\infty$. Other bits are ignored. So, if we try to represent 1×10^{1000} with a *decimal64* number, which is too large since the largest allowed exponent value is 369, we instead get infinity,

```
78 00 00 00 00 00 00 00
```

with,

$$S = \texttt{0}$$
$$CF = \texttt{11110}$$

and *BXCF* and *CCF* all zero. Negative infinity is the same with *S* set to 1.

The IEEE 754 standard for binary floating-point defines Not-a-Number values to represent attempts to encode values that are not legitimate numbers. NaNs are also defined for decimal floating-point by the special *CF* field value of `11111`. For example, attempting to calculate the logarithm of a negative number will result in a NaN. In addition to the bits of the *CF* field, the leftmost bit of the *BXCF* field is used to determine whether or not the NaN is signaling (bit set to 1) or quiet (bit set to 0). A signaling NaN will throw a floating-point exception while a quiet NaN will not.

When a NaN is generated the IEEE 754 standard allows implementations to use the *CCF* field to encode information about the source of the NaN. This is called its payload. Furthermore, the standard also ensures that subsequent operations involving a NaN will propagate the payload to the result of the operation. It appears, in practice, that this ability is seldom used. Instead, the appearance of the NaN is taken to mean there is an error in the expression or code without using the payload to provide any additional information. See Chap. 3 for an example using the payload for a binary floating-point number.

Rounding Modes The IEEE standard defines multiple rounding modes. The standard assumes that operations are performed with "infinite" precision and are then fit into the storage format at hand. This will require rounding of results so that the proper number of output digits can be set. Typically, users seldom modify these modes but there are times when it is beneficial to do so (interval arithmetic is one such instance). There are five rounding modes that can apply to decimal floating-point numbers. These are,

Name	Description
roundTowardsPositive	Round the result towards $+\infty$.
roundTiesToAway	Round the result to the closest representable value. Round ties to the larger in magnitude.
roundTiesToEven	Round the result to the closest representable value. Round ties towards the even value.
roundTowardsZero	Round the result towards zero.
roundTowardsNegative	Round the result towards $-\infty$.

with the default action to *roundTiesToEven*.

Storage Order For the examples in this section we have implicitly assumed that the bytes of the DFP number are stored in memory with the most significant byte first. This is called big-endian format and matches the way the bits are labeled in the diagrams of the standard. Naturally, the concept of big-endianness is more general than just this one instance. The library we will be using later in the chapter uses little-endian format. The difference is simply reversing the order of the bytes in memory. For convenience, our examples will continue to use big-endian format.

An Example Before we move on and start using the open source decimal floating-point library let's use what we've covered in this section to write a simple routine to parse a DFP number, stored as eight bytes, and output the value as an ASCII string. We will assume the storage order to be big-endian.

The main function, `dfp_parse()` is listed in Fig. 7.1. It accepts a big-endian *decimal64* number as an array of bytes and fills in a given string with the output value. The string array is assumed to be large enough.

In line 4 the sign of the output string is set. If the number is not negative the sign will be overwritten. Lines 5 through 7 extract the sign bit, *CF* field and *BXCF* field. The *BXCF* field is made up of the last two bits of the first byte and all of the second. At this point we can do two quick checks to see if the number is infinity or a NaN. If so, copy the proper string to the output and return (lines 9 and 10). Line 12 calls a helper function, `dfp_parse_exp()`, to extract the exponent and first output digit from the combined *CF* and *BXCF* fields. This function is,

```
void dfp_parse(unsigned char *b, unsigned char *s) {
    int sign, cf, bxcf, d0,d1,d2, dec, exponent, i=0;

    s[0] = '-';
    sign = (b[0] & 0x80) != 0;
    cf = (b[0] >> 2) & 0x1f;
    bxcf = ((b[0] & 0x3) << 6) + (b[1] >> 2);

    if (cf==0x1e) { strcpy(&s[sign],"inf"); return; }
    if (cf==0x1f) { strcpy(s,"NaN"); return; }

    exponent = dfp_parse_exp(cf, bxcf, &d0);
    i += sign;
    s[i++] = d0 + '0';

    dec = ((b[1] & 0x3) << 8) + b[2];
    dfp_parse_declet(dec, &d0, &d1, &d2);
    s[i++] = d0 + '0';
    s[i++] = d1 + '0';
    s[i++] = d2 + '0';
    dec = (b[3] << 2) + ((b[4] >> 6) & 0x3);
    dfp_parse_declet(dec, &d0, &d1, &d2);
    s[i++] = d0 + '0';
    s[i++] = d1 + '0';
    s[i++] = d2 + '0';
    dec = ((b[4] & 0x3f) << 4) + ((b[5] & 0xf0) >> 4);
    dfp_parse_declet(dec, &d0, &d1, &d2);
    s[i++] = d0 + '0';
    s[i++] = d1 + '0';
    s[i++] = d2 + '0';
    dec = ((b[5] & 0xf) << 6) + ((b[6] & 0xfc) >> 2);
    dfp_parse_declet(dec, &d0, &d1, &d2);
    s[i++] = d0 + '0';
    s[i++] = d1 + '0';
    s[i++] = d2 + '0';
    dec = ((b[6] & 0x3) << 8) + b[7];
    dfp_parse_declet(dec, &d0, &d1, &d2);
    s[i++] = d0 + '0';
    s[i++] = d1 + '0';
    s[i++] = d2 + '0';

    s[i++] = 'E';
    sprintf(&s[i], "%d", exponent);
}
```

Fig. 7.1 A function to parse a *decimal64* number

```
1  int dfp_parse_exp(int cf, int bxcf, int *d) {
2      static int digits[] = {0,1,2,3,4,5,6,7,
3                             0,1,2,3,4,5,6,7,
4                             0,1,2,3,4,5,6,7,
5                             8,9,8,9,8,9};
6      static int bits[] = {0,0,0,0,0,0,0,0,
7                           1,1,1,1,1,1,1,1,
8                           2,2,2,2,2,2,2,2,
9                           0,0,1,1,2,2};
10
11     *d = digits[cf];
12     return ((bits[cf] << 8) + bxcf) - 398;
13 }
```

which is an encoding of Table 7.1 so that the 5-bit *CF* value can be used to look up the proper digit and bits. The bits are combined with the *BXCF* to produce the biased exponent which is then reduced by subtracting the bias term (line 12), which is returned.

Once we have the exponent and first significand digit we can put it in the output (line 14) after setting up our index (line 13) which will cause the initial minus sign to be overwritten if the number is positive. Lines 16 through 40 appear daunting at first but they are really five repetitions of the same pattern. First we extract the 10 bits of a declet (lines 16, 21, 26, 31, and 36). Next, we recover the three decimal digits from this declet by calling the helper function `dfp_parse_declet()`. We then place the digits in the output string. The lines defining the declet are somewhat cryptic. They are pulling the 10 bits of the declet from the bytes of the input DFP number using standard bit manipulation to piece the declet together. A helpful exercise would be to write out all the bits for one of the hexadecimal examples above, say the one for 0.1, and then mark off the five declets. This will make clear the origin of the particular incantations used in this function. The helper function itself is an implementation of Table 7.2 and is given in Fig. 7.2 with `dfp_parse_bits()` defined as,

```
int dfp_parse_bits(int a, int b, int c, int d) {
    return 8*a + 4*b + 2*c + d;
}
```

The helper function assigns the bits from the input declet (lines 4 through 8) and then compares them to the values from the table in lines 10 through 44. When there is a match, the three decimal numbers are extracted by setting the proper bits according to Table 7.2.

After line 40 in Fig. 7.1 all that remains is to copy the exponent to the output string which is done in lines 42 and 43.

This example is only the simplest of ways to convert the DFP number to a decimal string. The output is given without any attempt to remove leading zeros of the significand or to scale the number to a more standard range. For example,

```
void dfp_parse_declet(int dec, int *d0, int *d1, int *d2) {
    int p,q,r,s,t,u,v,w,x,y;

    y = (dec & 0x1) != 0;      x = (dec & 0x2) != 0;
    w = (dec & 0x4) != 0;      v = (dec & 0x8) != 0;
    u = (dec & 0x10) != 0;     t = (dec & 0x20) != 0;
    s = (dec & 0x40) != 0;     r = (dec & 0x80) != 0;
    q = (dec & 0x100) != 0;    p = (dec & 0x200) != 0;

    if (v==0) {
        *d0 = dfp_parse_bits(0,p,q,r);
        *d1 = dfp_parse_bits(0,s,t,u);
        *d2 = dfp_parse_bits(0,w,x,y);
    }
    if ((v==1) && (x==0) && (w==0)) {
        *d0 = dfp_parse_bits(0,p,q,r);
        *d1 = dfp_parse_bits(0,s,t,u);
        *d2 = dfp_parse_bits(1,0,0,y);
    }
    if ((v==1) && (x==0) && (w==1)) {
        *d0 = dfp_parse_bits(0,p,q,r);
        *d1 = dfp_parse_bits(1,0,0,u);
        *d2 = dfp_parse_bits(0,s,t,y);
    }
    if ((v==1) && (x==1) && (w==0)) {
        *d0 = dfp_parse_bits(1,0,0,r);
        *d1 = dfp_parse_bits(0,s,t,u);
        *d2 = dfp_parse_bits(0,p,q,y);
    }
    if ((v==1) && (x==1) && (w==1) && (s==0) && (t==0)) {
        *d0 = dfp_parse_bits(1,0,0,r);
        *d1 = dfp_parse_bits(0,p,q,u);
        *d2 = dfp_parse_bits(1,0,0,y);
    }
    if ((v==1) && (x==1) && (w==1) && (s==1) && (t==0)) {
        *d0 = dfp_parse_bits(0,p,q,r);
        *d1 = dfp_parse_bits(1,0,0,u);
        *d2 = dfp_parse_bits(1,0,0,y);
    }
    if ((v==1) && (x==1) && (w==1) && (s==1) && (t==1)) {
        *d0 = dfp_parse_bits(1,0,0,r);
        *d1 = dfp_parse_bits(1,0,0,u);
        *d2 = dfp_parse_bits(1,0,0,y);
    }

    return;
}
```

Fig. 7.2 A function to parse a DPD declet into three BCD digits

if the function `dfp_parse()` is given `22 18 00 00 19 43 65 65` as input (in hexadecimal) it produces

```
0000000314159265E-8
```

as output. Clearly, a more complete implementation would change this to something closer to the more familiar 3.14159265 format. We leave implementing a function for *decimal32* DFP numbers to the exercises.

7.3 Decimal Floating-Point in Software

In this section we look at DFP implementations in C and Python along with some examples of their use. DFP operations are complex enough that beyond the example above to parse a DFP number it does not make sense to attempt an implementation of our own in this case.

DFP in C DFP implementations in hardware are not yet common. However, IBM has released, as open source, a DFP library which is compatible with the IEEE 754 standard. It is this library that we will be using in this section. The *decNumber* library is available here,

http://speleotrove.com/decimal/decnumber.html

along with some documentation on its use. We are using version 3.68 but later versions should operate in much the same way. Download the library and expand it. There is no Makefile or installation, the library is simply a set of source code files in C along with documentation and a few examples. The IEEE compliant portion is implemented in the `decimal32.c`, `decimal64.c` and `decimal128.c` files. We will focus on the *decimal64* implementation. Explore the other options the library provides as we will only scratch the surface here.

The library uses little-endian storage order. A useful example is `example5.c` which we can trivially modify to output in big-endian order. In a text editor, open `example5.c` and change line 30 from,

```
sprintf(&hexes[i*3], "%02x ", a.bytes[i]);
```

to,

```
sprintf(&hexes[i*3], "%02x ", a.bytes[7-i]);
```

so that bytes are output with the highest order byte first. With this change the program will output hex digits as we have been using them above. Compile the program with,

```
gcc example5.c -o example5 decimal64.c decNumber.c decContext.c
```

then test the program with,

```
./example5 0.1
```

which should produce,

```
0.1 => 22 34 00 00 00 00 00 01 => 0.1
```

matching the hexadecimal numbers we decoded in the previous section. This confirms that the *decNumber* library is working properly.

The full *decNumber* library is far more complete than what we will need here. We will work with IEEE 754 compliant functions as defined in the `decDouble.c` file. These functions ensure that we get results matching IEEE 754 decimal floating-point hardware. In order to compare results and performance between DFP and binary floating-point we will again make use of the logistic map which we first saw in Chap. 4. To review, this map iterates values between [0, 1] and can be used to demonstrate the bifurcation route to chaotic dynamic behavior, among other things. For the present, it is sufficient to know that this map has the form,

$$x_{i+1} = rx_i(1 - x_i)$$

for some initial x_0, which we set to 0.25 and a fixed r of 3.8 which is in the chaotic region.

First we will implement this map in C using variables of type `double`. We will then implement the map using the *decNumber* library and compare outputs and execution time. This will also illustrate how to use the library functions to set up a context, initialize variables, and perform arithmetic.

Figure 7.3 shows the plain C version. This version will initialize x and then output the first eight iterates. It then runs for 10 million more iterates and outputs the final eight. Note that we have broken the calculation into individual operations (lines 9–11) so that the DFP implementation matches in terms of the order in which the expressions are evaluated.

Figure 7.4 shows the *decNumber* library version. This requires some explanation. Before the main program we define a helper function, `pp()`, which will print a *double64* number by first converting it to a string. The function uses the `decDoubleToString()` library function. The constant `DECDOUBLE_String` is large enough to hold the biggest string that might be produced. The argument is a pointer to a `decDouble`. This is the base type for a *decimal64* number. The type is a structure which does not use any heap memory which means that it can be created, used, and ignored like a normal C variable.

The program starts in line 12 by defining the temporary variables we will use to store the logistic map calculation. Line 13 declares a context. This will be initialized to properly handle a *decimal64* number in line 17. This context must be passed to many of the library functions to get the expected behavior. The context includes the rounding mode. Here we leave it as the default. We will describe how to work with rounding modes below. Line 14 defines our variable, x, as well as r and a constant *one*. Initial values are set from strings in lines 19 through 21. Since this is base ten,

Fig. 7.3 A program to iterate the logistic map, $x_{i+1} = rx_i(1 - x_i)$, using IEEE 754 binary floating-point

```
#include <stdio.h>

int main(int argc , char *argv[]) {
    double a,b;
    double x=0.25, r=3.8;
    int i;

    for(i=0; i < 8; i++) {
        a = 1.0 - x;
        b = x * a;
        x = r * b;
        printf("%0.16f\n", x);
    }

    for(i=0; i < 10000000; i++) {
        a = 1.0 - x;
        b = x * a;
        x = r * b;
    }

    printf("\n");

    for(i=0; i < 8; i++) {
        a = 1.0 - x;
        b = x * a;
        x = r * b;
        printf("%0.16f\n", x);
    }

    return 0;
}
```

the value set is exact. Lines 23 through 28 iterate the equation eight times printing the value of x after each iteration. These will be compared to the initial values generated by the binary floating-point version. The equation itself is implemented in a way that mirrors the binary implementation but using the `decDoubleSubtract()` and `decDoubleMultiply()` library functions. There are similarly named functions for addition and division plus many other operations. The final eight iterates are printed with lines 38 through 43. Compile the programs with,

```
gcc logistic.c -o logistic
gcc logistic64.c -o logistic64 decContext.c decDouble.c decQuad.c
```

where `logistic64.c` needs both `decDouble.c` and `decQuad.c` so that computations can be performed with 128-bits and then rounded to 64-bits.

Let's compare the output of the two programs, Figs. 7.3 and 7.4. The first eight iterates for `logistic` and `logistic64` are,

```
#include "decDouble.h"
#include <stdio.h>

void pp(decDouble *x) {
    char s[DECDOUBLE_String];

    decDoubleToString(x, s);
    printf("%s", s);
}

int main(int argc , char *argv[]) {
    decDouble a,b;
    decContext set;
    decDouble x,r,one;
    int i;

    decContextDefault(&set, DEC_INIT_DECDOUBLE);

    decDoubleFromString(&x, "0.25", &set);
    decDoubleFromString(&r,  "3.8", &set);
    decDoubleFromString(&one, "1", &set);

    for(i=0; i < 8; i++) {
        decDoubleSubtract(&a, &one, &x, &set);
        decDoubleMultiply(&b, &x, &a, &set);
        decDoubleMultiply(&x, &r, &b, &set);
        pp(&x); printf("\n");
    }

    for(i=0; i < 10000000; i++) {
        decDoubleSubtract(&a, &one, &x, &set);
        decDoubleMultiply(&b, &x, &a, &set);
        decDoubleMultiply(&x, &r, &b, &set);
    }

    printf("\n");

    for(i=0; i < 8; i++) {
        decDoubleSubtract(&a, &one, &x, &set);
        decDoubleMultiply(&b, &x, &a, &set);
        decDoubleMultiply(&x, &r, &b, &set);
        pp(&x); printf("\n");
    }

    return 0;
}
```

Fig. 7.4 A program to iterate the logistic map, $x_{i+1} = rx_i(1 - x_i)$, using IEEE 754 decimal floating-point

`logistic`	`logistic64`
0.7124999999999999	0.71250
0.7784062500000002	0.77840625000
0.6554618478515620	0.6554618478515625
0.8581601326777955	0.8581601326777949
0.4625410135688510	0.4625410135688527
0.9446679324750937	0.9446679324750942
0.1986276333476368	0.1986276333476352
0.6048638471497435	0.6048638471497399

where we already see that after a few iterations the two sequences begin to diverge. Note that the decimal result is exact at least for the first two iterations while the binary result is already showing rounding error. The final eight iterations are,

`logistic`	`logistic64`
0.6245962562304587	0.9494112872063385
0.8910079371467448	0.1825120807398593
0.3690286137353397	0.5669654002706923
0.8848166847236699	0.9329594136330268
0.3872812528014777	0.2376753553568733
0.9017190393139833	0.6885059670888577
0.3367628911200712	0.8149689014131991
0.8487438558811845	0.5730194463417375

where there is no longer any relationship between the two sequences. This is after $10,000,000$ iterations. In this case, the departure is not fatal due to the shadowing theorem which states that for any chaotic trajectory the numerically computed trajectory using that initial condition (here 0.25) ultimately follows another chaotic trajectory with a slightly different initial condition. This theorem means that chaotic trajectories generated by computers are real.

If we use the Python rational arithmetic class developed in Chap. 5 we see that the first few iterates of the logistic map are exactly,

$$
\begin{aligned}
&57/80 &&= 0.7125000000000000\\
&24909/32000 &&= 0.7784062500000000\\
&3355964661/5120000000 &&= 0.6554618478515625\\
&112480764910343946501/ &&= 0.8581601326777950\\
&131072000000000000000 &&
\end{aligned}
$$

which matches the decimal floating-point output more closely than the binary.

Naturally, there is a large difference in performance between `logistic` and `logistic64` since the former is using hardware (on systems using Intel instructions). `Logistic64` is approximately 45 times slower than `logistic` so DFP's precision must be balanced by the runtime hit. For some applications it may not be feasible to use DFP.

```
int main(int argc , char *argv[]) {
    decDouble a;
    decContext set;

    decContextDefault(&set, DEC_INIT_DECDOUBLE);

    decContextSetRounding(&set, DEC_ROUND_HALF_EVEN);
    decDoubleFromString(&a, "0.12345678901234565", &set);
    pp(&a); printf("\n");
    decContextSetRounding(&set, DEC_ROUND_CEILING);
    decDoubleFromString(&a, "0.12345678901234565", &set);
    pp(&a); printf("\n");
    decContextSetRounding(&set, DEC_ROUND_UP);
    decDoubleFromString(&a, "0.12345678901234565", &set);
    pp(&a); printf("\n");
    decContextSetRounding(&set, DEC_ROUND_DOWN);
    decDoubleFromString(&a, "0.12345678901234565", &set);
    pp(&a); printf("\n");
    decContextSetRounding(&set, DEC_ROUND_FLOOR);
    decDoubleFromString(&a, "0.12345678901234565", &set);
    pp(&a); printf("\n\n");
    decContextSetRounding(&set, DEC_ROUND_HALF_EVEN);
    decDoubleFromString(&a, "-0.12345678901234565", &set);
    pp(&a); printf("\n");
    decContextSetRounding(&set, DEC_ROUND_CEILING);
    decDoubleFromString(&a, "-0.12345678901234565", &set);
    pp(&a); printf("\n");
    decContextSetRounding(&set, DEC_ROUND_UP);
    decDoubleFromString(&a, "-0.12345678901234565", &set);
    pp(&a); printf("\n");
    decContextSetRounding(&set, DEC_ROUND_DOWN);
    decDoubleFromString(&a, "-0.12345678901234565", &set);
    pp(&a); printf("\n");
    decContextSetRounding(&set, DEC_ROUND_FLOOR);
    decDoubleFromString(&a, "-0.12345678901234565", &set);
    pp(&a); printf("\n");
}
```

Fig. 7.5 A program to illustrate the effect of IEEE 754 decimal rounding modes

We conclude this introduction to the *decNumber* library by looking at rounding modes and their effect on calculations. Rounding modes are set using `decContextSetRounding()`. A simple example will show the effect. Figure 7.5 illustrates all the IEEE rounding modes using the *decNumber* library. The value is chosen so that the first digit beyond the limit of a *decimal64* number is 5 to enable a tie between representable numbers. The output of Fig. 7.5 is,

Mode	Value
roundTiesToEven	0.1234567890123456
roundTowardsPositive	0.1234567890123457
roundTiesToAway	0.1234567890123457
roundTowardsZero	0.1234567890123456
roundTowardsNegative	0.1234567890123456
roundTiesToEven	-0.1234567890123456
roundTowardsPositive	-0.1234567890123456
roundTiesToAway	-0.1234567890123457
roundTowardsZero	-0.1234567890123456
roundTowardsNegative	-0.1234567890123457

where the last digit of the converted number shows the choice made according to the IEEE rounding rules.

DFP in Python Python supports decimal floating-point through its `decimal` module. This module is really a wrapper on the *decNumber* library but is easier to use because of operator overloading. A quick example shows it in action,

```
1  >>> from decimal import *
2  >>> "%0.16f" % 0.1
3      '0.1000000000000000'
4  >>> "%0.16f" % (1.95-1.85)
5      '0.0999999999999999'
6  >>> setcontext(Context(prec=16))
7  >>> print Decimal("0.1")
8      0.1
9  >>> print Decimal("1.95") - Decimal("1.85")
10     0.10
```

After importing the DFP library the first two examples (lines 2 and 4) show why decimal floating-point might be desirable. The output should be the same, 0.1, but it is not because of rounding and the fact that 0.1 is a repeating value in binary.

Line 6 sets the context of the `decimal` module to use 16 digits. This matches the `decDouble` library used above and corresponds to an IEEE *decimal64* number. Note that both libraries support arbitrary precision but we restrict ourselves to IEEE sized numbers here.

The `Decimal` class defines a DFP number, typically by using a string as we did above. Lines 7 and 9 repeat lines 2 and 4 but show that there is no rounding error but only an exact result. Additionally, line 9 returns “0.10” instead of “0.1” to indicate that the result is accurate to two digits.

Figure 7.6 shows a Python implementation of the logistic example given above. The output is identical to that of `logstic64` as expected since the underlying library is the same and we have initialized the context to 16 digits of precision. However, the

Fig. 7.6 A Python version of the logistic map using the `decimal` module

```
from decimal import *

def main():
    setcontext(Context(prec=16))

    x = Decimal("0.25")
    r = Decimal("3.8")
    one = Decimal("1")

    for i in xrange(8):
        a = one - x
        b = x * a
        x = r * b
        print x

    for i in xrange(10000000):
        a = one - x
        b = x * a
        x = r * b

    print

    for i in xrange(8):
        a = one - x
        b = x * a
        x = r * b
        print x

main()
```

run time is significantly worse, taking 82× longer than the C *decNumber* version. This is due to the Python interpreter itself and the overhead of object creation and not the performance of the C library itself.

7.4 Thoughts on Decimal Floating-Point

Decimal floating-point has not increased in use as might have been expected. Humans work in base ten almost universally so making calculations in base ten seems quite reasonable. One reason for the slow adoption of DFP is the fact that hardware floating-point in binary is so much faster. The few hardware implementations of DFP are slower than their binary cousins but not terribly so. Still, even a 10% reduction in performance may not be acceptable in some circumstances.

The exactness of the results of DFP calculations makes this a perfect number format for commercial or financial work. Many, if not most, databases are using decimal storage for monetary values, often as character strings in base ten, so that the possible application of DFP to these databases is obvious. It would probably also result in smaller storage requirements and would certainly result in more accuracy or fewer checks for edge cases that cannot be allowed in financial calculations.

Is DFP useful for scientific programming? I would think so, if it were fast enough. The logistic example above clearly shows that DFP results are sometimes "truer" than binary floating-point. The preservation of knowledge that there are meaningful trailing zeros in a value is also potentially of significance. The Python example returning "0.10" as the answer instead of "0.1" is helpful as it tells users that the digit in the hundredths place is truly zero and not simply unknown. A scientific result might require a great deal of effort to know that that digit is exactly zero so it would be good if the computer representation preserved it.

While writing this chapter two potentially highly useful exercises came to mind. As far as I know as of this writing neither has been done but I present them as a challenge to motivated individuals,

- Python is widely used in scientific programming because of its well-supported array-processing libraries, particularly NumPy and libraries built on top of it (like SciPy). These array-processing libraries are highly optimized and performant. It would be perhaps quite useful to implement a version of NumPy using decimal floating-point in such a way that if compiled on a system with actual DFP hardware it could take advantage of it automatically.

- Interval arithmetic adds bounds to numbers by specifying a range in which the actual number exists rather than simply a single value. This requires proper control of rounding modes (see Chap. 8). A decimal floating-point based interval arithmetic package would perhaps be useful and could be combined with a DFP based NumPy for Python.

Hopefully this chapter has illustrated that decimal floating-point is of value and should be of primary concern to developers working with financial data. The thought perhaps should not be "should I use decimal floating-point?" but rather, "Is there any valid reason why I should not use decimal floating-point?"

7.5 Chapter Summary

In this chapter we reviewed the IEEE standard for storing decimal floating-point numbers using the format that seems to be in widest use. We learned how to decode such numbers and produce the corresponding decimal value as a string. We then examined an IEEE compliant C library for decimal floating-point and learned the basics of how to use it. We used the logistic map from chaos theory as an example of how decimal floating-point can be more accurate than binary floating-point for iterated calculations. We then examined a Python core module that uses the C library to implement a decimal floating-point class and showed that it produces the same results as before for the logistic map. We concluded by presenting, as food for future thought, two possible ways in which decimal floating-point might grow in its use which in turn would hopefully lead to more accessible hardware implementations.

Exercises

7.1 Parse the following *decimal64* numbers:

```
29 ff 98 42 d2 e9 64 45
25 ff 18 0c ce ef 26 1f
25 fe 14 44 ee 27 cc 5b
```

7.2 The function `dfp_parse()` in Fig. 7.1 works with *decimal64* numbers only. Create a version that works with *decimal32* numbers. Test it with the following inputs,

`A2 10 C6 15` → -3.1415
`F8 00 00 00` → -inf
`22 40 00 01` → 0.1

7.3 Encode the number 123.456 as a *decimal64* number. First decide on the significand and exponent to use, then encode the exponent and first digit of the significand to create the *CF* and *BXCF* fields. Last, use Table 7.3 to encode the significand mapping the remaining 15 decimal digits to five declets.

References

1. IEEE Standards Association. Standard for Floating-Point Arithmetic. IEEE 754-2008 (2008).
2. Duale, A., *et al.* Decimal floating-point in z9: An implementation and testing perspective. IBM Journal of Research and Development 51.1.2 (2007): 217–227.

Chapter 8
Interval Arithmetic

Abstract Floating-point numbers do not map exactly to the real numbers. Also, sometimes there are uncertainties in our knowledge of the true value for a quantity. Both of these situations can be addressed by using interval arithmetic which keeps bounds on the possible value of a number while a calculation is in progress. In this chapter we will describe interval arithmetic, implement basic operations for interval arithmetic in C, discuss functions as they pertain to intervals, examine interval implementation for C and Python, and finally offer some advice on when to use interval arithmetic.

8.1 Defining Intervals

A number like π is exact in the sense that there is no uncertainty as to its numeric value even though in this case it is impossible to write it. However, in the real world, there is almost never complete certainty as to the value of a quantity since most are measurements made by some finite device which has limitations. A scale may only be accurate to the ounce or gram. A ruler may only be accurate to a millimeter. Scientists associate uncertainties with measured quantities as a way to mathematically account for the inexactness of our dealings with reality. Similarly, since computers use a finite number of bits to store floating-point numbers, we know that it is impossible to store just any real number in a computer, we can only store the floating-point number we deem to be "nearest to" the actual real number.

Mathematics has ways to deal with uncertainty, be it through probability and statistics or through error analysis and propagation. The latter are mathematical concepts which involve a function, f, of input variables, say, x and y, and how to determine the uncertainty in $f(x, y)$, denoted as σ_f, given x, y and their uncertainties, σ_x and σ_y. It is the propagation of errors that decides how to calculate σ_f. This is an essential technique for science as a whole and the reader is directed to the very useful presentation in Bevington and Robinson [1] for examples and advice on how to calculate the uncertainty of expressions involving variables with uncertainties. For our purposes we need only consider what σ_x means for a particular x and how to represent that meaning in a computer using floating-point numbers. We will always assume that any floating-point numbers we use are IEEE 754 compliant.

R.T. Kneusel, *Numbers and Computers*, DOI 10.1007/978-3-319-50508-4_8

Scientific literature is full of measured quantities which are often given along with some uncertainty as to their actual value. This uncertainty may be estimated, the standard deviation of a set of measurements of a quantity, the mean of a set of measurements of a quantity along with the standard error of the mean, etc. For example, it might be reported that the temperature of a reaction was 74.0 ± 1.3 C where the ± 1.3 portion is the uncertainty. This means that, with some level of confidence, usually 95%, we know that the temperature of the experiment was in the interval $[72.7, 75.3]$ C because this is 1.3 C less and more than the first value reported. Instead of tracking mean values and uncertainties we could just track the interval itself and work with it. This is the essence of *interval arithmetic* and we will write intervals in the form $t = [72.7, 75.3]$ where the first number is the lower bound on the interval and the second is the upper bound (the bounds are included in the interval). Symbolically we will write $t = [\underline{t}, \overline{t}]$ where $\underline{t}$ is the lower bound and $\overline{t}$ is the upper.

We should slow down a bit here. The previous paragraphs mentioned the concept of "propagation of errors" without describing it and without discussing why one might choose interval arithmetic over propagating uncertainty. Without deriving it, we give the general formula for determining the uncertainty of the value of a function (σ_f) from the uncertainties of the arguments to that function. We assume $f(x, y)$ with argument uncertainties of σ_x and σ_y but the formula holds for any number of arguments, just use a similar term for each additional variable. The formula is,

$$\sigma_f^2 = \sigma_x^2 \left(\frac{\partial f}{\partial x}\right)^2 + \sigma_y^2 \left(\frac{\partial f}{\partial y}\right)^2$$

under the assumption that x and y are independent (i.e., not correlated with each other so that the covariance terms disappear) and that x and y represent random variables from Gaussian distributions.

Do not be concerned if the last sentence is not clear. The point to remember is that the formula used for propagation of errors is based on assumptions about the way in which the values behave, or, rather, the environment in which they were produced. And, as is always the case, an assumption may be incorrect. With intervals there is no assumption beyond the statement that the true value lies within the interval.

We will see below how to calculate intervals but let's quickly look at an example comparing intervals to propagation of errors. In this example we simply want to calculate $f(x, y) = x + y$ and find the uncertainty or interval containing $f(x, y)$. We use a C program which uses interval arithmetic to add x and y and to calculate σ_f from the formula above. When we do this we get the following answers,

$$\begin{aligned} \textit{propagation of errors} &= [23.344, 23.564] \\ \textit{interval arithmetic} &= [23.337, 23.571] \end{aligned}$$

for $x = [12.12, 12.34]$ and $y = [11.217, 11.231]$. Notice that while not exact the two methods produce comparable answers so we know at least that interval arithmetic might be a useful way to account for uncertainty.

In some sense the origin of interval arithmetic stretches back into antiquity. Archimedes, in his *On the measurement of the circle* estimated the value of π by using a two-sided bound which is in effect an interval since he showed that $\underline{x} \leq \pi \leq \bar{x}$ for the bounds $[\underline{x}, \bar{x}]$ which he placed on the value of π.

In modern times intervals were discussed by Young [2] and extended by Dwyer [3]. Sunaga again extended the algebra of intervals [4]. Dwyer discussed computing with approximate numbers but it was Moore who first wrote about implementing interval arithmetic on digital computers [5]. Moore refers to Dwyer and then extends intervals to floating-point numbers representable on the computer. Indeed, [5] is more than simply interval arithmetic and is an interesting historical work which discusses how to represent floating-point numbers on digital computers. Note that Moore calls interval arithmetic "range arithmetic" after Dwyer and then extends this to "digital range arithmetic" by which is meant the interval arithmetic that is the subject of this chapter.

The remainder of this chapter discusses interval arithmetic over basic operations (+,−,×,/, comparison) with examples in C and Python. Next comes a discussion of functions and intervals. This includes elementary functions like sine, cosine, and exponential along with general functions made up of expressions involving variables which are themselves intervals. We will see that general functions lead to some undesirable consequences and discuss the "dependency effect" which is the biggest drawback of interval arithmetic. After this we look at intervals in programming, examining some implementations. We conclude the chapter with thoughts on intervals and their use.

8.2 Basic Operations

Basic arithmetic operations are well-suited to working with intervals but before we start our library of functions we need to consider carefully how we will go about defining intervals using floating-point numbers. Mathematically, the bounds of an interval are real numbers and as such are exact. Unfortunately, computers do not operate on real numbers but instead use floating-point numbers which are by necessity a finite set. So, when an operation produces a result, and this is true even when the operands are already floating-point numbers, the result is seldom a floating-point number but must be rounded in some way to be expressed as a floating-point number. How should the result be rounded? IEEE 754 floating-point, which is the standard, has multiple rounding modes. By default, IEEE 754 states that results should be rounded to the nearest floating-point value. This is a completely reasonable thing to do but completely wrong when it comes to interval arithmetic.

If we write our interval arithmetic functions so that they round results to the nearest floating-point number we will quickly see that this introduces a bias that will make the intervals change so that we can no longer be certain that the actual number lies in the specified interval. We can fix this issue, however, by making use of the non-default rounding modes supported by IEEE 754. One of these modes is to always round down towards $-\infty$. Another is to round up towards $+\infty$. These two modes are what we need. When computing the lower bound of an interval we will always round towards $-\infty$ while for an upper bound we will round towards $+\infty$ (see [6]). This will ensure that our interval does not shrink during a calculation.

We begin our library by including required C libraries and by defining the data structure we will use to represent an interval,

```
#include <fenv.h>
#include <stdio.h>
#include <math.h>

typedef struct {
    double a;
    double b;
} interval_t;

char *pp(interval_t x) {
    static char s[100];
    sprintf(s, "[%0.16g, %0.16g]", x.a, x.b);
    return s;
}

interval_t interval(double a, double b) {
    interval_t ans;
    ans.a = a;
    ans.b = b;
    return ans;
}
```

where we will use the function `pp()` to display an interval which is of type `interval_t`. To maximize precision we use C `doubles` with `a` the lower bound and `b` the upper. The function `interval()` builds an interval from two floating-point numbers representing the bounds.

Note the inclusion of not only `stdio.h` and `math.h` but also `fenv.h`. This latter include is the floating-point library and it defines constants and functions related to floating-point including those necessary to change the rounding mode. Including `fenv.h` is not enough on its own, however. We must also compile our programs with the `-frounding-math` compiler switch (for `gcc`) otherwise the function calls for changing the rounding mode will be ignored.

Addition and Subtraction Now, let's take a look at how to add two intervals $x = [\underline{x}, \bar{x}]$ and $y = [\underline{y}, \bar{y}]$. In this case, the answer is not too surprising,

$$x + y = [\underline{x}, \bar{x}] + [\underline{y}, \bar{y}] \equiv [\underline{x} + \underline{y}, \bar{x} + \bar{y}]$$

where we must remember in our implementation to use rounding towards $-\infty$ for the lower bound and rounding towards $+\infty$ for the upper. Intuitively it makes sense that the lower bound would simply be the sum of the two lower bounds and likewise for the upper bounds. So far, so good. Here's the C code,

```
1  interval_t int_add(interval_t x, interval_t y) {
2      interval_t sum;
3
4      fesetround(FE_DOWNWARD);
5      sum.a = x.a + y.a;
6      fesetround(FE_UPWARD);
7      sum.b = x.b + y.b;
8      fesetround(FE_TONEAREST);
9      return sum;
10 }
```

where lines 5 and 7 add the lower and upper bounds. Line 4 sets the floating-point rounding mode towards $-\infty$ just before adding the lower bounds. Line 6 then sets the rounding mode towards $+\infty$ before adding the upper bounds in line 7. Lastly, line 8 returns to rounding towards the nearest floating-point number just before returning the new sum in line 9.

Subtraction is, as might be expected, similar in form to addition but with a twist. Interval subtraction is defined to be,

$$x - y = [\underline{x}, \bar{x}] - [\underline{y}, \bar{y}] \equiv [\underline{x} - \bar{y}, \bar{x} - \underline{y}]$$

which at first glance might seem a bit odd. One way to see why subtraction is defined like this is to consider what negation of an interval might mean. To negate a value is to change its sign, $x \rightarrow -x$. For an interval the operation is the same but not only are the signs of the bounds changed but also their order. If the order were not changed the lower bound would suddenly be larger than the upper bound. So, negation is,

$$-[\underline{x}, \bar{x}] \equiv [-\bar{x}, -\underline{x}]$$

where we have swapped the order of the bounds. How does this help in understanding subtraction? By remembering that subtraction is simply addition of a negative value. So,

$$[\underline{x}, \bar{x}] - [\underline{y}, \bar{y}] = [\underline{x}, \bar{x}] + [-\bar{y}, -\underline{y}] \equiv [\underline{x} - \bar{y}, \bar{x} - \underline{y}]$$

which is exactly the definition of interval subtraction. In code this looks like,

```
1  interval_t int_sub(interval_t x, interval_t y) {
2      interval_t diff;
3
4      fesetround(FE_DOWNWARD);
5      diff.a = x.a - y.b;
6      fesetround(FE_UPWARD);
7      diff.b = x.b - y.a;
8      fesetround(FE_TONEAREST);
9      return diff;
10 }
```

where we set the proper rounding mode as in addition and lines 5 and 7 set the lower and upper bounds using the definition of interval subtraction.

Multiplication The goal of interval arithmetic is to ensure that the interval represents the entire range in which the actual value may be found. This means that the lower bound must be the smallest such value and the upper bound must be the largest. For the addition of two intervals it is clear that the lower bound must be simply the sum of the lower bounds of the operands and similarly for the upper bound. With subtraction the same holds after we see that it is addition of the negation of the second operand. For multiplication the situation is more general, however, because the interval may include bounds of opposite signs. Therefore, we define multiplication of two intervals as,

$$[\underline{x}, \bar{x}] \times [\underline{y}, \bar{y}] \equiv [\min\{\underline{x}\underline{y}, \underline{x}\bar{y}, \bar{x}\underline{y}, \bar{x}\bar{y}\}, \max\{\underline{x}\underline{y}, \underline{x}\bar{y}, \bar{x}\underline{y}, \bar{x}\bar{y}\}]$$

which reduces to,

$$[\underline{x}, \bar{x}] \times [\underline{y}, \bar{y}] = [\underline{x}\underline{y}, \bar{x}\bar{y}]$$

if the lower bounds of x and y are greater than zero. This definition covers all cases in order to find the smallest and largest product of the components of x and y.

With multiplication defined we can now implement it in our library,

```
1  interval_t int_mul(interval_t x, interval_t y) {
2      interval_t prod;
3      double t;
4
5      fesetround(FE_DOWNWARD);
6      prod.a = x.a * y.a;
7      t = x.a * y.b;
8      if (t < prod.a) prod.a = t;
9      t = x.b * y.a;
10     if (t < prod.a) prod.a = t;
11     t = x.b * y.b;
12     if (t < prod.a) prod.a = t;
13
14     fesetround(FE_UPWARD);
15     prod.b = x.a * y.a;
16     t = x.a * y.b;
17     if (t > prod.b) prod.b = t;
18     t = x.b * y.a;
19     if (t > prod.b) prod.b = t;
20     t = x.b * y.b;
21     if (t > prod.b) prod.b = t;
22
23     fesetround(FE_TONEAREST);
24     return prod;
25 }
```

where we accept two input intervals in `x` and `y` and return `prod` as their product. Lines 5 through 12 set the lower bound so we first set the IEEE rounding mode to $-\infty$ (line 5). Line 6 sets the lower bound of `prod` to $\underline{x}\underline{y}$. This will be our lower bound for the product unless one of the other component products is smaller. We set the auxiliary variable `t` to the next candidate (line 7) and in line 8 check to see if this is smaller than our current lower bound. If so, we update the lower bound. Lines 9 through 12 continue this process for the remaining component products so that after line 12 we know `prod.a` is set to the smallest value. Line 14 sets the rounding mode to $+\infty$ and repeats this process for the upper bound keeping the largest component product in `prod.b`. Line 23 then restores the default IEEE rounding mode and the final product interval is returned in line 24.

Reciprocal and Division Just as subtraction is really the addition of a negative value so it is that division is really multiplication by the reciprocal. Therefore, in order to discuss division, we need to take a look at reciprocal in interval arithmetic. Our first thought would be to simply take the reciprocal, $x \rightarrow 1/x$, of each component. However, just as negation of an interval required swapping the bounds so will the reciprocal because $1/\bar{x} < 1/\underline{x}$, like so,

$$1/[\underline{x}, \bar{x}] \equiv [1/\bar{x}, 1/\underline{x}]$$

where we have swapped the bounds as needed. However, there is an important caveat that we must keep in mind. Division by zero is mathematically undefined so we must be careful of the case when the interval we are taking the reciprocal of contains zero. If the interval does not contain zero, which means that $\underline{x} > 0$ or $\bar{x} < 0$, we can proceed normally because the reciprocal is defined over the entire interval. This means we must add a qualifier to our original definition,

$$1/[\underline{x}, \bar{x}] \equiv [1/\bar{x}, 1/\underline{x}], \underline{x} > 0 \text{ or } \bar{x} < 0$$

which makes it explicit that we are not allowed to consider intervals which contain zero.

For scalars, we simply say that division by zero is undefined and leave it at that. Can we say something more in the case of intervals? One approach is to split the interval into two parts at zero,

$$1/[\underline{x}, 0] \rightarrow [-\infty, 1/\underline{x}]$$
$$1/[0, \bar{x}] \rightarrow [1/\bar{x}, \infty]$$

following the definition above and writing $\pm\infty$ for $1/0$. If we do this we get two overlapping intervals representing the reciprocal of our original interval, x, which contained zero. Because of this we are really saying that there is no information in writing the reciprocal of an interval containing zero, but we could continue to work through an expression involving such an interval if we use both the intervals above. This is one reason why interval arithmetic is not more widely used, though it isn't the main one (we will see that below). For our convenience, and because we are trying to be pedagogical and not obfuscate things, we will declare reciprocals of intervals which contain zero to be illegal and ignore them.

Armed with a definition of the reciprocal of an interval we are now ready to calculate x/y. We do this by first calculating the reciprocal, $y' \equiv 1/y$, and then find $x \times y'$. The C code we need is,

```
interval_t int_recip(interval_t x) {
    interval_t inv;

    if ((x.a < 0.0) && (x.b > 0.0)) {
        inv.a = -exp(1000.0);
        inv.b =  exp(1000.0);
    } else {
        fesetround(FE_DOWNWARD);
        inv.a = 1.0 / x.b;
        fesetround(FE_UPWARD);
        inv.b = 1.0 / x.a;
        fesetround(FE_TONEAREST);
    }

    return inv;
}

interval_t int_div(interval_t x, interval_t y) {
    return int_mul(x,int_recip(y));
}
```

where we call the `int_mul()` function defined above to do the actual multiplication with proper rounding modes.

The function `int_div()` is straightforward, simply multiply the first argument by the reciprocal of the second as in line 19 above. All the action is in `int_recip()` which first checks to see if zero is in the interval (line 4). If zero is in the interval we call the reciprocal undefined and return $[-\infty, +\infty]$ by attempting to calculate the exponential of a large number (lines 5 and 6). Line 8 sets rounding towards $-\infty$ and then calculates the lower bound in line 9. This is repeated by lines 10 and 11 with rounding set towards $+\infty$. Line 12 resets the rounding mode and line 15 returns the new interval.

Powers What it means to raise a number to a positive integer power is clear, simply multiply the number by itself that many times so that $x^4 = x \times x \times x \times x$. For intervals we must be more careful. The result of raising an interval to an integer power depends both upon the even or oddness of the exponent and the signs of the components of the interval.

Let's present the formulas and then look at why they work. For an odd exponent, $n = 2k + 1,\ \ k = 0, 1, \ldots$, things work as expected,

$$[\underline{x}, \bar{x}]^n = [\underline{x}^n, \bar{x}^n],\ \ n = 2k + 1,\ \ k = 0, 1, \ldots$$

It is only when the exponent is even that things become interesting. In this case we have,

$$\begin{aligned}[\underline{x}, \bar{x}]^n &= [\underline{x}^n, \bar{x}^n],\ \ \underline{x} \geq 0,\ \ n = 2k,\ \ k = 1, 2, \ldots \\ &= [\bar{x}^n, \underline{x}^n],\ \ \bar{x} < 0 \\ &= [0, \max\{\underline{x}^n, \bar{x}^n\}],\ \ \text{otherwise}\end{aligned}$$

For an odd exponent it is straightforward to see why the formula takes that form. If the interval has bounds that are negative then an odd power of those bounds will still be negative and the order of the original will not change. This means that for an odd exponent $\underline{x}^n < \bar{x}^n$ so the answer is simple, no special cases. When the exponent is even, however, we have to pay attention to the signs of the bounds of x. In the first case, if $\underline{x} \geq 0$ we know that both bounds are positive and that $\underline{x} < \bar{x}$. Therefore, the order of the bounds need not change since $\underline{x}^n < \bar{x}^n$ as well. The second case tells us that both bounds are negative. Since $a^n = |a|^n$ if n is even and $a < 0$ it follows that $\underline{x}^n > \bar{x}^n$ when $\underline{x}, \bar{x} < 0$ so we must swap the order of the bounds. The last case covers situation when the interval includes zero. As x^n is always positive when n is even (and a positive integer) we know that the lower bound of the result must be zero. The upper bound is set to the larger of the two original bounds after raising them to the nth power.

What if n is a negative integer? In that case we use the rule that $x^{-n} = 1/x^n$. To do this, first calculate $y = x^n = [\underline{x}, \bar{x}]^n$ and then calculate the result as $[\underline{x}, \bar{x}]^{-n} = [1/\bar{y}, 1/\underline{y}]$ with suitable warnings about y containing zero.

Adding `int_pow()` to our library is straightforward, if a little tedious in checking all the special cases. The C code is,

```
1  interval_t int_pow(interval_t x, int n) {
2      interval_t ans;
3
4      if ((n%2) == 1) {
5          fesetround(FE_DOWNWARD);
6          ans.a = pow(x.a,n);
7          fesetround(FE_UPWARD);
8          ans.b = pow(x.b,n);
9      } else {
10         if (x.a >= 0) {
11             fesetround(FE_DOWNWARD);
12             ans.a = pow(x.a,n);
13             fesetround(FE_UPWARD);
14             ans.b = pow(x.b,n);
15         } else {
16             if (x.b < 0) {
17                 fesetround(FE_DOWNWARD);
18                 ans.a = pow(x.b,n);
19                 fesetround(FE_UPWARD);
20                 ans.b = pow(x.a,n);
21             } else {
22                 ans.a = 0.0;
23                 fesetround(FE_UPWARD);
24                 ans.b = pow(x.a,n);
25                 if (pow(x.b,n) > ans.b) ans.b = pow(x.b,n);
26             }
27         }
28     }
29
30     fesetround(FE_TONEAREST);
31     return ans;
32 }
```

where line 4 checks to see if the exponent n is even or odd. If the remainder after dividing by 2 is 1, then n is odd and we calculate the lower bound in line 6 and the upper in line 8 after setting the floating-point rounding mode appropriately. If n is even we move to line 10 and check the three cases given above. If the lower bound is greater than or equal to zero (line 10) we calculate the new bounds in lines 12 and 14. If the upper bound is less than zero (line 16) we swap the results (lines 18 and 20). Finally, if the interval includes zero we set the lower bound to zero (line 22) and set the upper bound to the larger of the two exponentiated initial bounds (lines 24 and 25). Line 30 restores the default rounding mode and line 31 returns our answer.

When using intervals it is important to remember that, unlike real numbers, exponentiation is not simply repeated multiplication. Let's look at some examples using our library routines above. We will give an example of an even and odd power of intervals that are positive, negative, and include zero. We will use `int_pow()` for the powers and `int_mul()` for the multiplications.

First, an all negative interval,

x = `[ -2.7182818284590451,  -2.6182818284590454]`
x^3 = `[-20.0855369231876679, -17.9493685483622478]`
$x \times x \times x$ = `[-20.0855369231876679, -17.9493685483622443]`
x^4 = `[ 46.9965055024911891,  54.5981500331442291]`
$x \times x \times x \times x$ = `[ 46.9965055024911749,  54.5981500331442433]`

Second, an all positive interval,

x = `[ 2.7182818284590451,  2.8182818284590452]`
x^3 = `[20.0855369231876644, 22.3848022077206359]`
$x \times x \times x$ = `[20.0855369231876608, 22.3848022077206359]`
x^4 = `[54.5981500331442220, 63.0866812956689813]`
$x \times x \times x \times x$ = `[54.5981500331442149, 63.0866812956689884]`

Lastly, an interval including zero,

x = `[ -2.7182818284590451,  2.8182818284590452]`
x^3 = `[-20.0855369231876679, 22.3848022077206359]`
$x \times x \times x$ = `[-21.5905309612583913, 22.3848022077206359]`
x^4 = `[  0.0000000000000000, 63.0866812956689813]`
$x \times x \times x \times x$ = `[-60.8482010748969273, 63.0866812956689884]`

The examples above are displayed using a format specifier of `%0.16f` which always displays 16 digits after the decimal point. Note that the repeated multiplications lead to result intervals that are a little too large compared to the powers. Especially note that the even power of the interval including zero is completely wrong in the sense that it is far too large by including a negative lower bound.

Other Operations There are many other operations we could add to our library. Here we add only a few of these. The new operations are negation, absolute value, comparison for equality, comparison for less than or equal, and a concept of "distance" between two intervals. We will describe the C code first and then give some examples. The code for negation is,

```
interval_t int_neg(interval_t x) {
    interval_t ans;
    fesetround(FE_DOWNWARD);
    ans.a = -x.b;
    fesetround(FE_UPWARD);
    ans.b = -x.a;
    fesetround(FE_TONEAREST);
    return ans;
}
```

which implements $-[\underline{x}, \bar{x}] = [-\bar{x}, -\underline{x}]$. This is always true because the lower bound of an interval is, by definition, less than (or equal) to the upper bound. Therefore, changing the sign of these values means that their order must be reversed to keep the lower bound less than the upper.

The absolute value of an interval depends on the signs of the bounds. Formally, the absolute value is defined to be,

$$\begin{aligned} |x| &= [\min\{|\underline{x}|, |\bar{x}|\}, \max\{|\underline{x}|, |\bar{x}|\}], \text{ if } \underline{x}\bar{x} \geq 0 \\ &= [0, \max\{|\underline{x}|, |\bar{x}|\}], \text{ if } \underline{x}\bar{x} < 0 \end{aligned}$$

where we maximize the interval if the signs of the bounds are the same ($\underline{x}\bar{x} \geq 0$) or we set zero to be the lower bound if the interval includes zero ($\underline{x}\bar{x} < 0$).

The C code for this operation is,

```
1  interval_t int_abs(interval_t x) {
2      interval_t ans;
3      if (x.b * x.a >= 0) {
4          fesetround(FE_DOWNWARD);
5          ans.a = (fabs(x.b) < fabs(x.a)) ? fabs(x.b) :
                       fabs(x.a);
6          fesetround(FE_UPWARD);
7          ans.b = (fabs(x.b) < fabs(x.a)) ? fabs(x.a) :
                       fabs(x.b);
8      } else {
9          ans.a = 0.0;
10         fesetround(FE_UPWARD);
11         ans.b = (fabs(x.b) < fabs(x.a)) ? fabs(x.a) :
                       fabs(x.b);
12     }
13     fesetround(FE_TONEAREST);
14     return ans;
15 }
```

where line 3 looks to see if the signs of the bounds are the same. If so, the minimum is used for the lower bound (line 5) and the maximum is used for the upper bound (line 7) with proper rounding modes set. If the interval includes zero we move to line 9 and set the output lower bound to zero and the upper bound to the maximum (line 11). We reset to the default rounding mode and return the answer in line 14.

To this point the interval library consists entirely of operations. Here we define the concepts of equality and less than or equal for intervals. Equality is straightforward: two intervals are equal if their bounds are equal. Symbolically this is represented as,

$$[\underline{x}, \bar{x}] = [\underline{y}, \bar{y}] \text{ iff } (\underline{x} = \underline{y}) \wedge (\bar{x} = \bar{y})$$

where $\wedge$ is conjunction or "logical and".

For less than or equal we compare both the bounds for less than or equal like so,

$$[\underline{x}, \bar{x}] \leq [\underline{y}, \bar{y}] \text{ iff } (\underline{x} \leq \underline{y}) \wedge (\bar{x} \leq \bar{y})$$

Why equal and less than or equal? These relations are enough to define the others as needed and are called out explicitly in [7] which works towards defining a standard for interval arithmetic.

The last operation we are including in our interval library is the concept of the distance between two intervals. This is a scalar measuring the difference between the bounds of two intervals returning the largest difference. Symbolically this is,

$$\text{dist}(x, y) = \max\{|\underline{x} - \underline{y}|, |\bar{x} - \bar{y}|\}$$

which measures the degree to which two intervals "overlap". The larger this value, the more dissimilar the intervals are.

In C code these operations are defined to be,

```
1  unsigned char int_eq(interval_t x, interval_t y) {
2      return (x.a==y.a) && (x.b==y.b);
3  }
4
5  unsigned char int_le(interval_t x, interval_t y) {
6      return (x.a <= y.a) && (x.b <= y.b);
7  }
8
9  double int_dist(interval_t x, interval_t y) {
10     double a,b;
11     a = fabs(x.a - y.a);
12     b = fabs(x.b - y.b);
13     return (a>b) ? a : b;
14 }
```

where line 2 simply compares the bounds for equality and line 6 compares them for less than or equal. For `int_dist()` we first compute $|\underline{x} - \underline{y}|$ (line 11) and $|\bar{x} - \bar{y}|$ (line 12) and then return the larger of the two in line 13.

As a quick example of these functions consider two intervals,

```
x = [-2.7182818284590451, 2.8182818284590452]
y = [ 2.5182818284590449, 2.9182818284590453]
```

which overlap. With these, then, the function calls to our new operations return,

```
int_neg(x,y)  = [-2.8182818284590452, 2.7182818284590451]
int_abs(x)    = [ 0.0000000000000000, 2.8182818284590452]
int_eq(x,y)   = 0(false)
int_le(x,y)   = 1(true)
int_dist(x,y) = 5.2365636569180900
```

The Python Version Before we move on with our investigation of interval arithmetic let's quickly implement an interval class in Python. For this class we will only implement the basic arithmetic operations and leave extending the class to the exercises. The key to interval arithmetic is the ability to control the floating-point rounding mode. This is easily accomplished in C by including the `fenv.h` header file and compiling with the `-frounding-math` option. However, the Python interpreter is already compiled and does not offer direct access to the constants and functions defined in `fenv.h`. Fortunately, all is not lost, we are still able to access the underlying C library that contains these functions and use them to set the rounding mode as necessary. For this trick we thank Rafael Barreto for his blog post [8] which describes how to do this to set the rounding mode.

Jumping in, then, we start with,

```
1  from math import exp
2  from ctypes import cdll
3  from ctypes.util import find_library
4
5  class Interval:
6      libc = cdll.LoadLibrary(find_library("m"))
7      FE_TOWARDZERO = 0xc00
8      FE_DOWNWARD = 0x400
9      FE_UPWARD = 0x800
10     FE_TONEAREST = 0
11
12     def __init__(self, a,b=None):
13         if (b == None):
14             self.a = a
15             self.b = a
16         else:
17             self.a = a
18             self.b = b
```

where lines 2 and 3 import standard Python libraries which will give us access to the C library so we can call the rounding functions. Line 1 imports the exponential function which we will use to generate infinity. The class starts on line 5. Line 6 sets a class-level variable, `libc`, to the C library which on Linux systems can be found by searching for "m" (hence `-lm` when compiling). We also define the rounding modes as class-level variables. The specific values were determined by a short C program which displayed them after including `fenv.h`. The constructor for the class starts on line 12 and takes up to two arguments. If only one is given the argument is assumed to be a real number and an interval is created from this number. Doing this simplifies the class by removing the need to consider mixed interval and float operations. If both arguments are given they are assumed to be the lower (`a`) and upper (`b`) bounds of the interval being defined. These are stored in `self.a` and `self.b`, respectively.

In order to round properly we need methods which set the rounding mode. These are defined next,

```
    def __RoundDown(self):
        self.libc.fesetround(self.FE_DOWNWARD)

    def __RoundUp(self):
        self.libc.fesetround(self.FE_UPWARD)

    def __RoundNearest(self):
        self.libc.fesetround(self.FE_TONEAREST)
```

where `__RoundDown()`, `__RoundUp()`, and `__RoundNearest()` are wrappers on calls to the C `fesetround()` function. Note the use of the double underscore prefix as Python notation for "private". We will call these methods whenever we need to change the rounding mode.

Next comes addition and subtraction. Instead of methods with those names we use the flexibility built into Python to define methods that work directly with the $+$ and $-$ operators. This means we can define expressions with intervals as if they were any other number or variable. The code is,

```
    def __add__(self, y):
        self.__RoundDown()
        a = self.a + y.a
        self.__RoundUp()
        b = self.b + y.b
        self.__RoundNearest()
        return Interval(a,b)

    def __sub__(self, y):
        self.__RoundDown()
        a = self.a - y.b
        self.__RoundUp()
        b = self.b - y.a
        self.__RoundNearest()
        return Interval(a,b)
```

where these methods, `__add__()` and `__sub__()`, will be called when objects of this class are encountered in arithmetic expressions involving + and −. Since we are not implementing mixed operations both operands of the + or − must be of this class. Just as we did for the C library we set the rounding modes to down (lower bound) or up (upper bound) before calculating the new value. We return the new value as a new `Interval` object.

Multiplication is next (`__mul__`). This is directly analogous to the C version where we consider each case for the new lower and upper limit with proper rounding. The result is returned as a new interval object,

```
1    def __mul__(self, y):
2        self.__RoundDown()
3        a = min([self.a*y.a,self.a*y.b,self.b*y.a,self.b*y.b])
4        self.__RoundUp()
5        b = max([self.a*y.a,self.a*y.b,self.b*y.a,self.b*y.b])
6        self.__RoundNearest()
7        return Interval(a,b)
```

with line 2 setting rounding towards $-\infty$ before calculating all the products between the two intervals. It then selects the smallest of these and sets the lower bound to it. Line 4 changes to rounding towards $+\infty$ and repeats the calculation this time selecting the maximum value. We must do the multiplications twice because we changed the rounding mode. Line 6 returns to default rounding and the new object is returned in line 7.

Division, with reciprocal, and powers are all that remain to be implemented. Division is multiplication of the reciprocal, as before,

```
1    def recip(self):
2        if (self.a < 0.0) and (self.b > 0.0):
3            return Interval(-exp(1000.0), exp(1000.0))
4        self.__RoundDown()
5        a = 1.0 / self.b
6        self.__RoundUp()
7        b = 1.0 / self.a
8        self.__RoundNearest()
9        return Interval(a,b)
10
11   def __div__(self, y):
12       return self.__mul__(y.recip())
```

where division is called in line 12. The reciprocal method, not associated with an operator but not private, is used to return the reciprocal of the interval. If the interval contains zero (line 2) we return $[-\infty, +\infty]$ (line 3). Otherwise, we round down and calculate the lower bound, round up, calculate the upper bound, and finally return the new reciprocal object.

The last method we will implement is raising an interval to a positive integer power. This will mimic the C code as,

```
    def __pow__(self, n):
        if (n%2) == 1:
            self.__RoundDown()
            a = self.a**n
            self.__RoundUp()
            b = self.b**n
        elif (self.a >= 0):
            self.__RoundDown()
            a = self.a**n
            self.__RoundUp()
            b = self.b**n
        elif (self.b < 0):
            self.__RoundDown()
            a = self.b**n
            self.__RoundUp()
            b = self.a**n
        else:
            a = 0.0
            self.__RoundUp()
            b = self.a**n
            t = self.b**n
            if (t > b):
                b = t
        self.__RoundNearest()
        return Interval(a,b)
```

with each condition checked as before. Note that `**` is exponentiation in Python.

Let's look at some examples using the class. We give a running interactive Python session with >>> the Python prompt followed by user input. Python output does not have the prompt and has been slightly intended for clarity,

```
>>> from Interval import *
>>> from math import sqrt
>>> x = Interval(-sqrt(2), sqrt(2)+0.1)
>>> x; x*x; x**2
    [-1.4142135623730951, 1.5142135623730952]
    [-2.1414213562373101, 2.2928427124746200]
    [0.0000000000000000, 2.2928427124746200]
>>> x*x*x; x**3
    [-3.2425692603699230, 3.4718535316173851]
    [-2.8284271247461903, 3.4718535316173846]
>>> x = Interval(sqrt(2), sqrt(2)+0.01)
>>> y = x**2
>>> x/y
    [0.6972118559681739, 0.7121067811865476]
>>> y/x
    [1.4042837765619229, 1.4342842730512142]
```

We first import the library and the square root function. We then define an interval, x, and show it along with $x \times x$ and x^2. Notice that the former gives too large an interval in this case because the interval contains zero. The more correct interval is the one returned by x^2. The same comparison is made between multiplying x by itself three times and calling the exponential. The multiplication leads to too large an interval. Re-defining x to be positive lets us demonstrate division thereby illustrating all the basic operations of the class.

In this section we explored and implemented basic interval arithmetic in C and Python. Next we consider functions of intervals. This is where we will run into the biggest reason, apart from performance issues due to repeated calculations, why interval arithmetic is not more widely used.

Properties of Intervals We summarize this section by discussing some of the mathematical properties of intervals. The first few are familiar and mimic those of normal arithmetic. Things become more interesting when we get to the distributive property.

For any three intervals, $x = [\underline{x}, \bar{x}]$, $y = [\underline{y}, \bar{y}]$, and $z = [\underline{z}, \bar{z}]$, the following is,

$$\begin{aligned}
(x + y) + z &= x + (y + z) && (\textit{associative}) \\
(x - y) - z &= x - (y - z) && \\
x + y &= y + x && (\textit{commutative}) \\
x \times y &= y \times x && \\
x + [0, 0] &= x && (\textit{additive identity}) \\
x \times [1, 1] &= x && (\textit{multiplicative identity})
\end{aligned}$$

which comes as no surprise. However, the distributive property causes problems. In general, for intervals,

$$x \times (y + z) \neq xy + xz$$

For example if $x = [1, 2]$, $y = [1, 2]$, and $z = [-2, -1]$,

$x \times (y + z)$	$xy + xz$
$[1, 2] \times ([1, 2] + [-2, -1])$	$[1, 2] \times [1, 2] + [1, 2] \times [-2, -1]$
$[1, 2] \times [-1, 1]$	$[1, 4] + [-4, -1]$
$[-2, 2]$	$[-3, 3]$

which is at first glance unexpected. This property is called *subdistributivity* and implies that for intervals we have,

$$x \times (y + z) \subseteq x \times y + x \times z$$

which is seen clearly in the example above because $[-2, 2] \subseteq [-3, 3]$ as it is a tighter bound. This means that in practice one needs to be careful with the form of an expression.

8.3 Functions and Intervals

Now that we know how to implement intervals we are ready to use them and be sure of accurate and tight bounds on our calculations. Or not. We must take a look at general functions of intervals and when we do we will see that a large elephant is indeed in the living room.

If a function of one variable is monotonic over a range this means that the function, $f(x)$, is either increasing or decreasing over the range. If this is the case, then interval arithmetic will work as expected because for the range of the interval the function will maintain (or reverse) the ordering of the bounds. For example, consider the square root function for positive intervals. Since the square root is monotonically increasing over any positive interval the output of $\sqrt{x}$ for an interval x will also be increasing as shown in Fig. 8.1.

Because of this consistent increase the tightest output interval will be simply $[\sqrt{\underline{x}}, \sqrt{\bar{x}}]$. If the function is monotonically decreasing over the interval the result will be similar but with the bounds flipped to maintain a proper ordering. So, in general, for monotonically increasing functions of one variable ($f(x)$) we have,

$$f(x) = f([\underline{x}, \bar{x}]) = [f(\underline{x}), f(\bar{x})]$$

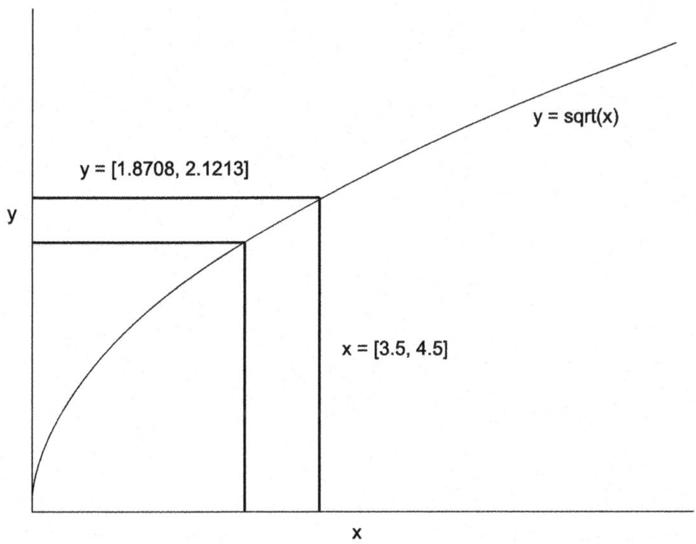

Fig. 8.1 An example of a monotonic function. Since the function $y = \sqrt{x}$ is always increasing for a positive input the bounds on any output interval are simply the function values at the lower and upper bounds of the input interval as shown for the interval $x = [3.5, 4.5]$. This leads to the proper output interval of approximately $[1.8708, 2.1213]$. If the function were monotonically decreasing over this interval the output would be the same but with the upper and lower bounds swapped so that the interval has the right ordering

while for monotonically decreasing functions ($g(x)$) we have,

$$g(x) = g([\underline{x}, \bar{x}]) = [g(\bar{x}), g(\underline{x})]$$

where it should be noted that many elementary functions are of this form. This includes the exponential function, roots (e.g. cube root), and logarithms. For exponentials and logarithms we can define the output explicitly,

$$a^{[\underline{x}, \bar{x}]} = [a^{\underline{x}}, a^{\bar{x}}],\ a > 1$$

$$\log_a [\underline{x}, \bar{x}] = [\log_a \underline{x}, \log_a \bar{x}]$$

What about functions that are not monotonically increasing or decreasing? In this case things become a bit more difficult. For example, what about the sine function? We assume the reader is familiar with the shape of the sine function. Clearly, it is sometimes increasing and sometimes decreasing. How to work with this and by extension the cosine?

One possibility is to use a Taylor series expansion. This is the solution presented in Daumas et al. [9] in terms of bounded functions using a parameter n to control the bounds where in the limit $n \to \infty$ the bounds tighten to the actual value of the function. Specifically, for the sine function [9] defines the output interval to be,

$$[\sin(x)]_n = [\underline{\sin}(\underline{x}, n), \overline{\sin}(\bar{x}, n)],\ x \subseteq [\frac{-\pi}{2}, \frac{\pi}{2}]$$

for a specified value of n. We'll see shortly what a good value of n might be. The functions $\underline{\sin}(\underline{x}, n)$ and $\overline{\sin}(\bar{x}, n)$ are themselves defined in terms of Taylor series parametrized by n and operating on scalar values (the bounds of x). The series expansions are,

$$\underline{\sin}(x, n) = \sum_{i=1}^{m} (-1)^{i-1} \frac{x^{2i-1}}{(2i-1)!}$$

$$\overline{\sin}(x, n) = \sum_{i=1}^{m+1} (-1)^{i-1} \frac{x^{2i-1}}{(2i-1)!}$$

where $m = 2n + 1, x \geq 0$, otherwise, $m = 2n$. In [9] the value of π itself is defined as an interval but for our purposes we will treat it as an exact quantity and note that the argument to our interval sine be restricted to the range given above.

Looking at the series expansions and the dependency upon the summation limit as a function of n we see where the cleverness lies in the definition. The bounds of the sine come from the number of terms used in the summation with the lower bound set by $m(n)$ and the upper set by $m(n) + 1$. Also, as $n \to \infty$ it is clear that the series will converge more and more to the sine of x.

Let's add this function to our interval library. The implementation is straightforward requiring a main function,

```
1  interval_t int_sin(interval_t x) {
2      int n=8;
3      interval_t ans;
4
5      fesetround(FE_DOWNWARD);
6      ans.a = int_sin_lo(x.a, n);
7      fesetround(FE_UPWARD);
8      ans.b = int_sin_hi(x.b, n);
9      fesetround(FE_TONEAREST);
10     return ans;
11 }
```

which follows our standard format of setting the rounding mode, first towards $-\infty$ for the lower bound and then towards $+\infty$ for the upper bound. Lines 6 and 8 set the output value by calling helper functions to compute the series value for the lower and upper bound. Note that line 2 sets n to eight. This value can be adjusted but eight gives good results. The helper functions are,

```
double fact(int n) {
    return (n==1) ? 1.0 : n*fact(n-1);
}

double int_sin_lo(double x, int n) {
    int i, m = (x < 0) ? 2*n : 2*n+1;
    double ans=0.0;

    for(i=1; i <= m; i++) {
        ans += pow(-1,i-1) * pow(x,2*i-1) / fact(2*i-1);
    }
    return ans;
}

double int_sin_hi(double x, int n) {
    int i, m = (x < 0) ? 2*n : 2*n+1;
    double ans=0.0;

    for(i=1; i <= m+1; i++) {
        ans += pow(-1,i-1) * pow(x,2*i-1) / fact(2*i-1);
    }
    return ans;
}
```

where `int_sin_lo()` and `int_sin_hi()` call `fact()` to compute the factorial. We use the recursive algorithm for simplicity noting that we will only need less than a dozen terms in the series. The functions themselves are a literal translation of the formulas with m set appropriately when declared (lines 6 and 16). Note also that for `int_sin_hi()` we add one to m in the loop limit. Why do we not combine these two functions? If `int_sin_hi()` needs to calculate only one more term in the series expansion than `int_sin_lo()` why calculate everything twice? The answer lies in the guarantee implicit in intervals implemented in floating-point, namely, that the bounds on the interval will be certain to include the correct answer. When `int_sin_lo()` is called from `int_sin()` the floating-point rounding mode has been set towards $-\infty$. Therefore, all the calculations of the series terms will round in that direction thereby assuring us that rounding error will not shrink the interval we expect. Likewise, when `int_sin_hi()` is called the rounding mode it towards $+\infty$ for the same reason. This is why the calculation must be done twice.

Let's look at the output of our sine function and compare it to the "gold standard" present in the C math library. For,

$$x = [0.785\underline{2}981625, 0.785\underline{4}981625]$$

where the difference between the bounds is underlined, we get,

$$\sin(x) = [0.7070360663383\underline{567}, 0.7071774876944\underline{870}] \quad (\textit{interval sine})$$

while the `sin()` function from `math.h` produces the interval,

$$\sin(x) = [0.7070360663383\underline{576}, 0.7071774876944\underline{862}] \quad (\textit{math.h sine})$$

Looking at the underlined digits of each interval we see that our sine function does indeed produce an interval which will be sure to encompass the "true" value.

The formulas for the interval version of cosine are,

$$[\cos(x)]_n = [\underline{\cos}(\bar{x}, n), \overline{\cos}(\underline{x}, n)], \; x \subseteq [0, \pi]$$

and,

$$\underline{\cos}(x, n) = 1 + \sum_{i=1}^{m+1} (-1)^i \frac{x^{2i}}{(2i)!}$$

$$\overline{\cos}(x, n) = 1 + \sum_{i=1}^{m} (-1)^i \frac{x^{2i}}{(2i)!}$$

again where $m = 2n + 1, x \geq 0$, otherwise, $m = 2n$. We leave the implementation, in C and Python, to the exercises.

So far in this section we have examined interval versions of elementary functions. What about general functions of x? Let's take a look and we will soon see that this is the elephant mentioned above. Applying interval arithmetic to general functions leads to the *dependency problem* (or *dependency effect*) which is the bane of intervals.

This problem shows itself when applying intervals to general functions where the interval variable appears more than once in the expression. The most commonly cited example of this is the evaluation of the expression $x - x^2$ for $x = [0, 1]$. Another way to write this expression is $x(1 - x)$. Let's look at how these two compare when we evaluate them,

$x(1-x)$	$x-x^2$
$[0,1]\times(1-[0,1])$	$[0,1]-[0,1]^2$
$[0,1]\times[0,1]$	$[0,1]-[0,1]$
$[0,1]$	$[-1,1]$

where we clearly have differing answers. We also have that the interval for $x(1-x) \subseteq x-x^2$ which is exactly what we would expect due to the subdistributivity of intervals. However, while neither answer is technically incorrect, since the proper answer lies inside of each interval, neither answer is satisfactory because of the dependency problem.

The expression we are evaluating is a downward facing parabola with a maximum value of 0.25 at $x = \frac{1}{2}$ and minimum values of zero at $x = 0$ and $x = 1$. Therefore, over the given range, $x = [0, 1]$, we would expect the output interval to be $[0, \frac{1}{4}]$. While this interval is within the range of both of the intervals above we do not find the narrow interval unless we remove the dependency problem by rewriting the expression so that x appears only once. If we complete the square of $x - x^2$ we can rewrite the original expression as $\frac{1}{4} - (\frac{1}{2} - x)^2$ in which x appears only once. If we evaluate this expression over $x = [0, 1]$ we get,

$\frac{1}{4} - (\frac{1}{2} - x)^2$
$0.25 - (0.5 - [0,1])^2$
$0.25 - [0, 0.25]$
$[0, 0.25]$

which is the narrowest interval in this case.

The dependency problem is likely the primary reason interval arithmetic is not more widely used. It is often impossible to rewrite an expression so that the variable appears only once. When this is the case it is possible to introduce auxiliary intervals whose union results in the original interval and then apply the expression to each of these intervals. This *interval splitting* will result in less of a dependency problem. Let's look at an example of this from [9] where the expression under consideration is $2x - x$ with $x = [0, 1]$. If we simply evaluate the expression we get $[-1, 2]$ as the answer when the narrowest interval is naturally $[0, 1]$. The observation key to interval splitting is that while $x = [0, 1]$ does not lead to the narrowest interval we can split the interval into two parts, $x_1 = [0, 0.5]$ and $x_2 = [0.5, 1]$, and evaluate each of these in turn. In that case we get $[-0.5, 1]$ for x_1 and $[0, 1.5]$ for x_2. The union of these two intervals is $[-0.5, 1.5]$ which is closer to the expected answer of $[0, 1]$. If the intervals are divided further and further the resulting answer will approach $[0, 1]$ which we can see if we evaluate $x_1 = [0, 0.05]$ to $x_{20} = [0.95, 1]$ which gives $[-0.05, 0.1]$ up to $[0.9, 1.05]$ the union of which is $[-0.05, 1.05]$. While interval splitting can help deal with the dependency problem it introduces a significant amount of overhead to the calculations. This is why our simple interval C library and Python class do not implement interval splitting even though more complete libraries do.

8.4 Implementations

In this section we look at two implementations of interval arithmetic, one for C and the other for Python. For C perhaps the most widely used library is the Multiple Precision Floating-point Interval library or MPFI for short [10]. This library is built on top of the widely used GNU GMP and MPFR libraries for multiple precision floating-point numbers. This means that in addition to interval arithmetic the MPFI library supports more than the standard IEEE 754 32-bit and 64-bit floating-point. For our purposes we will ignore this important and useful feature and instead focus on the interval support MPFI provides. Using MPFI requires us to install GMP (*http://gmplib.org/*) and MPFR (*http://www.mpfr.org/*). Once these are installed with the typical "`configure; make; make install`" sequence of Unix instructions we can download MPFI (*http://perso.ens-lyon.fr/nathalie.revol/software.html*). Building MPFI follows the same sequence of steps as GMP and MPFR.

Let's start with a simple example that defines two intervals and then prints them on standard out. This will teach us how to define and initialize intervals and how to print them. Please note that there is an extensive API for this library and that we are only using a small part of it. Our initial program, then, is,

```
#include <stdio.h>
#include <mpfr.h>
#include <mpfi.h>
#include <mpfi_io.h>

int main() {
    mpfi_t x, y;

    mpfr_set_default_prec(64);

    mpfi_init(x);
    mpfi_set_str(x, "[-1.0, 1.0]", 10);

    mpfi_init_set_str(y, "[0,2]", 10);

    printf("x = ");
    mpfi_out_str(stdout, 10, 0, x);
    printf("\n");

    printf("y = ");
    mpfi_out_str(stdout, 10, 0, y);
    printf("\n");

    mpfi_clear(x);
    mpfi_clear(y);
}
```

where lines 1 through 4 include necessary header files. Note that the documentation for MPFI recommends including `stdio.h` before the remaining header files. Line 7 declares two interval variables, x and y. This declares the variables but does not initialize them. They must be initialized before they are used. Line 9 sets the default precision to 64 bits to mimic a `double`. Note that this is a call to the MPFR library. Lines 11 and 12 show one way to initialize an interval variable and set its value from a string. We will use strings to set values though there are many options in the API. For `mpfi_set_str()` the first argument is the variable, the second is a string which defines the interval in the way we have been displaying them throughout this chapter, and the last argument is the base in which the interval is defined (2 through 36). Line 14 combines lines 11 and 12 into a single initialize and set function. The arguments are the same as line 12. We then display the intervals using `mpfi_out_str()` in lines 17 and 21. The first argument is the output file stream, the second is the base, the third is the number of digits (0 says use the precision of the variable) and the fourth argument is the interval variable itself. Lastly, lines 24 and 25 clear the memory used by x and y. While not necessary in this case because the program is about to exit it is necessary before re-assigning these variables.

This program should be compiled with,

```
gcc mpfi_ex.c -o mpfi_ex -lmpfi -lmpfr
```

assuming `mpfi_ex.c` to be the file name. Note that MPFI installs itself in `/usr/local/lib` which might require adding that path to the `LD_LIBRARY_PATH` environment variable.

Now that we know how to initialize, set, and display intervals in MPFI let's look at basic arithmetic functions. These are,

```
int mpfi_add(mpfi_t ROP, mpfi_t OP1, mpfi_t OP2)
int mpfi_sub(mpfi_t ROP, mpfi_t OP1, mpfi_t OP2)
int mpfi_mul(mpfi_t ROP, mpfi_t OP1, mpfi_t OP2)
int mpfi_div(mpfi_t ROP, mpfi_t OP1, mpfi_t OP2)
```

where `ROP` is the variable to receive the result and `OP1` and `OP2` are the operands. Note that division using an interval containing zero will result in $[-\infty, +\infty]$. The return value indicates whether or not endpoints of the interval are exact.

MPFI is powerful but a bit cumbersome to use because of its C language heritage. There is a C++ version as well which we are not exploring here. Let's look at our dependency problem example again and implement it using MPFI. This example computes $x(1-x)$ and $x-x^2$ for $x=[0,1]$. The C code is,

```
1 int main() {
2     mpfi_t x, one, t, y;
3
4     mpfr_set_default_prec(64);
5     mpfi_init_set_str(x, "[0, 1]", 10);
6     mpfi_init_set_str(one, "[1,1]", 10);
7
8     mpfi_init(t);
9     mpfi_sub(t, one, x);
10    mpfi_init(y);
11    mpfi_mul(y, t, x);
12
13    printf("y = ");
14    mpfi_out_str(stdout, 10, 0, y);
15    printf("\n");
16
17    mpfi_clear(y);
18    mpfi_clear(t);
19    mpfi_clear(one);
20
21    mpfi_init(t);
22    mpfi_mul(t, x, x);
23    mpfi_init(y);
24    mpfi_sub(y, x, t);
25
26    printf("y = ");
27    mpfi_out_str(stdout, 10, 0, y);
28    printf("\n");
29 }
```

where we see that it is necessary to preserve the intermediate temporary variables so that they can be released properly without leaking memory. Specifically, line 5 initializes x while line 6 initializes the constant interval, `one`. The expression $x(1-x)$ is calculated in lines 8 through 11. First, $t = 1-x$ is calculated in line 9 and then the final answer is found as $y = t \times x$ in line 11. This is the first y output (line 14). Lines 17 through 19 release memory so that we can reuse y and t to calculate $x - x^2$. This expression is calculated in lines 21 through 24 and output in line 27. The output,

```
y = [0,1.00000000000000000000]
y = [-1.00000000000000000000,1.00000000000000000000]
```

is precisely what we would expect it to be for $x(1 - x)$ and $x - x^2$.

Interval arithmetic is available in Python via the `mpmath` [11] library. This library is implemented in pure Python and is quite powerful. As with MPFI, we will only look at the simplest of uses with intervals. To install mpmath on Ubuntu use the command,

```
sudo apt-get -y install python-mpmath
```

with suitable repository access for other Linux distributions or build it from the website.

Once the library is installed we can access it easily from the Python command prompt. The following console session shows some of mpmath's basic interval abilities,

```
1  >>> from mpmath import mpi
2  >>> x = mpi('0','1')
3  >>> print x
4        [0.0, 1.0]
5  >>> print x*(1-x)
6        [0.0, 1.0]
7  >>> print x-x**2
8        [-1.0, 1.0]
9  >>> y = mpi('-1','1')
10 >>> print y**2
11       [0.0, 1.0]
12 >>> print y*y
13       [-1.0, 1.0]
14 >>> print 1/y
15       [-inf, +inf]
16 >>> print 1/x
17       [1.0, +inf]
```

where the Python prompt is `>>>` and Python output is indented for clarity.

In line 1 we import the interval module (`mpi`) from the larger `mpmath` library. Line 2 defines a new interval and stores it in x. Note that the bounds are given as two separate arguments and may be given as strings or as numbers. If given as numbers they are interpreted as Python `floats` (64-bits) while strings are interpreted as exact decimals. Line 5 prints $x(1-x)$ with the expected output while line 7 prints $x-x^2$, again with the expected output. Line 9 defines y to be an interval containing zero. Lines 10 through 15 display the results of several simple calculations on y. Note that y^2 (line 10) returns an interval different than $y \times y$ (line 12) but positive, as in our C library above. Note that line 14 returns $[-\infty, +\infty]$ as the reciprocal of y since it contains zero while line 16 prints the reciprocal of x which has zero as a lower bound. In this case the resulting interval is $[0, +\infty]$.

The bounds of an interval are acquired by the `a` and `b` properties on the object. Additionally, there are also `mid` and `delta` properties for the midpoint of an interval and for the difference between the endpoints,

```
>>> from mpmath import mpi
>>> x = mpi('0.41465', '0.5')
>>> x.a
    mpi('0.41464999999999996', '0.41464999999999996')
>>> x.b
    mpi('0.5', '0.5')
>>> x.mid
    mpi('0.45732499999999998', '0.45732499999999998')
>>> x.delta
    mpi('0.085350000000000037', '0.085350000000000037')
```

with `x.a` returning the lower bound, `x.b` returning the upper, `x.mid` returning the middle position between the lower and upper bounds, and `x.delta` to calculating the range of the interval. Note, the return values are themselves intervals even if the lower and upper bounds are the same.

This section took a quick look at two common interval libraries to extend C and Python. Intervals have been added to many other languages and sometimes the compiler used will support intervals. For example, Fortran has had interval packages for some time. More recent languages like R and MATLAB® also support intervals.

8.5 Thoughts on Interval Arithmetic

Interval arithmetic is most useful when uncertainty is part of the data. With intervals, and proper rounding, we have assurances that the correct answer lies within a particular range. For basic operations intervals work well even if they may involve significantly more low-level calculations than floating-point numbers on their own. For that reason, unless truly necessary, intervals are generally not recommended in situations where high-performance is required.

Floating-point arithmetic is standardized via the IEEE 754 standard. IEEE 754 defines the number of bits for each precision, including single (32-bit) and double (64-bit), as well as rounding modes, comparisons, etc. The IEEE organization, as of this writing (Fall 2016) is developing IEEE P1788 (Standard for Interval Arithmetic) as a project [12]. This standard will address the fact that currently intervals are implemented in many different ways and without a consistent set of algorithms. While not yet complete and adopted, the standard is worth remembering as it may go a long way towards increasing the use of interval arithmetic. A good summary of interval arithmetic, along with a discussion of its uses in scientific and mathematical research, can be found in [13].

Because of the difficulty involved in properly applying interval arithmetic to general functions and expressions one must be careful and thoughtful when using intervals to avoid falling into a situation where the uncertainty grows too large. Recall that unlike standard floating-point computation, the interval $[-\infty, +\infty]$ is a valid return value for an interval expression, albeit, not a particularly useful one. However, if uncertainty and bounds on it are crucial to the task at hand, then interval arithmetic may be the ideal choice.

8.6 Chapter Summary

In this chapter we took a practical look at interval arithmetic. We defined what we mean by intervals and interval arithmetic, summarized the history of intervals, and explained basic operations on intervals. We implemented an interval library in C and Python which included some elementary functions. We then discussed the dependency problem, with examples, and mentioned interval splitting as a technique for minimizing the dependency problem. Lastly, we looked at some implementations of interval arithmetic and offered some thoughts on interval arithmetic and its utility.

Exercises

8.1 Add `int_lt()`, `int_gt()`, `int_ge()`, and `int_ne()` functions to test for $<$, $>$, $\geq$, and $\neq$ using `int_eq()` and `int_le()`. **

8.2 Add exponential (e^x), square root ($\sqrt{x}$) and logarithm functions (any base) to the C interval library. **

8.3 The Python interval class only contains basic operations. Complete the class by adding the missing operations found in the C library. **

8.4 Extend the Python interval library to support mixed scalar and interval arithmetic for $+$, $-$, $\times$, and $/$. (Hint: look at how methods like `__radd__()` might be used.) **

8.5 Extend the Python `__pow__()` method to proper handle negative integer exponents.

8.6 Extend the C library by adding cosine using the formulas in Sect. 8.3. *

8.7 Extend the Python class by adding sine and cosine using the formulas in Sect. 8.3. *

8.8 Add interval splitting to the Python interval class in order to minimize the dependency problem. Automatically split intervals into n parts and evaluate a function on each of those parts reporting the union of them when complete. ***

References

1. Bevington, PR., Robinson, DK. Data reduction and error analysis for the physical sciences. Vol. 336. McGraw-Hill (1969).
2. Young, RC., The algebra of multi-valued quantities, Mathematische Annalen, 1931, Vol. 104, pp. 260–290.
3. Dwyer, PS., Linear Computations, J. Wiley, N.Y., 1951.

4. Sunaga, T., Theory of interval algebra and its application to numerical analysis, In: Research Association of Applied Geometry (RAAG) Memoirs, Ggujutsu Bunken Fukuy-kai. Tokyo, Japan, 1958, Vol. 2, pp. 29–46 (547–564); reprinted in Japan Journal on Industrial and Applied Mathematics, 2009, Vol. 26, No. 2–3, pp. 126–143.
5. Moore, RE. Automatic error analysis in digital computation. Technical Report Space Div. Report LMSD84821, Lockheed Missiles and Space Co., 1959.
6. Hickey, T., Ju, Q. and van Emden, MH., Interval Arithmetic: From principles to implementation. Journal of the ACM (JACM) 48.5 (2001): 1038–1068.
7. Bohlender, G., Kulisch, U., Definition of the Arithmetic Operations and Comparison Relations for an Interval Arithmetic Standard. Reliable Computing 15 (2011): 37.
8. Barreto, R., Controlling FPU rounding modes with Python, http://rafaelbarreto.wordpress.com/2009/03/30/controlling-fpu-rounding-modes-with-python/ (accessed 07-Nov-2014).
9. Daumas, M., Lester, D., Muoz, C. Verified real number calculations: A library for interval arithmetic. Computers, IEEE Transactions on 58.2 (2009): 226–237.
10. Revol, N., Rouillier, F., Multiple Precision Floating-point Interval Library. http://perso.ens-lyon.fr/nathalie.revol/software.html (retrieved 15-Nov-2014) (2002).
11. Johansson, F. et al. mpmath: a Python library for arbitrary-precision floating-point arithmetic (version 0.14), February 2010. http://code.google.com/p/mpmath (accessed 16-Nov-2014).
12. Kearfott, R. An overview of the upcoming IEEE P-1788 working group document: Standard for interval arithmetic. IFSA/NAFIPS. 2013.
13. Kearfott, R. Interval computations: Introduction, uses, and resources. Euromath Bulletin 2.1 (1996): 95–112.

Chapter 9
Arbitrary Precision Floating-Point

Abstract Floating-Point numbers using the IEEE standard have a fixed number of bits associated with them. This limits the precision with which they can represent actual real numbers. In this chapter we examine arbitrary precision, also called multiple precision, floating-point numbers. We look at how they are represented in memory and how basic arithmetic works. We end with an examination of arbitrary precision floating-point libraries that are currently available and offer some thoughts on when to consider arbitrary precision floating-point.

9.1 What is Arbitrary Precision Floating-Point?

By "arbitrary precision floating-point" we mean a floating-point representation that is able to vary the level of precision to suit any calculation. Unlike arbitrary sized integers, which can grow without bound (as long as there is memory), arbitrary precision floating-point generally fixes the precision before calculations begin. The "arbitrary" label is appropriate, however, because we can set the precision as necessary to maintain any level of accuracy (again, assuming memory allows). This is what we mean by the phrase "arbitrary precision floating-point".

9.2 Representing Arbitrary Precision Floating-Point Numbers

There are two main approaches to representing arbitrary precision floating-point numbers: extend the sign-significand-exponent approach to a large number of digits in the significand or use a fixed-point representation with a large number of digits in the fractional part. In both cases the precision can be set by the user, either the number of digits in the significand or in the fractional part. The base can be binary (most efficient) or decimal (better accuracy, slower). The libraries that we will examine below use the exponent approach and represent numbers as,

$$s \times b_{n-1}b_{n-2}\dots b_0 \times 2^e$$

R.T. Kneusel, *Numbers and Computers*, DOI 10.1007/978-3-319-50508-4_9

where s is the sign, e is an integer exponent, potentially large, and b is the binary significand of a preset precision (n bits) with an implied leading "1" digit. In this case, arithmetic proceeds as in IEEE 754 floating-point but in software so that the significand can be arbitrarily large.

Our small implementation of arbitrary precision floating-point, in Python, will use a fixed-point representation and make copious use of the fact that Python natively supports infinite precision integers. We will also freely abuse the term "floating-point" and use it in an expanded sense that includes fixed-point representations as a form of floating-point.

As we saw in Chap. 6, fixed-point representations make particularly good use of integer arithmetic. Specifically, we will represent floating-point numbers in base 10 with a given number of digits in the decimal part. The integer part can be arbitrarily large and will be handled automatically by Python. We will set the number of decimal digits in the fractional part before beginning calculations, to simplify the code, and then proceed using that precision. So, if we set the precision to eight digits we will represent numbers as,

$$\begin{array}{rr} 10.93210032 \rightarrow & 10\ 93210032 \\ -10.9320122356 \rightarrow & -10\ 93201224 \\ 0.003423 \rightarrow & 342300 \end{array}$$

where we actually store the sign separately from the magnitude and we have added a space on the right to differentiate between the integer portion and the scaled fractional part. Note also that since the precision is eight digits after the decimal that the second example, which has ten digits after the decimal, is rounded when stored. To return to the actual floating-point number we need to divide the stored value by 10^d where d is the precision of the fractional part. For the example above, $d = 8$. In practice, we will convert to a float by using the string representation of the integer value, with sign, and insert the decimal point in the proper place (d digits from the end of the string representation).

Let's start our Python class implementing arbitrary precision fixed-point numbers (`APFP.py`),

```
from types import *

class APFP:
    dprec = 23

    def __toFP(self, arg):
        sign = 1
        arg = arg.strip()
        if (arg[0] == "."):
            arg = "0" + arg
        if (arg[0] == "-"):
            sign = -1
            arg = arg[1:]
```

```
14        if (arg[0] == "+"):
15            arg = arg[1:]
16        idx = arg.find(".")
17        ipart = arg[0:idx]
18        fpart = arg[(idx+1):]
19        flen = len(fpart)
20        rounding = 0
21
22        if (flen <= APFP.dprec):
23            fpart += "0"*(APFP.dprec-flen)
24        else:
25            if (int(fpart[APFP.dprec]) >= 5):
26                rounding = 1
27            fpart = fpart[0:APFP.dprec]
28
29        self.n = sign*(int(ipart) * 10**APFP.dprec +
              int(fpart) + rounding)
30
31    def __init__(self, x="0"):
32        if (type(x) == StringType) or (type(x) ==
              UnicodeType):
33            if (x.find(".") != -1):
34                self.__toFP(str(x))
35            else:
36                self.__toFP(str(x)+".0")
37        elif (type(x) == FloatType):
38            self.__toFP("%0.18f" % x)
39        elif (type(x) == IntType) or (type(x) ==
              LongType):
40            self.__toFP("%d.0" % x)
41        else:
42            raise ValueError("unsupported value")
```

where we store the decimal precision in the class variable `dprec` (line 4). Set this once to the desired precision and then use the class. A more sophisticated library would track the precision per value and scale accordingly. Here the default precision is randomly set to 23 decimal digits. We start with two methods, `__init__` and `__toFP`, the constructor which accepts a value to store in `x`, and a private method to process that value. The `__init__` method (line 31) uses the type of the argument to make a string representation and passes that to `__toFP` (line 6). The code in `__init__` has the task of creating a string of the argument with a decimal point in the proper place. Note that the conversion on line 38 uses 18 digits for the decimal so that a Python float (internally a C double) is represented with complete precision.

The `__toFP` method performs string checks and processing (lines 8 through 15) and sets the sign (`sign`) accordingly where `1` is used for a positive number and `-1` is used for a negative number. Finally, lines 17 and 18 extract the integer and fractional parts of the input string (i.e., left of the decimal point and right of the decimal point). Note that these are kept as strings for now. The variable `rounding` (line 20) is set to zero. It will be set to one if it is necessary to truncate the fractional part of the input and the $d + 1$th digit is ≥ 5. Adding one to the scaled input is equivalent to adding 10^{-d}, the smallest value that can be represented.

Lines 22 through 27 manipulate the fractional part to handle the case when the fractional part is less than d digits in which case trailing zeros are added (line 23) to make it exactly d digits long. If the fractional part is longer than d digits we check if the first digit to be dropped is ≥ 5 and if it is, increment `rounding` (line 26). We then extract the first d digits of the fractional part.

Lastly, we can set the scaled integer value that represents the input (line 29). We scale the integer part, as an integer, by 10^d and add in the integer value of the fractional part. We also add in `rounding` to handle the rounding case of line 25 and multiply all by `sign`. We now have the scaled value in `self.n`.

To display an arbitrary precision number we will overload the Python methods for generating a string version of an object. This will let us get the value with the `print` statement as well as with the `str` and `repr` functions. We will also overload the methods corresponding to the `float` and `int` functions so that they will work as expected. This is done with,

```
def __str__(self):
    if (self.n == 0):
        return "0.0"

    s = str(self.n)
    if (len(s) <= APFP.dprec):
        s = "0"*(APFP.dprec-len(s)) + s
        ans = "0." + s
    else:
        ip = len(s) - APFP.dprec
        ans = s[0:ip] + "." + s[ip:]

    k = len(ans)-1
    while (ans[k] == "0"):
        k -= 1
    ans = ans[0:(k+1)]
    if (ans[-1] == "."):
        ans += "0"
    if (self.n < 0):
        ans = "-" + ans

    return ans

def __repr__(self):
    return self.__str__()

def __float__(self):
    return float(self.__str__())

```

```
30 def __int__(self):
31     ans = self.__str__()
32     idx = ans.find(".")
33     return int(ans[0:idx])
```

where the main method is `__str__` which returns a string representing the number. If zero, return zero (line 3). Otherwise, convert the scaled integer to a string (line 5). If this string is less than the precision the integer part is zero so return a value starting with zero after prepending the fractional part with enough leading zeros to match the precision (lines 7 and 8). Otherwise, insert the decimal point before the fractional part (line 11). When line 13 is reached `ans` contains a string representation of the number but it also has all the trailing zeros necessary to reach the fractional precision. To make the return value look nicer lines 13 through 15 locate the first nonzero digit scanning right to left. Line 16 then truncates the answer to drop the trailing zeros. Lines 17 and 18 add a final zero after the decimal point if the number has no fractional part. Lines 19 and 20 prepend the sign if the number is negative. Finally, the exact representation is returned (line 22).

The `__repr__` method, called by the `repr` function, simply returns the string representation (line 25). Similarly, to convert to a Python float the `__float__` method simply attempts to interpret the string representation as a floating-point number. If it is too large in magnitude, it will return positive or negative infinity. Lastly, `__int__` finds the integer portion of the string representation and returns it as an integer. This integer, because of Python's big integer support, will be exact (without the fractional part, which is dropped, not rounded).

Here is an example session with the class as it stands now,

```
>>> from APFP import *
>>> APFP.dprec = 6
>>> x = APFP("3.14159265")
>>> x
    3.141593
>>> print x
    3.141593
>>> str(x)
    '3.141593'
>>> repr(x)
    '3.141593'
>>> float(x)
    3.141593
>>> int(x)
    3
```

where we set the decimal precision to six. In this case, π is truncated with rounding to use exactly six digits, 3.141593. The overloaded `__str__` method then ultimately handles all the calls that follow. Note, the `float` call returns the exact representation in this case because six digits fits easily in a C `double`. Naturally, the integer part is simply three.

9.3 Basic Arithmetic with Arbitrary Precision Floating-Point Numbers

Let's add basic arithmetic to our library. We follow the methods used in Chap. 6 for addition, subtraction, multiplication and division of fixed-point numbers. Addition and subtract are particularly simple. Since each number is stored internally as a a signed integer that is properly scaled by 10^d where d is the decimal precision we simply need to add or subtract as normal,

```
1 def __add__(self, b):
2     z = APFP()
3     z.n = self.n + b.n
4     return z
5
6 def __sub__(self, b):
7     z = APFP()
8     z.n = self.n - b.n
9     return z
```

where we create a new object (lines 2 and 7) and then set it to the sum or difference of the current number and the given number (lines 3 and 8). We return the new object and are done. Why does this work? A simple example will illustrate,

$$\begin{array}{rr|rr} & 123 & & 123\ 0000000 \\ + & 34 & + & 34\ 0000000 \\ \hline & 157 & & 157\ 0000000 \end{array}$$

for $d = 7$ and added space after the integer part on the right side. Similarly, for subtraction we have,

$$\begin{array}{rr|rr} & 123.549 & & 123\ 5490000 \\ - & 34.067 & - & 34\ 0670000 \\ \hline & 89.482 & & 89\ 4820000 \end{array}$$

where it is plain that the scaled versions of the numbers lead to exact results since there is no shifting of the implied decimal point. Overflow and underflow are not an issue in this case because we are using infinite precision integers.

What about multiplication and division? They are almost as simple but we do need to be aware of what is happening in order to rescale the results properly. For multiplication we have,

$$XY = (x \times 10^d) \times (y \times 10^d) = (xy) \times 10^{2d}$$

where X and Y are the scaled versions of x and y. This means that if we simply multiply the scaled integers we will end up with the product we want (xy) but instead of being scaled by 10^d it will be scaled by 10^{2d}. We can fix this by dividing (integer division) by 10^d but if we simply do that we will introduce a bias by truncating digits that do not fit without properly rounding. We can fix this by adding 5 to the highest digit that will be truncated and then dividing. The smallest value we can represent, when the entire number is scaled by 10^{2d}, is 10^d so if we add $5 \times 10^{d-1}$ we will be adding five to the first truncated digit which will round the next digit, the smallest digit we are keeping, to the proper value. For example, what we will do is equivalent to rounding $\pi^2 = 9.8696043785340225$ to nine decimal places ($d = 9$) and getting 9.86960437**9** instead of 9.86960437**8** because we added 5 to the digit after the final 8 to round up (the next digit is 5).

So, after all the words above, the implementation is straightforward,

```
1 def __mul__(self, b):
2     z = APFP()
3     z.n = (self.n * b.n + 5*10**(APFP.dprec-1)) /
              10**APFP.dprec
4     return z
```

where z holds the answer and we set it in line 3 by first multiplying the two numbers (`self.n * b.n`), adding our bias correction (`5*10**(APFP.dprec-1)`), and then dividing by 10^d (`10**APFP.dprec`) to rescale. The signs of the two numbers are handled automatically by the multiplication.

Division is similarly straightforward once we pay attention to what is actually happening and account for rounding. Following the derivation of Chap. 6 we see that division of decimal fixed-point numbers is best represented with,

$$\frac{x \times 10^{2d} + 5 \times y \times 10^{d-1}}{y \times 10^d} = \frac{x}{y} \times 10^d + \frac{5}{10}$$

where $\frac{5}{10} = \frac{1}{2}$ is added to round properly and avoid bias due to truncation effects, just as was done for multiplication. For example, if $d = 8$ and the bias correction is present $1/1120 = 0.0008928\mathbf{6}$ because the actual result is closer to $0.0008928\mathbf{5}\mathbf{7}14\ldots$ where the next digit, the 7, requires rounding to 6. Without the bias correction the $d = 8$ answer would end in a 5, not a 6.

With this in mind, division is easily implemented,

```
1 def __div__(self, b):
2     z = APFP()
3     z.n = (self.n * 10**APFP.dprec + (5*b.n) / 10) / b.n
4     return z
```

where the result (line 2) is set with the expression that is a properly scaled and bias corrected quotient (line 3). This line is a direct translation of the left-hand side of the equation above defining division. As with multiplication, the signs are properly handled by the expression as well. Note that `(5*b.n)/10` is equivalent to $5 \times y \times 10^{d-1}$ since `b.n` is already scaled by 10^d so $10^d/10 = 10^{d-1}$.

Let's look at a sample session using our arbitrary precision floating-point class as it currently is,

```
1  >>> from APFP import *
2  >>> x = APFP(11.09); y = APFP(1120.7768)
3  >>> x+y
4      1131.86680000000093718
5  >>> x = APFP("11.09"); y = APFP("1120.7768")
6  >>> x+y
7      1131.8668
8  >>> x-y
9      -1109.6868
10 >>> x*y
11     12429.414712
12 >>> x/y
13     0.00989492287848927636618
14 >>> APFP.dprec
15     23
16 >>> APFP.dprec = 64
17 >>> x = APFP("11.09"); y = APFP("1120.7768")
18 >>> x/y
19     0.0098949228784892763661774583485311259119567785485
           92369149682613
```

where we added line numbers for reference. The module is imported in line 1. Line 2 defines two variables and line 3 adds them producing line 4. This seems mysterious at first, why the extra bit at the end? It has to do with the way the variables were defined in line 2. The values given to the constructor were Python floats, C doubles, and as such were subject to the imprecision IEEE 754 numbers have. In line 5 we redefine x and y using exact strings as arguments. In this case the sum shown in line 7 is exact, as expected. The same is true for subtraction (line 9), multiplication (line 11), and division (line 13) which uses all the available decimal digits for the default precision of 23 digits (shown in line 15). Line 16 changes the precision to 64 digits, line 17 redefines x and y in this new precision, and line 19 shows the division accurate to 64 digits.

So far, so good. The library works and is already useful but we can improve it. We are missing comparison operators and a few other "nice to have" functions to say nothing of trigonometric and other transcendental functions.

9.4 Comparison and Other Methods

Python kindly offers us many methods that we can overload to work with existing operators. This includes all the expected comparison operators: $>$, $<$, $\geq$, $\leq$, $==$, and $!=$ which become one-liners,

```
def __lt__(self, b):
    return self.n < b.n

def __gt__(self, b):
    return self.n > b.n

def __le__(self, b):
    return self.n <= b.n

def __ge__(self, b):
    return self.n >= b.n

def __eq__(self, b):
    return self.n == b.n

def __ne__(self, b):
    return self.n != b.n
```

courtesy of the magic of fixed-point arithmetic. Since each number is (assumed!) to be scaled by the same value direct comparison of the integer representation is equivalent to a floating-point comparison. So, we simply compare the integer representations directly and return the result.

Absolute value and unary negation are also easily implemented,

```
def __abs__(self):
    z = APFP()
    z.n = abs(self.n)
    return z

def __neg__(self):
    z = APFP()
    z.n = -self.n
    return z
```

where in each case we return a new instance and apply the operation to the integer representation directly.

The operation of these new methods is self-explanatory so we will dispense with the examples and move directly to the more interesting issue of trigonometric and transcendental functions. We'll throw square root in there as well, just for fun.

9.5 Trigonometric and Transcendental Functions

A complete arbitrary precision library will include the basic trigonometric functions, sin, cos, and tan, as well as transcendental functions line log (natural log), exp (e^x), and a fast square root. We will add these functions here by either adapting the cookbook examples given in the Python decimal module documentation [1] which are already designed for arbitrary precision or by using Newton's method to iteratively determine the proper value. We use the Python cookbook examples for the trigonometric functions and exp which add terms to a series representation until the value is no longer changing. We use Newton's method for square root and log as we did in Chap. 6 for fixed-point arithmetic.

Sine and cosine are as follows, with tangent defined as $\tan\theta = \frac{\sin\theta}{\cos\theta}$,

```
def sin(self):
    i, lasts, fact, sign = [APFP(j) for j in [1,0,1,1]]
    s = self
    num = self

    while (s != lasts):
        lasts = s
        i += APFP("2")
        fact *= i * (i - APFP("1"))
        num *= self * self
        sign *= APFP("-1")
        s += num / fact * sign
    return s

def cos(self):
    i, lasts, s, fact, num, sign = [APFP(j) for j in
        [0,0,1,1,1,1]]

    while (s != lasts):
        lasts = s
        i += APFP("2")
        fact *= i * (i - APFP("1"))
        num *= self * self
        sign *= APFP("-1")
        s += num / fact * sign
    return s

def tan(self):
    return self.sin() / self.cos()
```

where `sin` and `cos` are implementations of the Taylor series expansions for sin and cos,

$$\sin x = x - \frac{x^3}{3!} + \frac{x^5}{5!} - \frac{x^7}{7!} + \frac{x^9}{9!} - \cdots$$

$$\cos x = 1 - \frac{x^2}{2!} + \frac{x^4}{4!} - \frac{x^6}{6!} + \frac{x^8}{8!} - \cdots$$

which add terms to the result until it no longer changes (`s == lasts`). This approach adapts the methods to any precision at the expense of potentially requiring hundreds of terms. The tangent is simply defined in line 28.

Adding e^x is similarly accomplished by a Taylor's series expansion which sums terms until they no longer affect the result,

```
def exp(self):
    i, lasts, s, fact, num = [APFP(j) for j in [0,0,1,1,1]]

    while s != lasts:
        lasts = s
        i += APFP("1")
        fact *= i
        num *= self
        s += num / fact
    return s
```

where the series is,

$$e^x = 1 + x + \frac{x^2}{2!} + \frac{x^3}{3!} + \frac{x^4}{4!} + \frac{x^5}{5!} + \cdots$$

To implement square root and natural logarithm we switch from summing a series to iterating via Newton's method. As shown in Chap. 6, iterating,

$$x_{i+1} = \frac{1}{2}(x_i + \frac{n}{x_i})$$

will result in $x \to \sqrt{n}$. In code this becomes,

```
def sqrt(self):
    if (self < APFP(0)):
        return APFP(-1)
    if (self == APFP(1)):
        return APFP(1)

```

```
7        s = self / APFP(2)
8        lasts = APFP(0)
9
10       while (s != lasts):
11           lasts = s
12           s = APFP("0.5") * (s + self / s)
13       return s
```

where we check for negative and zero arguments (lines 2 through 5) and then iterate an initial guess, $x_0 = n/2$, in line 7 until it no longer changes (`s == lasts`). It is known that this method for approximating the square root converges very rapidly which we can see if we plot the number of iterations required as a function of the precision,

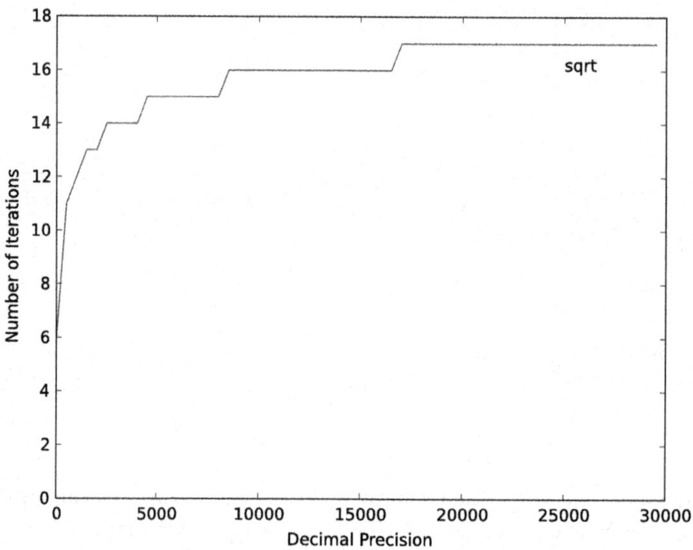

where the number of iteration necessary for convergence levels off quickly.

Our last transcendental function is log which, again following the derivation in Chap. 6, is found by iterating,

$$x_{i+1} = x_i - \frac{f(x_i)}{f'(x_i)} = x_i - \frac{e^{x_i} - c}{e^{x_i}} = x_i - 1 + ce^{-x_i}$$

with $x_0 = c/2$ and $x \to \log(c)$. Unlike square root, this function might, for certain precisions, result in a set repeating values so that simply checking for equality with the last estimate is not sufficient. So, we add a maximum iteration term to ensure that the function will end and return eventually. In code this becomes,

```
1  def log(self, imax=20):
2      if (self <= APFP(0)):
3          return APFP(-1)
4      if (self == APFP(1)):
5          return APFP(0)
6
7      s = self / APFP(10)
8      lasts = APFP(0)
9      i = 0
10
11     while (s != lasts) and (i < imax):
12         lasts = s
13         s += self * (-s).exp()
14         s -= APFP(1)
15         i += 1
16     return s
```

where unlike the other methods we allow a keyword to set the maximum number of iterations before returning (`imax`). Lines 2 through 5 check for bad values or simple values. The loop (line 11) runs until we do get convergence or max out the number of iterations.

Before we move on to the next section, let's take a look at the performance of sin, cos, and exp by number of terms required as a function of precision. Plotting these gives,

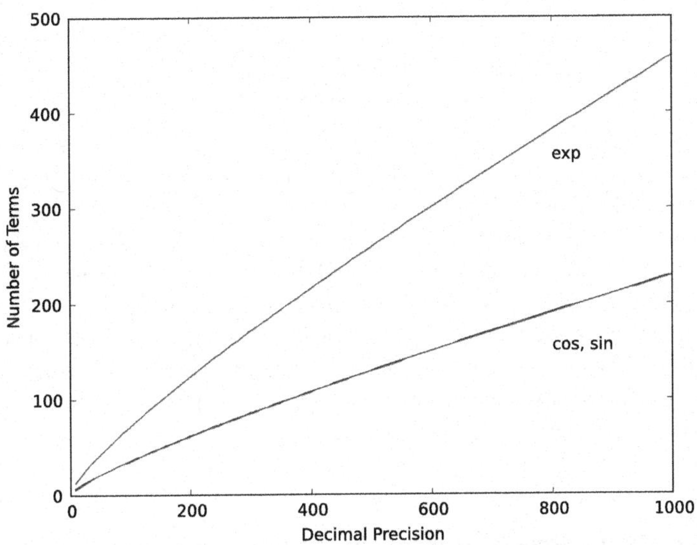

where we see that sin and cos are virtually indistinguishable but that exp requires approximately twice as many terms for any given precision.

We have not attempted to optimize this simple library but it is instructive to take a look at performance in terms of time. The arithmetic operations ($+$, $-$, $\times$, $\div$) will be quite performant relative to the trigonometric and transcendental functions because they are using the optimized, C-based, routines of the Python interpreter. Python makes use of Karatsuba multiplication for large integers, which is a good choice offering $O(n^{\log_2 3})$ performance, better than the grade school method which is $O(n^2)$. See Chap. 5, Sect. 5.4 for a description of the Karatsuba algorithm.

If we plot the evaluation time of sin, cos, and exp as a function of precision we get,

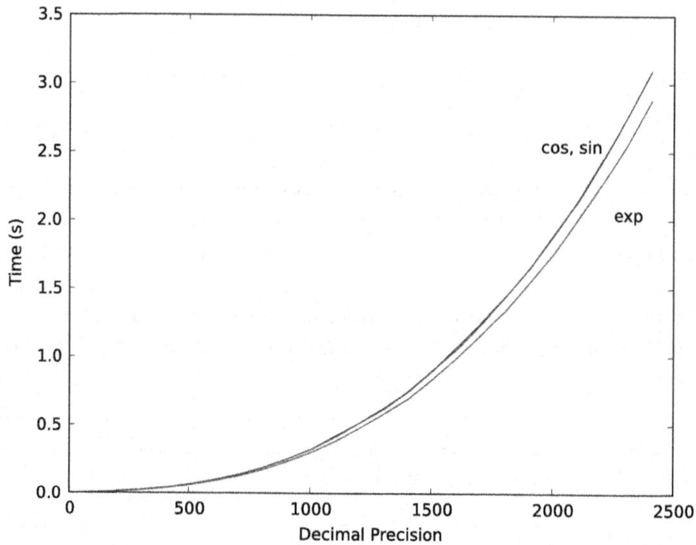

where we see that our implementations are far from stellar and that they give the appearance of a run time like $O(n^2)$.

We have created a simple but effective implementation of arbitrary precision floating-point numbers in Python, via fixed-point numbers, and have added more than just basic arithmetic. We definitely want to remember that the library is meant for illustration purposes only and understand, especially when we consider high precision performance, that we want to look to more advanced implementations for serious work. In the next section we take a look at some powerful arbitrary precision floating-point libraries.

9.6 Arbitrary Precision Floating-Point Libraries

Most major programming languages support arbitrary precision floating-point extensions via libraries. Since we are focusing here on Python and C we will look at two such libraries, one for Python (mpmath) and one for C (GNU MPFR).

The former is implemented in pure Python, like the APFP class we defined above, while the latter is pure C and has been wrapped many times to make it available to different programming languages.

We will discuss how to install the libraries and summarize and demonstrate the main functionality of each including the way in which the libraries store and manipulate arbitrary precision numbers.

mpmath Our first library is `mpmath` [2] which describes itself as *"a free (BSD licensed) Python library for real and complex floating-point arithmetic with arbitrary precision. It has been developed by Fredrik Johansson since 2007, with help from many contributors."* We can install `mpmath` easily (assuming a 32-bit Linux-based system which is the standard assumption for this book) using `pip`,

```
sudo pip install mpmath
```

and we can check that it installed inside of Python with,

```
>>> import mpmath
>>> mpmath.__version__
    '0.19'
```

where your version should be at least 0.19 and will probably be higher.

The `mpmath` library provides, in pure Python without requiring external libraries, all the basic functionality of our `APFP` class along with much more. We can test that `mpmath` installed correctly and get an idea of the plethora of functionality it provides by running its internal tests,

```
>>> import mpmath
>>> mpmath.runtests()
```

where the test results should all be `'ok'` across all the tested areas. We will do a great disservice to this elegant library by only considering here the functionality that matches our `APFP` class.

Let's look at how `mpmath` sets the precision for calculations. These examples illustrate how to do it,

```
>>> from mpmath import mp
>>> mp.mpf('1') / mp.mpf('1120')
    mpf('0.00089285714285714283')
>>> mp.pretty = True
>>> mp.mpf('1') / mp.mpf('1120')
    0.000892857142857143
>>> mp.dps = 20
>>> mp.mpf('1') / mp.mpf('1120')
    0.00089285714285714285714
>>> mp.dps = 50
>>> mp.mpf('1') / mp.mpf('1120')
    0.000892857142857142857142857142857142857142857142857142
             5714285714
```

where we have added line numbers to aid our description of what is going on. Line 1 imports mpmath. In particular, it imports the mp context. Several different kinds of contexts are supported, we will only use the mp context. Notice that this line is different than the typical NumPy import line of `import numpy as np` but the effect is similar, we prefix mpmath calls with mp. Line 2 evaluates the expression 1/1120 by defining two mpmath numbers using `mp.mpf()` which creates a mpmath floating-point scalar. Line 3 shows the response. Since we are using the Python command line we will get the output of `repr()`. Line 4 sets an internal flag so that repeating the calculation (line 5) gives us a prettier output (line 6).

The default decimal precision for mpmath is 15 digits which is what we get in line 6. Ignore the leading zeros—we'll see why when we talk about how mpmath stores numbers internally below. Line 7 ups the decimal precision to 20 digits which we see in line 9. Lines 10 and 12 do the same for 50 decimal digits. One other thing to notice in the example above is that the `mpf` constructor accepts strings, integers and floats, just like the `APFP` class. Similarly, use strings to most accurately specify numbers.

Naturally, if division works as in line 2, the other basic arithmetic operations will also work in a similar fashion. Lastly, mpmath supports free mixing of its own numbers with native Python integers and floats. This is something we didn't add to the `APFP` class in order to keep the implementation simple.

If we want to compare mpmath with `APFP`, we should be aware of the difference in meaning between the precision each library uses. For example,

```
>>> mp.dps = 15
>>> APFP.dprec = 15
>>> mp.mpf("1") / mp.mpf("1120")
    0.000892857142857143
>>> APFP("1") / APFP("1120")
    0.000892857142857
```

where we set both mpmath and `APFP` to use a precision of 15 digits. We then calculate the same quantity and see that the output is different. This is because when mpmath uses 15 digits it means 15 significant digits. When `APFP` uses 15 digits it means, literally, that there are 15 digits after the decimal point, including leading zeros. With this in mind, a few examples of arithmetic will show that mpmath and `APFP` are giving identical results but with some interesting differences,

```
>>> mp.dps = 23
>>> APFP.dprec = 23
>>> mp.mpf("12.9432") * mp.mpf("-0.043225")
    -0.55946982
>>> APFP("12.9432") * APFP("-0.043225")
    -0.55946982
>>> mp.mpf("12.9432") / mp.mpf("-0.043225")
    -299.43782533256217466744
```

```
9  >>> APFP("12.9432") / APFP("-0.043225")
10     -299.43782533256217466743783
11 >>> mp.mpf("12.9432") * mp.mpf("43225094543.33")
12     559471043693.228856
13 >>> APFP("12.9432") * APFP("43225094543.33")
14     559471043693.228856
15 >>> mp.mpf("129435433.2095345332") * mp.mpf
         ("43225094543.33")
16     5594858837739005819.5525
17 >>> APFP("129435433.2095345332") * APFP("43225094543.33")
18     5594858837739005819.552518723556
```

Lines 1 and 2 set our precision to 23 decimals. So, lines 3 and 5 show that each library gives the same result for an operation that fits well within the allowed precision. Lines 9 and 11 show the difference between the libraries that we noticed above, namely, that `mpmath` is using 23 decimal digits for the entire number while `APFP` uses 23 digits in the fractional part so that where `mpmath` ends with 744 `APFP` ends with 743**783** where the bold digits are used to get to 23 decimal digits in the fractional part and are also correctly rounded in the `mpmath` answer.

Lines 11 and 13 show that, again, when everything fits nicely in the specified number of digits both libraries give identical results. However, lines 15 and 17 show something new. Namely, since `mpmath` uses a fixed number of digits for the *entire* number, the large product returns only four digits after the decimal point. However, `APFP` uses a fixed number of digits after the decimal point but an arbitrary number of digits in the integer part of the number so that it gives the same integer part but still has 23 digits available for the fractional part which is more than is needed to *exactly* express the full result of the multiplication. This difference between the two libraries is important to remember and could be seen as an advantage to the fixed-point representation that is missing in the floating-point representation (more below). For example, perhaps the smallest values used in science are the Planck units of which the Planck time is on the order of 10^{-44} s. So, naïvely, one might imagine setting `APFP.dprec` to 50, or better still 100 (to maintain full fractional precision during multiplication) to cover the smallest of fractions of a value and, with an "infinite" sized integer part, be able to represent any physically meaningful quantity with all the precision necessary to accurately carry out virtually any calculation.

The number of trigonometric and transcendental functions supported by `mpmath` exceeds several dozen. Here we will do quick comparisons between `APFP` and `mpmath` for *sin* and *exp* to see how an optimized implementation can improve over a naïve implementation. If we measure the computation time for $sin(1)$ and $e^{0.5}$ for each library we get numbers like these,

	sin(1)		$e^{0.5}$	
Precision	APFP	mpmath	APFP	mpmath
100	0.004277	0.000201	0.003466	0.000073
1000	0.301725	0.002486	0.245470	0.001851
2000	1.758381	0.008085	1.442545	0.006231
3000	5.005423	0.016561	4.142346	0.013261
4000	10.866939	0.028612	9.002207	0.023152
5000	20.118882	0.044388	16.673306	0.036576

which shows that our naïve `APFP` implementation, while accurate, could be greatly improved.

How does `mpmath` get such good performance? An examination of the code shows that it is finely tuned and employs a lot of tricks including caching, symmetries, and preset precision thresholds where different approaches are used. If the precision is high enough the Taylor series expansion is used but it is also finely tuned and far more sophisticated than the simple approaches used by `APFP`. Interestingly, one of the optimizations used by `mpmath` is to switch internally to a fixed-point representation in some cases. To explore these optimizations, see the file `libelefun.py` in the `mpmath/libmp` directory of the `mpmath` distribution.

As alluded to above, `mpmath` uses a different storage format than `APFP`. Where `APFP` uses a scaled arbitrary precision integer, a standard Python data type, `mpmath` uses a true floating-point format stored internally as a tuple. Specifically, `mpmath` stores a number as a 4-tuple: *(sign, significand, exponent, bit count)*. The *sign* is zero (positive) or one (negative). The significand is a Python integer, as in `APFP`, but in this case it represents the significand. The exponent is also a Python integer. The bit count is the size of the significand in bits. So, the number expressed by a 4-tuple is,

$$x = (-1)^{sign} \times significand \times 2^{exponent}$$

where `mpmath` ensures that the *significand* is always odd.

As we saw with the examples above, `mpmath` is a solid, well-optimized library, all the more impressive because it is pure Python. It is well recommended as mature, highly portable, software.

GNU MPFR We discussed the `mpmath` library in some detail because it was directly comparable to our own Python `APFP` class. Here we will take a quick look at the popular *MPFR* library [3] which forms the basis of many other libraries and tools.

The GNU *MPFR* arbitrary precision floating-point library is part of a loose suite of tools that implement arbitrary precision integers (*GMP*) and intervals (*MPFI*). *MPFR* is written in C but with easy access from C++. The library also usable from MATLAB®, Python (via the `bigfloat` module), and other languages. Here we will show simple examples in C.

First, install *MPFR* from http://www.mpfr.org/ by following the directions on that site. We will be using version 3.1.4. Then, to test the installation, let's run the sample program from the website,

```
#include <stdio.h>
#include <gmp.h>
#include <mpfr.h>

int main (void)
{
  unsigned int i;
  mpfr_t s, t, u;

  mpfr_init2 (t, 200);
  mpfr_set_d (t, 1.0, MPFR_RNDD);
  mpfr_init2 (s, 200);
  mpfr_set_d (s, 1.0, MPFR_RNDD);
  mpfr_init2 (u, 200);
  for (i = 1; i <= 100; i++)
    {
      mpfr_mul_ui (t, t, i, MPFR_RNDU);
      mpfr_set_d (u, 1.0, MPFR_RNDD);
      mpfr_div (u, u, t, MPFR_RNDD);
      mpfr_add (s, s, u, MPFR_RNDD);
    }
  printf ("Sum is ");
  mpfr_out_str (stdout, 10, 0, s, MPFR_RNDD);
  putchar ('\n');
  mpfr_clear (s);
  mpfr_clear (t);
  mpfr_clear (u);
  mpfr_free_cache ();
  return 0;
}
```

which can be compiled with,

```
$ gcc mpfr_ex1.c -o mpfr_ex1 -lmpfr -lgmp
```

assuming the code is stored in `mpfr_ex1.c`. Note the inclusion of the *MPFR* library via `-lmpfr`. We also include the *GMP* library via `-lgmp`. When executed the program produces,

```
Sum is 2.718281828459045235360287471352662497757247093699
          9595749669131
```

which is a reasonably accurate value for e since the program implements,

$$e = \sum_{k=0}^{\infty} \frac{1}{k!}$$

We say "reasonably" because the last three digits (131) are not correct, they should be, according to APFP, 676. This is exactly what we see if we change the program to increase the precision by replacing "200" in lines 10, 12, and 14 with "400".

A lot is happening in the sample program. Let's go over it carefully as this will introduce fundamental concepts associated with the *MPFR* library. First, notice that lines 2 and 3 import both `gmp.h` and `mpfr.h`. Line 8 creates several variables of `mpfr_t` type. These store *MPFR* numbers. Lines 10 through 14 are,

```
10  mpfr_init2 (t, 200);
11  mpfr_set_d (t, 1.0, MPFR_RNDD);
12  mpfr_init2 (s, 200);
13  mpfr_set_d (s, 1.0, MPFR_RNDD);
14  mpfr_init2 (u, 200);
```

which initialize and set *MPFR* variables. The variable `t` is initialized (line 10) by calling `mpfr_init2`. This function initializes the given variable setting its precision to the given number of *bits* (not decimal digits like APFP or `mpmath`). The number itself is set to NaN indicating that *MPFR* supports the expected floating-point concepts of NaN and $\pm\infty$. Line 11 sets `t` to 1.0 via `mpfr_set_d` which assigns an IEEE double precision value to an MPFR variable. The third argument of this call, `MPFR_RNDD`, rounds the value towards $-\infty$. So, *MPFR* supports rounding modes as well, just like IEEE 754. Lines 12 and 13 do the same for `s`. What if we want to assign a variable to a high precision constant? We can simply call `mpfr_set_str` instead which takes a string argument representing the value. Line 14 initializes `u` but does not assign it a value yet.

Since *MPFR* uses bits for precision, how do we know the number of decimal digits? The relationship is $digits \approx bits/3.32$ so 200 bits implies about 60 decimal digits.

The main loop implements the series summation with,

```
15  for (i = 1; i <= 100; i++)
16    {
17      mpfr_mul_ui (t, t, i, MPFR_RNDU);
18      mpfr_set_d (u, 1.0, MPFR_RNDD);
19      mpfr_div (u, u, t, MPFR_RNDD);
20      mpfr_add (s, s, u, MPFR_RNDD);
21    }
```

where different *MPFR* functions are called to perform basic arithmetic. Each function follows this form,

```
mpfr_add(c, a, b, rnd)
```

which implements $c \leftarrow a + b$ with rounding mode `rnd`. So, line 17 multiplies `t` by `i` rounding towards positive infinity. In this case `mpfr_mul_ui` is called to multiply an *MPFR* number by an unsigned integer. Likewise, lines 19 and 20 implement $u \leftarrow u/t$ and $s \leftarrow s + u$ respectively with rounding towards negative infinity. The loop runs for a set 100 iterations which is sufficient to converge when 200 bits of precision are used.

When complete, the result is displayed by calling `mpfr_out_str` (line 23) to write a decimal representation of `s` to `stdout`. The third argument (`0`) selects the number of digits to output based on the precision of `s`. Note here that a specific rounding mode is given, in this case towards negative infinity.

When we are through with an *MPFR* variable we need to clear it and release the memory. This is done above in lines 25 through 27 by calling `mpfr_clear`. Line 28 calls `mpfr_free_cache` to explicitly free memory associated with predefined constants.

Let's look at a second example using *MPFR*. In Chap. 4 we present an exercise where we iterate the logistic equation written in different algebraically-equivalent forms to show that round-off error quickly dominates and causes calculations that should produce identical results to diverge. We will repeat that exercise here using *MPFR* now that we have arbitrary precision at our disposal.

The logistic equation is often used to illustrate the onset of chaotic behavior. It is an iterated one-dimensional map,

$$x_{i+1} = rx_i(1 - x_i) = rx_i - rx_ix_i = x_i(r - rx_i)$$

where $x = [0, 1]$ and $r < 4$. If r is above a certain value the behavior of the logistic equation is chaotic. We will select $r = 3.8$ which is in the chaotic region and iterate each of the three forms above. In an ideal world the values will track identically but because of round-off error in floating-point, even with arbitrary precision, the values will eventually diverge after a sufficient number of iterations.

```
void pp(mpfr_t x0, mpfr_t x1, mpfr_t x2) {
    printf("    x0: "); mpfr_out_str(stdout,10,
            0,x0,MPFR_RNDN); printf("\n");
    printf("    x1: "); mpfr_out_str(stdout,10,
            0,x1,MPFR_RNDN); printf("\n");
    printf("    x2: "); mpfr_out_str(stdout,10,
            0,x2,MPFR_RNDN); printf("\n\n");
}

int main (int argc, char *argv[]) {
    unsigned int i, p, n;
    mpfr_t x0,x1,x2,u,v, r, one;

    p = (unsigned int)atoi(argv[1]);
    n = (unsigned int)atoi(argv[2]);
    mpfr_init2(x0, p);  mpfr_init2(x1, p);
    mpfr_init2(x2, p);  mpfr_init2(r,  p);
    mpfr_init2(one,p);  mpfr_init2(u,  p);
            mpfr_init2(v,  p);
    mpfr_set_d(one,1, MPFR_RNDN);
            mpfr_set_d(r, 3.8, MPFR_RNDN);
```

```
    mpfr_set_d(x0, 0.25, MPFR_RNDN); mpfr_set_d(x1, 0.25, MPFR_RNDN);
    mpfr_set_d(x2, 0.25, MPFR_RNDN);

    printf("Initial values:\n");
    for(i=0; i < 8; i++) {
        mpfr_sub(u, one, x0, MPFR_RNDN);  // x0 = r*x0*(1.0 - x0)
        mpfr_mul(x0, x0, u, MPFR_RNDN);
        mpfr_mul(x0, r, x0, MPFR_RNDN);
        mpfr_mul(u, r, x1, MPFR_RNDN);    // x1 = r*x1 - r*x1*x1
        mpfr_mul(v, r, x1, MPFR_RNDN);
        mpfr_mul(v, v, x1, MPFR_RNDN);
        mpfr_sub(x1, u, v, MPFR_RNDN);
        mpfr_mul(u, r, x2, MPFR_RNDN);    // x2 = x2*(r - r*x2)
        mpfr_sub(v, r, u, MPFR_RNDN);
        mpfr_mul(x2, x2, v, MPFR_RNDN);
    }
    pp(x0,x1,x2);
    for(i=0; i < n-16; i++) {
        mpfr_sub(u, one, x0, MPFR_RNDN);  // x0 = r*x0*(1.0 - x0)
        mpfr_mul(x0, x0, u, MPFR_RNDN);
        mpfr_mul(x0, r, x0, MPFR_RNDN);
        mpfr_mul(u, r, x1, MPFR_RNDN);    // x1 = r*x1 - r*x1*x1
        mpfr_mul(v, r, x1, MPFR_RNDN);
        mpfr_mul(v, v, x1, MPFR_RNDN);
        mpfr_sub(x1, u, v, MPFR_RNDN);
        mpfr_mul(u, r, x2, MPFR_RNDN);    // x2 = x2*(r - r*x2)
        mpfr_sub(v, r, u, MPFR_RNDN);
        mpfr_mul(x2, x2, v, MPFR_RNDN);
    }
    printf("\n\nFinal values:\n");
    for(i=0; i < 8; i++) {
        mpfr_sub(u, one, x0, MPFR_RNDN);  // x0 = r*x0*(1.0 - x0)
        mpfr_mul(x0, x0, u, MPFR_RNDN);
        mpfr_mul(x0, r, x0, MPFR_RNDN);
        mpfr_mul(u, r, x1, MPFR_RNDN);    // x1 = r*x1 - r*x1*x1
```

```
        mpfr_mul(v, r, x1, MPFR_RNDN);
        mpfr_mul(v, v, x1, MPFR_RNDN);
        mpfr_sub(x1, u, v, MPFR_RNDN);
        mpfr_mul(u, r, x2, MPFR_RNDN);     // x2 = x2*(r - r*x2)
        mpfr_sub(v, r, u, MPFR_RNDN);
        mpfr_mul(x2, x2, v, MPFR_RNDN);
    }
    pp(x0,x1,x2);
    return 0;
}
```

where we are not listing the `#include` statements to save space. While the listing seems long, it is really three repetitions of a simple block to implement the three different forms of the logistic equation,

```
1  for(i=0; i < n-16; i++) {
2      mpfr_sub(u, one, x0, MPFR_RNDN);  // x0 = r*x0*(1.0 - x0)
3      mpfr_mul(x0, x0, u, MPFR_RNDN);
4      mpfr_mul(x0, r, x0, MPFR_RNDN);
5      mpfr_mul(u, r, x1, MPFR_RNDN);    // x1 = r*x1 - r*x1*x1
6      mpfr_mul(v, r, x1, MPFR_RNDN);
7      mpfr_mul(v, v, x1, MPFR_RNDN);
8      mpfr_sub(x1, u, v, MPFR_RNDN);
9      mpfr_mul(u, r, x2, MPFR_RNDN);    // x2 = x2*(r - r*x2)
10     mpfr_sub(v, r, u, MPFR_RNDN);
11     mpfr_mul(x2, x2, v, MPFR_RNDN);
12 }
```

where lines 2 through 4 iterate `x0`, lines 5 through 8 iterate `x1`, and lines 9 through 11 iterate `x2`. The first eight iteration values are displayed along with the final eight. We compile the program and call it with two command line arguments: precision and number of total iterations. As we saw in Chap. 3, even after eight iterations with standard floating-point the values are already diverging significantly. Here we do not see this deviation but do with the final eight as we will see below. Let's fix the number of total iterations at 10,000. If we do this with 113 bits of precision (equivalent to an IEEE 754 quadruple precision float) we get the following output for the final iteration,

$$
\begin{aligned}
x_0&: \ 0.949404460849763073594113685574416764\\
x_1&: \ 0.457638502447496138812013997759474251\\
x_2&: \ 0.798660115295355269211918348596331193
\end{aligned}
$$

where the values are clearly no longer in sync with each other. This begs the question, for a fixed iteration limit of 10,000, what precision do we need to get final values are equal to some level of precision? Some experimentation using a manual binary search shows that when we set the precision to 6250 bits we get the following,

$$x_0: \quad 0.67946589407247434\,62$$
$$x_1: \quad 0.67264664286395118\,24$$
$$x_2: \quad 0.67387779797033019\,64$$

where the first two decimals are now equal and we are showing the precision we'd expect from an IEEE 754 double (53 bits). This implies that we need nearly *118* times the precision of a C double in order to even approach uniformity when calculating 10,000 iterations of the logistic map! What precision will let us calculate 10,000 iterates and maintain the 53 bit precision of a C double? Some more searching reveals that 6315 bits of precision is the value we seek. At that precision the final values are $x_0 = x_1 = x_2 = 0.7061911661282795065$. We begin to see how it is possible that mathematical research could require very high precision in order to be convinced results are legitimate. Fortunately for us, the real world seldom affords or requires such high precision but it is a bit disconcerting that the bulk of floating-point calculations performed by computers use so few bits.

Let's take a second look at the rounding modes supported by *MPFR*. These mirror the four rounding modes supported by IEEE 754 along with a mode not supported by IEEE,

Mode	Description
`MPFR_RNDU`	Round the result up towards $+\infty$.
`MPFR_RNDD`	Round the result down towards $-\infty$.
`MPFR_RNDZ`	Round the result up or down towards zero.
`MPFR_RNDN`	Round the result to the nearest representable number. Ties to even.
`MPFR_RNDA`	Round away from zero, up or down. Not in IEEE 754.

Note that in IEEE round to nearest is the assumed default while *MPFR* requires the rounding mode be explicitly given. In general, round to nearest (`MPFR_RNDN`) is usually sufficient.

How does *MPFR* store numbers internally? If we look at `mpfr.h` we see that `mpfr_t` is an instance of a specific structure,

```
typedef struct {
  mpfr_prec_t  _mpfr_prec;
  mpfr_sign_t  _mpfr_sign;
  mpfr_exp_t   _mpfr_exp;
  mp_limb_t   *_mpfr_d;
} __mpfr_struct;
```

where it is evident that *MPFR* is using a format similar to `mpmath` with separate sign, exponent and collection of "digits" (`_mpfr_d`). The `mpfr.h` header explicitly states how to generate the represented number,

$$x = s \times (\frac{d_{k-1}}{B} + \frac{d_{k-2}}{B^2} + \cdots + \frac{d_0}{B^k}) \times 2^e$$

where s is the sign (± 1), $k = \lceil p/g \rceil$, $B = 2^g$, p is x's precision in bits, e is the exponent, and g is a constant used by the underlying *GMP* library. On our test system $g = 32$ bits. Just as IEEE 754 requires that the highest order bit of the significand be one, *MPFR* requires that the highest order data value of the significand be $\geq B/2$.

How does *MPFR* stack up against `mpmath`? This is a somewhat unfair question as `mpmath` is pure Python and *MPFR* is written in C, but it is illustrative to do a quick comparison in order to understand when it might be acceptable to use `mpmath` for a calculation or when one might want to go through the extra work of using *MPFR*. To this end, we measured the runtime for 1000 calculations of $\sin(1)$ for both libraries as a function of the binary precision where the timing only measured the actual calculations and no other overhead. For `mpmath`, recall, we can set the binary precision with,

```
>>> from mpmath import mp
>>> mp.prec = 200
```

to set the binary precision to 200 bits. If we do the comparison and plot the results we get,

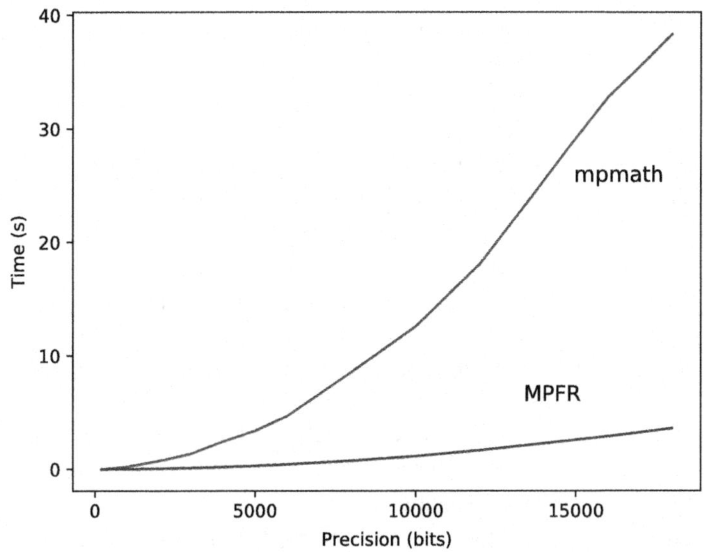

clearly showing that for precisions much above 1000 bits (about 300 decimal digits) we will want to use *MPFR* for serious calculations. Actually, this shows just how well optimized `mpmath` is in that it can effectively compete with a purely compiled library at precisions below 1000 bits. Clearly, a single comparison is not a fair evaluation but it does provide a ballpark estimate of relative performance.

This concludes our investigation into arbitrary precision floating-point libraries. There are others, to be sure, but `mpmath` and *MPFR* provide an adequate overview.

9.7 Thoughts on Arbitrary Precision Floating-Point

What to think of arbitrary precision floating-point? Clearly, if working simply, say as a sort of "desk calculator", almost any library, including the `APFP` library we developed in this chapter, will work nicely. However, if we are developing software that relies on many calculations we will want more optimized libraries. If working in a language like Python, and the precision need not be super high, something like `mpmath` will do the trick. For maximum performance, where speed is critical, or precision needs to be very high, we will want something like *MPFR*.

But, is arbitrary precision ever really needed? The cheeky answer is "it depends". High precision is helpful for some scientific calculations to avoid round-off effects or because the values really are known with more than the precision expressed by an IEEE 754 float. A glance at the list of software using *MPFR* reveals places where arbitrary precision is desirable. For example, paraphrasing select packages from the "Software Using *MPFR*" section of the main *MPFR* website gives the following list (as of July 2016),

ALGLIB.NET	Implements multiple-precision linear algebra using MPFR
BNCPack	The numerical analysis library which can be compiled with MPFR
CGAL	Computational Geometry Algorithms Library
DateTime-Astro	Functions for astronomical calendars
FLINT	Fast Library for Number Theory
FractalNow	A fractal generator
GCC	In GFortran, then in the middle-end phase as of GCC 4.3, to resolve math functions with constant arguments
libieeep1788	C++ implementation of the preliminary IEEE P1788 standard for interval arithmetic
Macaulay 2	A software system devoted to supporting research in algebraic geometry and commutative algebra
Magma	A computational algebra system
MATLAB®	Multiprecision Computing Toolbox for MATLAB®
Protea	Software devoted to protein-coding sequences identification
TIDES	To integrate numerically ordinary differential equations in arbitrary precision
TRIP	A general computer algebra system dedicated to celestial mechanics
ZKCM	A C++ library for multi-precision complex-number matrix calculations

indicating that many math packages and computer algebra systems are interested in high precision calculations along with certain scientific applications where large numbers are likely required.

9.8 Chapter Summary

In this chapter we developed a Python class which implements arbitrary precision floating-point calculations using fixed-point arithmetic. We demonstrated how to use the class and explored its limitations. We then introduced a sophisticated Python library for arbitrary precision floating-point and compared it to our own class revealing strengths and weaknesses of both. We then introduced a high-performance C library for arbitrary precision floating-point and explored the level of precision necessary to maintain complete precision for calculations that, naïvely, should be equivalent. We ended with some thoughts about arbitrary precision floating-point and when it might be useful.

Exercises

9.1 Add a `frac` method to the APFP class which returns the fractional part of a number.

9.2 Add `ceil` and `floor` methods to the APFP class which return $\lceil x \rceil$ and $\lfloor x \rfloor$ respectively. Then add `round` which returns the closest integer to x.

9.3 Add modulo to the APFP class by overloading the modulo operator (`%`) using the `__mod__` method.

9.4 Add x^y to the APFP class. Overload the `__pow__` method so `x**y` works properly. Distinguish between two cases: x^y where both x and y are arbitrary precision numbers and x^n where n is an ordinary Python integer, positive or negative. (Hint: recall that for real numbers, $x^y = e^{y \log x}$.)

9.5 Make a plot of the number of iterations for the following series approximations of e as a function of the precision. Which series converges in the fewest number of terms? Test precisions from 8 to 4500. Look at the implementation of the APFP `exp` method for hints on implementing the series. The series are,

$$e_0 = \frac{1}{2} \sum_{k=0}^{\infty} \frac{k+1}{k!}$$

$$e_1 = 2 \sum_{k=0}^{\infty} \frac{k+1}{(2k+1)!}$$

$$e_2 = \sum_{k=0}^{\infty} \frac{(3k)^2 + 1}{(3k)!}$$

$$e_3 = \sum_{k=1}^{\infty} \frac{k^7}{877(k!)}$$

References

1. https://docs.python.org/2/library/decimal.html. (retrieved 10 Jul 2016).
2. http://mpmath.org/. (retrieved 14 Jul 2016).
3. http://www.mpfr.org/. (retrieved 15 Oct 2014).

Chapter 10
Other Number Systems

Abstract In this chapter we take a look at a number of other somewhat unusual number systems. Most of these find use in hardware and are not encountered, typically, in software. Nonetheless, we will look at these number systems from a software perspective with the intention of understanding how they operate in order to see why and when they might be advantageous. Specifically, this chapter covers the logarithmic number system, double-base number system, residue number system, and redundant signed-digit number system.

10.1 Introduction

The other chapters in this book examined many of the key number systems used by computers. However, this list has not been exhaustive. Indeed, in this chapter we will briefly consider four other number systems which are intellectually interesting in their own right, each with strengths and weaknesses. These number systems are typically implemented in hardware but we will explore software implementations here.

First we will consider the logarithmic number system (LNS) which stores numbers in log form. We follow with the double-base number system (DBNS) which, as the name implies, uses more than one base to store numbers. These numbers are representable as a two-dimensional array or image. Lastly, we consider the residue (RNS) and redundant signed-digit (RSD) number systems which seek to improve arithmetic by sacrificing specificity.

10.2 Logarithmic Number System

The IEEE 754 floating-point storage format, discussed in detail in Chap. 3, stores real numbers in memory as an approximation using a base-2 floating-point significand times 2^i where i is an integer exponent. This directly preserves the magnitude of the number. While a completely natural and reasonable thing to do when faced with the question of "How do I store an arbitrary real number in a finite computer memory?" as Chap. 3 shows, manipulating these numbers requires considerable effort for certain operations, namely multiplication and division. So, the question

R.T. Kneusel, *Numbers and Computers*, DOI 10.1007/978-3-319-50508-4_10

could be asked, "Is there a better way to store numbers so that multiplication and division become simpler?". The answer is "yes" and logarithmic numbers are the solution. The logarithmic number system (LNS) stores numbers as 2^x where x is now a floating-point exponent. The significand is always an implied one.

To see why storing numbers this way helps with multiplication and division we need only remember high school algebra. For two numbers, X and Y,

$$\begin{aligned} A &= \log_2(X) \\ B &= \log_2(Y) \\ XY &= 2^A \times 2^B \\ &= 2^{A+B} \\ X/Y &= 2^A/2^B \\ &= 2^{A-B} \end{aligned}$$

where we see that if the numbers are stored in log form (A and B) multiplication and division are simply addition and subtraction. Of course, while very convenient, there is a price to pay. What about addition and subtraction of numbers stored in log form? This is where the difficulty lies and is why the logarithmic number system isn't widely used though it has been suggested that the advantage of casting multiplication and division as addition and subtraction outweighs the difficulty inherent in addition and subtraction of numbers stored in log form. Indeed, hardware projects have been undertaken to do just that. A good example is the *European Logarithmic Microprocessor* [1] which implemented logarithmic numbers in hardware. We will examine addition and subtraction of logarithmic numbers below but first let's look at how we will store the numbers in memory.

Representing and Storing LNS Numbers A 32-bit IEEE 754 number is stored in memory as,

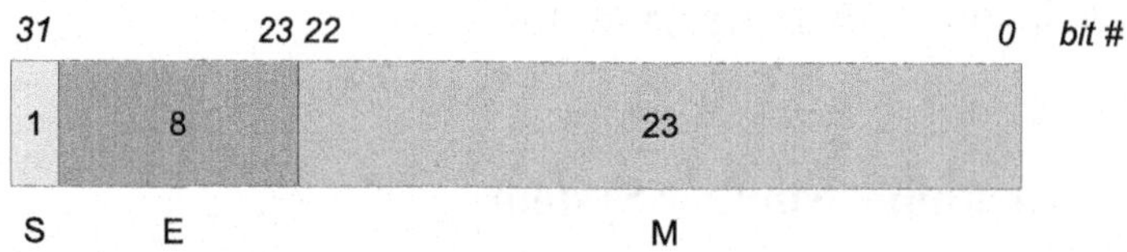

with a sign bit (S), 8 bits for the exponent (E), and 23 bits for the significand (M). The exponent is stored in excess-127 format with the special values of 0 and FF_{16} reserved. The implied base is 2 and the significand has an implied leading 1 bit for a normalized number. This results in a number stored as,

$$\pm 1.b_{22}b_{21}b_{20}\ldots b_0 \times 2^{E-127}$$

where b_n is the value of that bit of the significand. Note that this format scales the floating-point significand by 2 to an integer exponent ($2^i, i \in \mathbb{Z}$).

By way of contrast, a number stored in LNS format uses an implied integer significand of 1 and a floating-point exponent on the base of 2. This exponent, then, is the log base-2 of the actual number and it is the exponent that is stored, along with the sign. There is currently no standard defining how LNS numbers are stored in memory. We will adopt the format used in [2] for a 32-bit representation of an LNS number with the modification that the two bits reserved for flags corresponding to the IEEE 754 values of *zero*, *+inf*, *−inf*, and *NaN* will be replaced with a single bit flag to indicate *zero*. So, our numbers will be stored as,

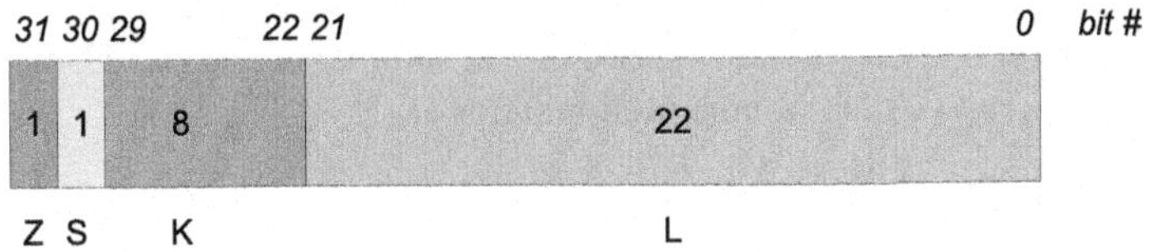

with one bit to indicate zero (Z), a sign bit (S), and floating-point exponent (E) stored in *fixed-point* format using eight bits for the integer part (K) and 22 bits for the fractional part (L), i.e., as a $Q7.22$ number. See Chap. 6 for a review of fixed-point numbers. We adopt this approach so that our LNS numbers can be represented as `uint32` numbers in C at the expense of some precision. This results in a number stored as,

$$\pm 1 \times 2^E$$

where the significand is always an implied "1" and E is a fixed-point number stored in two's-complement format so that numbers < 1.0 have a negative exponent.

What is the range of numbers we can represent with our logarithmic format? The answer depends on the range of the exponent. A Q7.22 number covers the range $[-2^7, 2^7 - 2^{-22}] = [-128, 127.9999998]$ which means that the smallest number, in magnitude, that can be represented is $2^{-128} \approx 2.9387359 \times 10^{-39}$ while the largest is $2^{2^7-2^{-22}} \approx 3.4028231 \times 10^{38}$ which compares favorably with the range of an IEEE 32-bit floating-point number: 1.17549×10^{-38} to 3.40282×10^{38}.

As with IEEE 754 floating-point numbers, the spacing between numbers in the LNS format is not constant. The minimum interval by which the exponent will change is $\Delta \equiv 2^{-22} \approx 2.3841858 \times 10^{-7} \approx 0.00000024$. Therefore, the difference between any number 2^N and the next largest number, $2^{N+\Delta}$ is,

$$2^{N+\Delta} - 2^N = 2^{\Delta}2^N - 2^N = (2^{\Delta} - 1)(2^N)$$

so that between the smallest representable number and the next smallest we have a difference of $2^{\Delta}2^{-128} \approx 4.9 \times 10^{-46}$ and between the largest representable number and the next largest we have $2^{\Delta}2^{2^7-2^{-22}} \approx 5.6 \times 10^{31}$.

For IEEE 754 floating-point the smallest change to the significand, for a 32-bit number, is $2^{-23} \approx 1.19 \times 10^{07}$ regardless of the exponent. Therefore, between any number and the next closest number the difference is $2^{-23}2^{\beta}$ where β is the exponent of the number (ignoring steps between exponents for now). Therefore, between the smallest number and the next smallest the difference is $2^{-23}2^{-126} \approx 1.4 \times 10^{-45}$ and between the largest and next largest we have $2^{-23}2^{127} \approx 2.0 \times 10^{31}$. This result indicates that LNS numbers are placed on the real number line in a manner similar to IEEE 754 floating-point. However, unlike IEEE 754, which has a fixed spacing between numbers until the exponent changes which then doubles the spacing, the spacing between LNS numbers is a smooth function which changes continuously.

We are now ready to start building C functions to handle input, output and storage of LNS numbers. Storage of LNS numbers is straightforward in that we arranged things so that a single LNS number uses the same number of bits as a 32-bit unsigned integer. Therefore, we define some preliminaries as,

```
#include <stdio.h>
#include <math.h>
#include <inttypes.h>

#define K 8
#define L 22
#define LNS_ZERO 0x80000000
#define LNS_SIGN 0x40000000
#define LNS_MASK 0x3FFFFFFF

typedef uint32_t lns_t;

double lg2(double x) {
    return log(x) / log(2.0);
}
```

where we include necessary C libraries, define a local version of log base-2 (in case not in `math.h`) and declare our LNS number type to be an unsigned 32-bit int. We also define K and L which are the integer and fractional parts of our fixed-point exponent. Last, we define three constants, `LNS_ZERO`, `LNS_SIGN`, and `LNS_MASK` which are used to check and set the zero and sign bits and to mask the bits for the exponent.

Conversion to and from LNS numbers will cheat and use existing C functionality. The process of doing the conversion in hardware, like in an FPGA, is more involved. Similarly, we will cheat when going from LNS to floating-point. Specifically,

```
1  lns_t to_lns(double x) {
2      double e = lg2(fabs(x));
3      uint8_t s = (x >= 0.0) ? 0 : 1;
4      lns_t ans;
5
6      if (x == 0.0) return LNS_ZERO;
7      ans = (lns_t)(LNS_MASK & (uint32_t)floor(pow(2.0,L)*e + 0.5));
8      if (s) { ans |= LNS_SIGN; }
9      return ans;
10 }
11
12 double to_double(lns_t x) {
13     uint32_t z = (uint32_t)(LNS_ZERO & x);
14     uint32_t s = (uint32_t)(LNS_SIGN & x);
15     int32_t p  = (uint32_t)(LNS_MASK & x);
16     double m;
17
18     p = (0x20000000 & p) ? 0xC0000000+p : p;
19     if (z != 0) { return 0.0; }
20     m = pow(2.0, (double)p * pow(2.0,-L));
21     return (double)(s == 0) ? m : -m;
22 }
```

Conversion to LNS (`to_lns`) first calculates the log base-2 of the input and stores it in `e` (line 2). It then captures the overall sign of the number in `s` (line 3). A quick special case check for zero is done in line 6. If so, return the constant for zero which is the highest bit set. The other bits need not be zero but it is simple enough to set them that way, so return `0x80000000`. Line 7 scales the log of the input by the number of bits in the fractional part of the exponent ($L = 22$) with an added 0.5 to round to the nearest integer value. Casting to `uint32_t` makes an integer of the scaled float. Line 7 also AND's the output with `0x3FFFFFFF` which sets the highest two bits to zero. Recall that the highest bit (bit 31) is our zero flag and the next highest (bit 30) is our sign. If the magnitude of the input is less than 1.0 the log is negative and the scaled integer output from the `pow` function will also be negative for all 32 bits. So, we need to mask the top two bits to make sure they are not set improperly. Line 8 OR's the sign bit, one if negative, to set the overall sign of the number. The LNS version is then returned in line 9.

Conversion from LNS (`to_double`) strips off the zero flag (`z`) and sign (`s`) in lines 13 and 14. Line 15 is where the fixed-point exponent is extracted. Note carefully that the extracted bits are stored in a *signed* integer, `p`. We need this integer to be signed so that the conversion back to double will properly account for the sign. Remember, `p` represents the exponent so it is negative only when the magnitude of the number is less than one. Line 18 adjusts `p` for the case when it is less than one. If bit 29 of `p` is set the exponent is negative. This is because bits 0 through 29 are reserved for the exponent using our LNS number definition. If it is set, we sign-extend by adding `0xC0000000` which sets all the high bits to make `p` a negative number. Line 19 checks for the zero flag and returns zero if set. Line 20 first "unscales" `p` by raising it to the negative `L` power and then uses that unscaled floating point value, which is the actual exponent of the LNS number, to calculate the final value by raising 2 to that power. Line 21 returns the number with the sign set based on the value in `s`.

Multiplication and Division Now that we have a way to represent LNS numbers we can implement basic operations with them. For the other number systems in this book we started with addition and subtraction and then developed multiplication and division. Here we switch the order and start with multiplication and division which are now easy because of the way we have chosen to represent LNs numbers in memory.

Since we know that $2^x 2^y = 2^{x+y}$ we can multiply LNS numbers by simply adding the exponents. We intentionally chose to store the exponents in two's-complement fixed-point format. Because of this, we can use normal integer addition to add the exponents which in turn multiplies the numbers. For example, $123.45 \times 10.0 = 1234.5$ can be calculated by simply adding the LNS representation of the numbers,

123.45	→	`1BCA87A`		
10.00	→	`0D49A78`		
		`29142F2`	→	1234.50

which works in this case because the numbers are positive and we need not consider the sign bit (bit 30) nor did we check for zero (bit 31). The exponent is stored in fixed-point format which has an implied radix point between bit 21 and bit 22 ($L = 22$). Addition of two numbers stored in this way works regardless of where the radix point is located so we need do nothing special beyond bit for bit addition.

Before defining multiplication we will introduce some helper functions,

```
unsigned char lns_isneg(lns_t x) {
    return (LNS_SIGN & x) != 0;
}

unsigned char lns_iszero(lns_t x) {
    return (LNS_ZERO & x) != 0;
}

lns_t lns_abs(lns_t x) {
    if (lns_iszero(x))
        return x;
    return LNS_MASK & x;
}

lns_t lns_negate(lns_t x) {
    if (lns_iszero(x))
        return LNS_ZERO;
    if (lns_isneg(x))
        return LNS_MASK & x;
    return LNS_SIGN | x;
}
```

where line 1 defines `lns_isneg` which returns whether or not its argument is negative. It does this by seeing if the sign bit is set or not. Line 5 performs a similar check to see if the number is zero. If it is, the zero bit is set which is checked in line 6. Line 9 defines `lns_abs` which returns a positive version of the input argument. It checks to see if the argument is zero in which case it returns zero. Line 12 makes the argument positive by masking off the zero and sign bits. Finally, line 15 defines `lns_negate` which returns its argument with the sign bit toggled. If zero (line 16) return zero. If the argument is negative return the absolute value (line 19), otherwise, set the sign bit to make the argument negative (line 20).

With these helper functions it becomes quite simple to implement multiplication of two LNS numbers,

```
1 lns_t lns_multiply(lns_t a, lns_t b) {
2     if (lns_iszero(a)) return LNS_ZERO;
3     if (lns_iszero(b)) return LNS_ZERO;
4 
5     if (lns_isneg(a) == lns_isneg(b))
6         return LNS_MASK & (lns_abs(a) + lns_abs(b));
7     else
8         return lns_negate(LNS_MASK & (lns_abs(a) + lns_abs(b)));
9 }
```

where the arguments are checked for zero in lines 2 and 3. If either is zero, return zero as the product. Line 5 checks if the signs of the arguments are the same. If so, the product is positive and line 6 adds the positive versions of the inputs. The AND with `LNS_MASK` keeps the highest two bits clear which is correct for a nonzero positive LNS number. If the signs differ the product is negative and line 8 negates the positive product to produce a negative result.

Division of LNS number, by analogy to multiplication, involves the subtraction of two exponents since $2^x/2^y = 2^{x-y}$. Again, the fixed-point exponent representation saves us trouble and as in multiplication simple subtraction with checks for the signs of the arguments will give us the proper results. Therefore, division of two LNS numbers is simply multiplication replacing addition with subtraction,

```
1 lns_t lns_divide(lns_t a, lns_t b) {
2     if (lns_iszero(a)) return LNS_ZERO;
3     if (lns_iszero(b)) return LNS_ZERO;
4 
5     if (lns_isneg(a) == lns_isneg(b))
6         return LNS_MASK & (lns_abs(a) - lns_abs(b));
7     else
8         return lns_negate(LNS_MASK & (lns_abs(a) - lns_abs(b)));
9 }
```

where lines 6 and 8 now subtract the arguments instead of adding them. However, line 3 shows a weakness of our chosen LNS representation in that instead of simply returning zero when the second argument is zero we should signal a division by zero error. To maximize precision we did not reserve any bits for infinity or NaN as we would in a more rigorous library.

Comparison Operators Before we define addition and subtraction for LNS numbers we will add basic logical comparison operators to our library. These are straightforward in this case because if $x > y$ then $2^x > 2^y$ so to decide on equality, greater than or less than we only need compare the exponents and signs. Also, since we are using a two's-complement fixed-point representation for the exponents we can compare magnitudes by using standard C comparison operators for signed integers. All the same, the implementation is a bit tedious because there are multiple cases to consider. Therefore, without further ado,

```
int32_t lns_exponent(lns_t x) {
    int32_t e = LNS_MASK & x;
    return (0x20000000 & e) ? 0xC0000000+e : e;
}

int8_t lns_compare(lns_t a, lns_t b) {
    int32_t ea = lns_exponent(a);
    int32_t eb = lns_exponent(b);

    if (lns_iszero(a) && lns_iszero(b))
        return 0;
    if (lns_iszero(a))
        return (lns_isneg(b)) ? 1 : -1;
    if (lns_iszero(b))
        return (lns_isneg(a)) ? -1 : 1;

    if ((lns_isneg(a)==1) && (lns_isneg(b)==0))
        return -1;
    if ((lns_isneg(a)==0) && (lns_isneg(b)==1))
        return 1;

    if (lns_isneg(a) && lns_isneg(b))
        return (ea == eb) ? 0
                          : (eb < ea) ? -1
                                      :  1;
    else
        return (ea == eb) ? 0
                          : (ea < eb) ? -1
                                      :  1;
}

unsigned char lns_equal(lns_t a, lns_t b) {
    return (lns_compare(a,b) == 0);
}

unsigned char lns_less(lns_t a, lns_t b) {
    return (lns_compare(a,b) == -1);
}

```

```
40 unsigned char lns_greater(lns_t a, lns_t b) {
41     return (lns_compare(a,b) == 1);
42 }
```

where we implement simple wrapper functions for `lns_equal` (line 32), `lns_less` (line 36), and `lns_greater` (line 40) which call `lns_compare` (line 6). This function does all the work and returns -1 if $a < b$, 0 if $a = b$, and $+1$ if $a > b$. Specifically, `lns_equal` returns 1 if $a = b$ because `lns_compare` returns 0 if its arguments are equal and `lns_equal` checks if the return value is 0 returning 1 if it is. Both `lns_less` and `lns_greater` work the same way checking for -1 and $+1$ respectively.

The `lns_compare` function uses several existing LNS library functions defined above, namely, `lns_isneg` which returns 1 if its argument is negative, 0 otherwise, and `lns_iszero` which acts identically if its argument is zero. Additionally, we define a new helper function, `lns_exponent`, which returns the exponent of the LNS number as a *signed* integer. This is defined in line 1. In line 2 the fixed-point exponent is returned by masking the zero and sign bits. Line 3 then returns this exponent after setting the two highest bits if the exponent is negative (`0x20000000 & e` is nonzero). The comparison asks if the highest bit of the exponent is set or not. If it is, the exponent is negative, since it is a two's-complement number, and we set the top two bits by AND-ing with `0xC0000000` to extend the sign before returning. With the exponent properly sign extended and stored in a signed integer we can perform normal integer comparisons to check the magnitude of the LNS numbers.

The `lns_compare` function extracts the exponents (lines 7, 8) and then begins a series of checks to cover the possible conditions that might be encountered. Lines 10 through 15 deal with numbers that are zero. If both are zero (line 10) return 0 to indicate that the numbers are equal. In line 12 we know that the second argument cannot be zero so we check to see if the first argument is zero. If it is, we return the proper value depending upon the sign of the second argument b (line 13). If b is negative we know to return 1 because a is zero and $0 > b$ when b is negative. If this isn't the case, return -1 to indicate that $0 < b$ when b is positive. Lines 14 and 15 make the same check for the case when the second argument is zero. These explicit checks against zero are necessary because we use a single bit to indicate that a number is zero.

Lines 17 through 20 handle the cases where the signs of the numbers are different. If they are different we already know the proper values to return regardless of the values of their exponents. If a is negative and b is positive (line 17) then we know $a < b$ and return -1. In line 19 we do the opposite check and if true return 1 because we know $a > b$ since a is positive and b is negative.

It is only when the signs are the same that we need look at the magnitude of the numbers. Line 27 handles the case where both arguments are positive. Here we simply compare the exponents in `ea` and `eb` and if equal return 0. If they are not equal we look at the outcome of $ea < eb$. If this is the case, $a < b$ so return -1 (line 28). Otherwise we know $a > b$ and return 1 in line 29. If both arguments are negative (line 22) similar comparisons are made but in this case we check $eb < ea$. If so, we know that the exponent of argument a is less negative than the exponent of b which means that $a < b$ since both a and b are negative and $-2^{|e_a|} > -2^{|e_b|}$ for e_a and e_b the exponents of arguments a and b, respectively. We return -1 in this case (line 24). Otherwise, the remaining case, $a > b$, is true and we return 1 (line 25).

Addition and Subtraction To this point we've seen some very simple implementations of operations that are typical difficult to implement (here multiplication and division). So, if LNS numbers are so simply implemented why are they not in wider use? The answer is that while multiplication and division are simple, addition and subtraction are now difficult. We can see that they will be troublesome because while simple formulas exist for multiplication and division of two numbers expressed as logs there are no simple formulas for addition and subtraction, $2^x + 2^y$ and $2^x - 2^y$ do not have simple solutions. However, all is not lost.

Following the notation in [2] we can represent an LNS number mathematically as $A = -1^{S_A} \cdot 2^{E_A}$ where S_A is the sign bit and E_A is the floating-point exponent. Similarly, we can write $B = -1^{S_B} \cdot 2^{E_B}$ and so forth. Therefore, to calculate $C = A \pm B$ where we assume $|A| \geq |B|$, we can write,

$$\begin{aligned}
S_C &= S_A \\
E_C &= \log_2 |A \pm B| \\
&= \log_2 |A(1 \pm \frac{B}{A})| \\
&= \log_2 |A| + \log_2 |1 \pm \frac{B}{A}| \\
&= E_A + f(E_B - E_A)
\end{aligned}$$

with,

$$f(E_B - E_A) = \log_2 |1 \pm \frac{B}{A}| = \log_2 |1 \pm 2^{E_B - E_A}|$$

where we have reduced the problem of addition and subtraction to function approximation since once we know the value of $f(E_B - E_A)$ we know the answer to our problem. Indeed, for actual hardware implementations of LNS numbers, the majority of the effort is placed in how one should deal with the function f. If we look at f graphically we get Fig. 10.1 where we have limited the domain to $[-10, 0]$. Our LNS implementation uses a $Q7.22$ fixed-point number to represent exponents in the range $[-128, 127.9999998]$ so that the difference, $E_B - E_A$, $E_B < E_A$, is closer to $[-128, 0]$ but as can be seen, for differences below about -10, the value of f, for addition or subtraction, is virtually zero. Some hardware implementations approximate f mathematically while others build large look-up tables. All use some form of interpolation to arrive at as accurate a value for f as possible which in turn makes $A \pm B$ as accurate as possible. We will take advantage of the closed functional

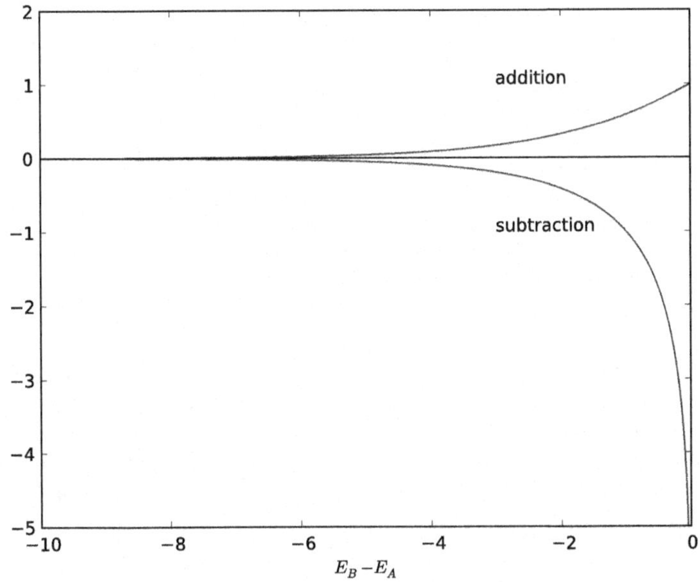

Fig. 10.1 The function f which must be approximated to implement LNS addition and subtraction. The argument to f is the difference between the two exponents where we require $|A| \geq |B|$ which implies that the exponent for B, E_B, is less than the exponent for A, E_A. Figure is a recreation of Fig. 1 in [2]

form for f and in our simple library use the C standard library functions to directly calculate the value. Yes, it is a cheat. One aspect of Fig. 10.1 is worth pointing out. If the two numbers being operated on, especially for subtraction, are close in value the difference between their exponents is small meaning the argument to f is close to zero. This is precisely the region where the function is changing most rapidly implying that interpolation in this region is crucial for accuracy to be maintained.

With the above in mind, then, we can implement addition and subtraction in stages. The first stage implements the function f for addition and subtraction of the magnitudes of two LNS numbers, A and B, where we already know that $|A| \geq |B|$ as,

```
double fminus(double ea, double eb) {
    return pow(2.0, ea + lg2(1.0 - pow(2.0, eb-ea)));
}

double fplus(double ea, double eb) {
    return pow(2.0, ea + lg2(1.0 + pow(2.0, eb-ea)));
}

```

```
9  lns_t lns_add_mag(lns_t a, lns_t b) {
10     double ea = (double)(LNS_MASK & a) * pow(2.0,-L);
11     double eb = (double)(LNS_MASK & b) * pow(2.0,-L);
12     return to_lns(fplus(ea,eb));
13 }
14
15 lns_t lns_sub_mag(lns_t a, lns_t b) {
16     double ea = (double)(LNS_MASK & a) * pow(2.0,-L);
17     double eb = (double)(LNS_MASK & b) * pow(2.0,-L);
18     return to_lns(fminus(ea,eb));
19 }
```

with the functions `lns_add_mag` and `lns_sub_mag` above called appropriately by `lns_add` and `lns_sub` (defined below) to handle all possible sign combinations of the arguments a and b.

The function f is implemented in `fminus` (line 1) and `fplus` (line 5) along with the rest of the calculation to return the sum or difference except for the sign. Specifically, these functions calculate,

$$E_C = E_A + log_2(1.0 \pm 2^{E_B - E_A}), \quad C = 2^{E_C}$$

with E_A and E_B the exponents of the inputs A and B. Note that these functions accept and return standard C double precision floats, not fixed-point floats. These helper functions are called by `lns_add_mag` (line 9) and `lns_sub_mag` (line 15). These functions perform the addition or subtraction ignoring the overall sign of the inputs and under the assumption that $|A| \geq |B|$ which was previously stated. These functions both extract the fixed-point exponents by AND'ing with `LNS_MASK` and then scale by 2^{-L} to create standard floating-point representations in `ea` and `eb`. They then call the proper helper function to calculate the sum or difference and return it as an LNS number.

Along with considering the signs of the arguments we must also consider the relative magnitudes of the arguments so that we call `lns_add_mag` and `lns_sub_mag` appropriately. For both addition and subtraction we have then a total of eight cases to consider: four for the possible signs of the arguments and for each of these two cases for the relative magnitudes. Therefore, we can define our top-level addition function as,

```
lns_t lns_add(lns_t a, lns_t b) {
    lns_t A = lns_abs(a);
    lns_t B = lns_abs(b);

    if (lns_iszero(a)) return b;
    if (lns_iszero(b)) return a;
    if (lns_equal(a,b))
        return lns_multiply(a, to_lns(2.0));

```

```
    if ((lns_isneg(a)==0) && (lns_isneg(b)==0)) {
        if (lns_greater(A,B))
            return lns_add_mag(A,B);
        else
            return lns_add_mag(B,A);
    }
    if ((lns_isneg(a)==1) && (lns_isneg(b)==1)) {
        if (lns_greater(A,B))
            return lns_negate(lns_add_mag(A,B));
        else
            return lns_negate(lns_add_mag(B,A));
    }
    if ((lns_isneg(a)==0) && (lns_isneg(b)==1)) {
        if (lns_greater(A,B))
            return lns_sub_mag(A,B);
        else
            return lns_negate(lns_sub_mag(B,A));
    }
    if ((lns_isneg(a)==1) && (lns_isneg(b)==0)) {
        if (lns_greater(A,B))
            return lns_negate(lns_sub_mag(A,B));
        else
            return lns_sub_mag(B,A);
    }
}
```

which for subtraction becomes,

```
lns_t lns_sub(lns_t a, lns_t b) {
    lns_t A = lns_abs(a);
    lns_t B = lns_abs(b);

    if (lns_iszero(a)) return lns_negate(b);
    if (lns_iszero(b)) return a;
    if (lns_equal(a,b)) return LNS_ZERO;

    if ((lns_isneg(a)==0) && (lns_isneg(b)==0)) {
        if (lns_greater(A,B))
            return lns_sub_mag(A,B);
        else
            return lns_negate(lns_sub_mag(B,A));
    }
    if ((lns_isneg(a)==1) && (lns_isneg(b)==1)) {
        if (lns_greater(A,B))
            return lns_negate(lns_sub_mag(A,B));
        else
            return lns_sub_mag(B,A);
    }
```

```
21     if ((lns_isneg(a)==0) && (lns_isneg(b)==1)) {
22         if (lns_greater(A,B))
23             return lns_add_mag(A,B);
24         else
25             return lns_add_mag(B,A);
26     }
27     if ((lns_isneg(a)==1) && (lns_isneg(b)==0)) {
28         if (lns_greater(A,B))
29             return lns_negate(lns_add_mag(A,B));
30         else
31             return lns_negate(lns_add_mag(B,A));
32     }
33 }
```

where each possible case is handled so that the proper final sign will be used and the relative magnitudes are respected. In particular, both functions store positive versions of their arguments in `A` and `B` (lines 2, 3) and then check if any of their arguments are zero. For addition, if either argument is zero, return the other one (lines 5, 6). For subtraction if the first argument is zero, negate the second one and return it (line 5) otherwise return the first (line 6). If the arguments are equal, using the comparison functions defined above, return zero for subtraction (line 7) or multiply the argument by two if adding (line 8). Each function then checks the signs of the inputs to find which of the four possible cases match. For each of these cases the magnitudes of the arguments are checked so that `lns_add_mag` or `lns_sub_mag` can be called appropriately and with the larger magnitude argument first so that the functions return meaningful values. In some cases this requires negating the return value to get the proper sign.

Discussion In this section we implemented a basic library of arithmetic and comparison routines for LNS numbers. We saw how easy it is to implement multiplication and division, operations that are difficult to implement in other representations. We also saw that comparison operations were simple as well because of the two's-complement fixed-point representation we chose for the exponents of our LNS numbers. Of course, little in life is free, and we saw that the power of LNS numbers comes with a definite cost when it comes to addition and subtraction. Still, since LNS numbers are typically implemented in hardware, especially in FPGAs, there are times when the speed of multiplication and division outweigh the pain of addition and subtraction. This is to say nothing of the improved energy efficiency of LNS versus IEEE floating-point when implemented in hardware. Indeed, a reasonable argument could be made that LNS numbers, if efficiently implemented in hardware, are actually a viable alternative to the IEEE 754 standard. Certainly, it is an entertaining intellectual exercise to consider how to work with numbers expressed as their exponents.

10.3 Double-Base Number System

The Double-Base Number System (DBNS) was developed by Dimitrov and Jullien [3] and offers a particularly compact way to represent large unsigned integers. As the name suggests, instead of the single base used by every number system we have examined in this book, be it binary, decimal, octal, or hexadecimal, the DBNS uses two bases, in this case 2 and 3. Specifically, for a given base, B, and digits, $d \in \{0, 1, \ldots, B-1\}$, we can represent a number, x, as,

$$x = \sum_i d_i B^i$$

In DBNS we replace the single base B with two bases, 2 and 3. We also limit the choice of digits to $d \in \{0, 1\}$ so that we represent x as a sum of products of powers of 2 and 3, e.g., $2^i 3^j$,

$$x = \sum_{i,j} 2^i 3^j$$

where i, j are exponents ≥ 0. Terms of the form $2^i 3^j$ are called *2-integers*. Note that we need not explicitly represent the zero digits as is done in standard place notation meaning that DBNS is not a positional number system. An example will clarify. The number 127 can be represented in DBNS as,

$$\begin{aligned} 127 &= 2^2 3^3 + 2^1 3^2 + 2^0 3^0 \\ &= 2^2 3^3 + 2^4 3^0 + 2^0 3^1 \\ &= 2^5 3^1 + 2^0 3^3 + 2^2 3^0 \end{aligned}$$

which can be written as a set of tuples,

$$\begin{aligned} 127 &= \{(2,3), (1,2), (0,0)\} \\ &= \{(2,3), (4,0), (0,1)\} \\ &= \{(5,1), (0,3), (2,0)\} \end{aligned}$$

listing only the powers of the terms $2^i 3^j$. We will use this representation in our implementation below.

The example with 127 above reveals an important aspect of the DBNS: representations are not unique. A number may be represented in many different ways. However, there are "smallest" representations in which the set of tuples is as small as possible. These are called *canonical* representations. The three given above for 127 can be shown by exhaustive search to be the three canonical representations of 127 in DBNS.

How does one find the canonical representation of a number? Locating the canonical representation requires an exhaustive search implying that it is NP-complete. However, not all is lost. In [4] a simple greedy algorithm is given which locates a "near" canonical DBNS representation for any unsigned integer. The algorithm is easily implemented in Python,

```
while (x > 0):
    w,i,j = Largest2Integer(x)
    print (i,j)
    x -= w
```

where `Largest2Integer()` returns the largest 2-integer $\leq$ to x along with the exponents, i.e., it returns $w = 2^i3^j \leq x$. Print or otherwise store the tuple (i,j) and subtract w from x. Repeat as long as x is greater than zero. This simple algorithm will not always produce the canonical representation of x but it will produce a very good one.

Graphical Representation of DBNS Numbers We saw above that a DBNS number can be represented by the set of exponents of the 2-integers that make up the number. It is also possible and quite informative to represent DBNS numbers as a 2-D grid making DBNS the only 2-D number system we will look at in this book.

A natural way to think of a DBNS number is as a matrix with the rows the powers of 2 and the columns the powers of 3. It is then a simple leap to visualize the matrix as an image. For example, a canonical representation of 127 is $\{(1,2),(0,0),(2,3)\}$ which can be put into matrix form as,

$$\begin{pmatrix} 1\,0\,0\,0 \\ 0\,0\,1\,0 \\ 0\,0\,0\,1 \end{pmatrix}$$

where a 1 indicates that the corresponding 2-integer is present and a 0 indicates that the corresponding 2-integer is not present. In the matrix above the element at row 3, column 4 is present meaning that $2^23^3 = 108$ is present in the DBNS representation of 127. Likewise, there are elements present at row 1, column 1 and row 2, column 3 for terms $2^03^0 = 1$ and $2^13^2 = 18$. The sum of the three terms is $108+18+1 = 127$ as expected. This matrix representation is useful for understanding the sparsity of a DBNS representation but it is still more impressive when cast as an image. If we do this, 127 can be visualized as,

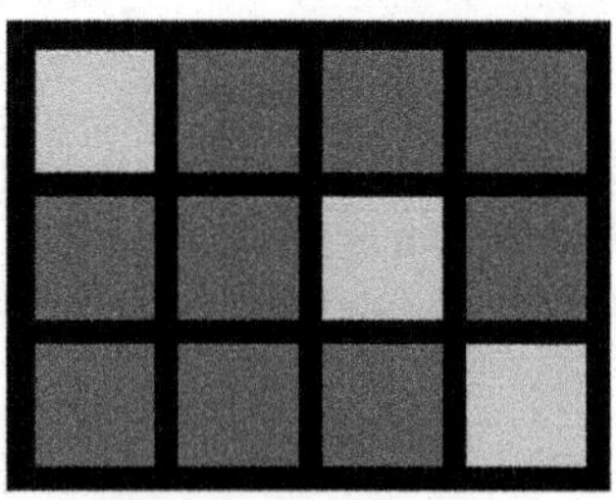

where light squares indicate the terms that are present in the DBNS representation. Not too impressive, to be sure, but consider a number like 785745187346896756987891,

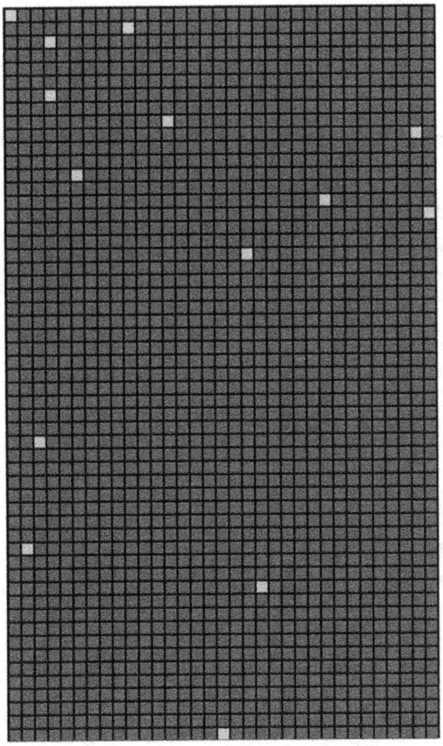

which has only a tiny handful of nonzero entries. We will see below that this 2-D representation is useful for illustrating operations on DBNS numbers.

A Python Class for DBNS Numbers We will develop a small Python class which implements the DBNS. We are using Python in this case for two reasons. First, the sparsity of DBNS numbers makes them an excellent choice for representing very large integers. Python supports big integers, see Chap. 5, so it makes sense to use it here. Second, it is natural to represent a DBNS number as a set of tuples, (i, j), and Python supports both sets and tuples as native data types.

The first thing our library needs to do is to accept an input number, x, and convert it to a set of DBNS tuples using the greedy algorithm. The code for this part is,

```
1  from math import log, ceil
2
3  class DBNS:
4      def Largest2Integer(self, x):
5          m = [1,0,0]
6
7          a = int(ceil(log(x)/log(2))) + 1
8          b = int(ceil(log(x)/log(3))) + 1
9
10         for i in range(a):
11             for j in range(b):
12                 n = [(2**i)*(3**j),i,j]
13                 if (n[0] > m[0]) and (n[0] <= x):
14                     m = n
15         return m
16
17     def Greedy(self, x):
18         while (x > 0):
19             w,i,j = self.Largest2Integer(x)
20             self.n.add((i,j))
21             x -= w
22
23     def Limits(self):
24         self.a = 0
25         self.b = 0
26
27         for i,j in self.n:
28             if (i > self.a):
29                 self.a = i
30             if (j > self.b):
31                 self.b = j
32
33     def __init__(self, x=0):
34         self.n = set()
35
36         if (x > 0):
37             self.Greedy(x)
38             self.Ready()
39             self.Limits()
```

where we will for the time being ignore the `Ready()` method of line 38. The `__init__` method for this class starts on line 33 and accepts an input number, `x`, which is the number we want to represent in the DBNS. We will store our representation as a set of tuples, (i, j), so define an empty set (line 34). The observant reader will have already realized that zero is not directly representable as the sum of 2^i3^j terms. So, we represent zero with an empty set. If `x` is not zero we apply the greedy algorithm (line 37) and then call `Limits()` (line 39).

The greedy algorithm is the simple `while` loop starting on line 18. As long as `x` is still greater than zero, subtract from it the largest possible 2-integer that is less than or equal to `x` and keep the exponents as a tuple in our DBNS representation. Specifically, line 19 calls `Largest2Integer()` passing the current value of `x`. This method returns the largest term, $w = 2^i 3^j$, which is the next tuple in the DBNS representation. The tuple is added to the set in line 20 and the value is subtracted from `x` in line 21. The `Largest2Integer` method starts on line 4. The goal is to locate the largest value and associated exponents that are less than or equal to the input, `x`. We store this set in `m` starting with $1 = 2^0 3^0$ or `[1,0,0]` (line 5). We locate the 2-integer by an exhaustive search of possible exponents. We define the limits of the search in lines 7 and 8, one each for powers of 2 and powers of 3. The limit is the corresponding logarithm of `x` plus one, $a = \lceil \log_2(x) \rceil + 1$, $b = \lceil \log_3(x) \rceil + 1$. We then search using the double loop of lines 10 and 11 testing to see if we've found a new maximum set of exponents (lines 13, 14). When done, return our maximum (line 15). A double loop is generally frowned upon but in this case the limits are small so the performance hit is minimal. There are alternative methods to find the largest 2-integer, see [5].

The last step in our conversion to DBNS format is to keep track of the largest powers of the bases. These values are used when generating the 2-D representations of the number which we will introduce below. The `Limits` method does this (line 23) using a simple loop over the tuples in the set (line 27) tracking the largest power of 2 (lines 28, 29) and 3 (lines 30, 31).

Let's add a simple method to calculate the decimal value of a DBNS number by summing the value of each tuple in the set,

```
def Value(self):
    m = 0
    for i,j in self.n:
        m +=  (2**i)*(3**j)
    return m
```

We now have the ability to turn a decimal integer into a DBNS number and back again.

If we want to see the DBNS representation of a number we can, ignoring best practices for object-oriented programming, simply look at the set by printing the value of the member variable, `n`. For example,

```
>>> from DBNS import *
>>> x = DBNS(127)
>>> print x.n
    set([(1, 2), (0, 0), (2, 3)])
```

which tells us that the greedy algorithm worked since, $127 = 2^1 3^2 + 2^0 3^0 + 2^2 3^3$. Note that there is no implied ordering in a set which is just fine in this case since we are summing terms.

The 2-D matrix representation is straightforward to add. We take advantage of Python's rich set of overloadable operators to let us see the 2-D representation by simply printing the object. Specifically,

```
1  def __str__(self):
2      m = ""
3      for i in range(self.a+1):
4          n = "|"
5          for j in range(self.b+1):
6              if ((i,j) in self.n):
7                  n += "1 "
8              else:
9                  n += "0 "
10         m += n[0:-1] + "|\n"
11     return m[0:-1]
12
13 def __repr__(self):
14     return self.__str__()
```

where `__str__()` returns a string that is the matrix representation of the number (lines 1–11). We overload the `__repr__()` method for completeness and simply call `__str__()` (line 14). Since we know the largest exponent used by any power of 2 (`self.a`) and 3 (`self.b`) we know the size of the output matrix, `self.a` rows by `self.b` columns. The loops at lines 3 and 5 cover all combinations. If the element, i.e., that power of 2 and 3, is present in the set of tuples (line 6), output a "1", otherwise, output "0" to indicate that element is not in the set. When done, return the string (line 11) dropping the final newline character. Testing at the command line gives,

```
>>> from DBNS import *
>>> x = DBNS(127)
>>> print x
    |1 0 0 0|
    |0 0 1 0|
    |0 0 0 1|
```

which matches the matrix and image representations of 127 above.

We are now ready to implement operations on DBNS numbers. We will implement two, addition and multiplication, and leave some others as straightforward exercises. Note that unsigned DBNS numbers represented in this way do not support subtraction or division. This, naturally, limits their use.

Addition of DBNS Numbers Addition of two DBNS numbers represented as sets of tuples is straightforward, simply take the union of the two sets accounting for any collisions, i.e., the same term, 2^i3^j, in both representations. For example, consider $23 + 11 = 34$. Using the matrix representation we have,

$$
\begin{array}{ccccc}
23 & + & 11 & = & 34 \\
\begin{pmatrix} 1\,0\,0 \\ 0\,0\,1 \\ 1\,0\,0 \end{pmatrix} & + & \begin{pmatrix} 0\,0\,1 \\ 1\,0\,0 \end{pmatrix} & = & \begin{pmatrix} 1\,0\,1 \\ 1\,0\,1 \\ 1\,0\,0 \end{pmatrix}
\end{array}
$$

where we simply overlay the two representations to generate a new representation. In this case there are no collisions between 23 and 11 so we have a final answer, 34, but, as we will see below, we will not keep the final answer in this form. Using the set of tuples notation we have,

$$
\begin{array}{ccccc}
23 & + & 11 & = & 34 \\
\{(0,0),(2,0),(1,2)\} & \cup & \{(1,0),(0,2)\} & = & \{(0,0),(1,0),(2,0),(0,2),(1,2)\}
\end{array}
$$

where we clearly see that addition is simply set union. So far, so good, but what if the two DBNS numbers share one or more terms? For example, consider $10 + 11 = 21$ which is, using matrix form and adding element by element,

$$
\begin{array}{ccccc}
10 & + & 11 & = & 21 \\
\begin{pmatrix} 1\,0\,1 \end{pmatrix} & + & \begin{pmatrix} 0\,0\,1 \\ 1\,0\,0 \end{pmatrix} & = & \begin{pmatrix} 1\,0\,\mathbf{2} \\ 1\,0\,0 \end{pmatrix}
\end{array}
$$

where we now have a problem, the bold 2, which is not allowed in the DBNS format. This is a collision and we need to deal with it as a "carry" of sorts. The solution is to realize that a carry is,

$$
2(2^i 3^j) = 2^{i+1} 3^j
$$

so we can replace the bold term in the answer with zero and increment $(i + 1, j)$ instead. If necessary, this process can be repeated if incrementing $(i + 1, j)$ results in a new carry. Then, we can rewrite the answer as,

$$
\begin{pmatrix} 1\,0\,\mathbf{2} \\ 1\,0\,0 \end{pmatrix} = \begin{pmatrix} 1\,0\,0 \\ 1\,0\,1 \end{pmatrix} = 21
$$

which is a valid form for a DBNS number. If we use set notation we have,

$$
\begin{array}{ccccc}
10 & + & 11 & = & 21 \\
\{(0,0),(0,2)\} & \cup & \{(1,0),(0,2)\} & \rightarrow & \{(0,0),(1,0),(0,2)\}
\end{array}
$$

which is a problem because there are really two $(0, 2)$ elements, one from each number. Conveniently, since we are using Python sets to implement DBNS numbers, to see if there are collisions when adding we simply look for a non-empty intersection between the numbers. If there are elements in the intersection, as is the case here, the element is a collision and we apply the carry reduction. If applying the carry reduction leads to a new carry we put that carry in the set of elements in the intersection and repeatedly pull elements until the intersection set is empty. For example, since $\{(0,0),(0,2)\} \cap \{(1,0),(0,2)\} = \{(0,2)\}$ we know we have at least one carry to deal with. The carry reduction replaces $(0,2)$ in the answer with $(1,2)$. As this does not result in a new carry, the term $(1,2)$ is not in the union, we are done and the final output set is $\{(0,0),(1,0),(1,2)\}$ which does indeed equal 21.

In Python this process can be added to our DBNS class by overloading the "+" operator,

```
1  def __add__(self, y):
2      z = DBNS()
3
4      z.n = self.n.union(y.n)
5      b = self.n.intersection(y.n)
6
7      while (b != set()):
8          i,j = b.pop()
9          z.n.remove((i,j))
10         if ((i+1,j) in z.n):
11             b.add((i+1,j))
12         else:
13             z.n.add((i+1,j))
14
15     z.Ready()
16     z.Limits()
17
18     return z
```

In line 2 we create a new output DBNS number. This will hold our answer. Nominally, the answer is the union of the two numbers, the object instance and the given object instance which is in `y`. This is line 4. This union is the set used by the answer, hence assigning to `z.n`. Yes, again we abuse object-oriented principles in the name of pedagogy. We check for collisions via the intersection which we assign to `b` (line 5). Then, lines 7 through 13 are a loop over the elements of the intersection, if any, in which we perform the carry reduction, possibly repeatedly. While there are still elements in the intersection (line 7) pull an element out of the set (line 8) and remove it from the answer (line 9). We know it is in `z` because it would be in the union between the two numbers. Line 10 asks whether the carry reduction element $(i,j) \rightarrow (i+1,j)$ is already in the answer. If so, we don't need to add it to the answer but we do need to add it to the intersection as a new element in need of the carry reduction (line 11). If the element is not already in the answer, add it (line 13). Continue this process until the intersection set is empty. Line 15 calls the mysterious `Ready()` method we saw above. Let's continue to ignore it for the moment. Line 16 sets the limits on the exponents of the answer so we can display it properly. Finally, line 18 returns the answer which is the sum of the number and the given number, `y`.

A Problem with Addition and Its Solution Our first addition example, $23+11=34$, produced solution of the form,

$$\begin{pmatrix} 1 & 0 & 1 \\ 1 & 0 & 1 \\ 1 & 0 & 0 \end{pmatrix}$$

which has quite a few nonzero elements. This means that the representation is no longer particularly sparse. We can improve this and change the representation back into a near canonical, sparse representation by repeatedly applying two reduction rules based on observations of sums of adjacent terms. Namely, we note that,

$$2^i3^j + 2^{i+1}3^j = 2^i3^{j+1}$$
$$2^i3^j + 2^i3^{j+1} = 2^{i+2}3^j$$

which we can visualize as,

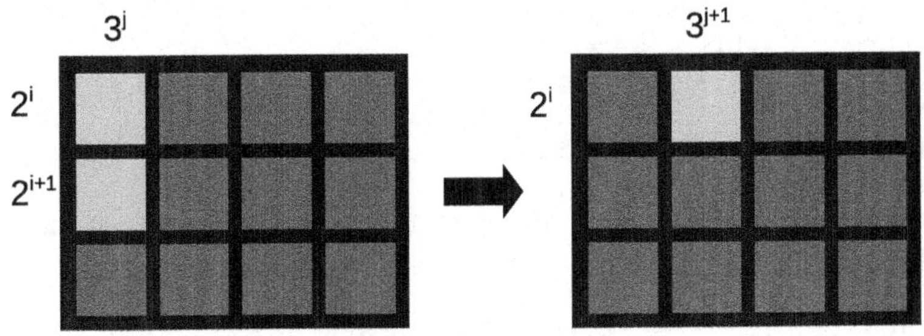

and,

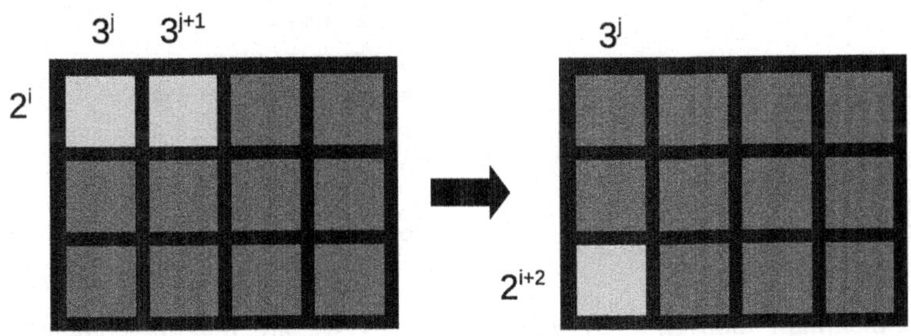

We also need to be aware that the reduction rules may result in new carries that also have to be dealt with. It is the job of the `Ready()` method in our Python class to apply these reduction rules and any needed carry reductions until the number is in this sparse, near canonical format. The `Ready()` method itself is straightforward,

```
def Ready(self):
    rows = cols = True

    while (rows or cols):
        cols = self.AdjacentColumn()
        rows = self.AdjacentRow()
```

where it repeatedly calls `AdjacentColumn()` and `AdjacentRow()` until both of these return `False` indicating that all instances of adjacent elements have been processed.

The method `AdjacentColumn()` appears to be a bit complicated but as we will see it isn't really so bad. The code is,

```
1  def AdjacentColumn(self):
2      modified = True
3      ever = False
4
5      while (modified):
6          modified = False
7          b = self.n.copy()
8          while (b != set()):
9              i,j = b.pop()
10             if ((i+1,j) in self.n):
11                 modified = True
12                 ever = True
13                 self.n.remove((i,j))
14                 self.n.remove((i+1,j))
15                 elem = (i,j+1)
16                 if (elem not in self.n):
17                     self.n.add(elem)
18                     b = self.n.copy()
19                 else:
20                     done = False
21                     while (not done):
22                         self.n.remove(elem)
23                         elem = (elem[0]+1,elem[1])
24                         if (elem not in self.n):
25                             self.n.add(elem)
26                             b = self.n.copy()
27                             done = True
28     return ever
```

where the main loop (line 5) runs as long as we are making changes to the representation. Each iteration of the loop starts by assuming no changes will be made (line 6) and making a duplicate of the current representation (line 7). We make a duplicate so we can look at each element (term, 2^i3^j) in the loop starting on line 8. Since we are looking for cases where for the current element at (i, j) the adjacent element in the same column, $(i + 1, j)$, is also present we ask if that element is in the representation (line 10). If it isn't we move to the next element. If it is, we know that we are going to make a change so we indicate that (line 11) and also indicate that we changed at least one element (line 12).

The reduction rule is $2^i3^j + 2^{i+1}3^j \rightarrow 2^i3^{j+1}$ so if both (i, j) and $(i + 1, j)$ are in the representation we remove them (lines 13 and 14) and try to add $(i, j + 1)$ (line 15). There are two possibilities at this point. If $(i, j + 1)$ is not already in the representation we simply add it (line 17), make a new copy of the representation (line 18) and continue looking at elements until b is empty (line 8). However, if $(i, j + 1)$ is already in the representation we have a new carry and pass to the code starting on line 20. This code loops through the representation removing the element that caused the carry and replacing it with a new element: $(i, j) \rightarrow (i + 1, j)$. The loop covers the case in which a chain of carries might happen.

Fortunately, the `AdjacentRow()` method is structurally identical to `AdjacentColumn()`, only the elements modified change. In particular, we can create a copy of `AdjacentColumn()`, change the name to `AdjacentRow()` and modify a few lines (in bold) to get,

```
1  def AdjacentRow(self):
2      modified = True
3      ever = False
4  
5      while (modified):
6          modified = False
7          b = self.n.copy()
8          while (b != set()):
9              i,j = b.pop()
10             if ((i,j+1) in self.n):
11                 modified = True
12                 ever = True
13                 self.n.remove((i,j))
14                 self.n.remove((i,j+1))
15                 elem = (i+2,j)
16                 if (elem not in self.n):
17                     self.n.add(elem)
18                     b = self.n.copy()
19                 else:
20                     done = False
21                     while (not done):
22                         self.n.remove(elem)
23                         elem = (elem[0]+1,elem[1])
24                         if (elem not in self.n):
25                             self.n.add(elem)
26                             b = self.n.copy()
27                             done = True
28     return ever
```

which acts exactly like `AdjacentColumn()` but implements the reduction $2^i3^j + 2^i3^{j+1} \rightarrow 2^{i+2}3^j$. Calling these methods repeatedly until neither changes the representation means that we will be in a suitable form to minimize carries on future operations.

Multiplication of DBNS Numbers The only other operation supported by DBNS numbers, at least in the form that we've been using them so far, is multiplication. It is particularly straightforward. To calculate $x \times y$ we multiply every term in y by every term in x. This is an $O(n^2)$ operation, true, but the very sparse nature of DBNS numbers means that in the end it is not all that bad. Multiplying two terms is particularly simple as well,

$$2^i3^j \times 2^a3^b = 2^i2^a3^j3^b = 2^{i+a}3^{j+b}$$

implying that the answer needs to contain the element $(i + a, j + b)$. If that element already exists we have a carry and do our usual transformation, $(i + a, j + b) \rightarrow (i + a + 1, j + b)$. In code, overloading the multiplication operator, we get,

```
1  def __mul__(self, y):
2      z = DBNS()
3  
4      for i,j in self.n:
5          for a,b in y.n:
6              elem = (i+a,j+b)
7              if (elem not in z.n):
8                  z.n.add(elem)
9              else:
10                 done = False
11                 while (not done):
12                     z.n.remove(elem)
13                     elem = (elem[0]+1,elem[1])
14                     if (elem not in z.n):
15                         z.n.add(elem)
16                         done = True
17 
18     z.Ready()
19     z.Limits()
20     return z
```

where our answer will be in `z`. We loop over every element of the first number (line 4) and for that element loop over every element of the second number (line 5). The product of these two elements (line 6) is looked for in the answer (line 7). If not present, it is added and the loops continue (line 8). Otherwise, we have a carry so we remove the element in the answer (line 12) and add the next element ($(i + 1, j)$) until carries are done (line 13). To finish the answer we make sure it is in proper form by calling `Ready()` (line 18) and set the limits on the exponents (line 19) before returning the new DBNS number as the answer (line 20).

How well does this algorithm perform? To multiply $1234567890 \times 9876543210$ we need to loop over the representations of each of these numbers. The representations are,

$$\begin{aligned} 1234567890 &= \{(8,14),(14,1),(9,9),(2,5),(1,1)\} \\ 9876543210 &= \{(0,0),(7,0),(5,5),(17,1),(11,14),(12,9),(0,2)\} \end{aligned}$$

so that multiplication requires $5 \times 7 = 35$ pairs of additions plus any occasional carry operations.

Representing Real Numbers Up to this point we have represented DBNS numbers as unsigned integers made by summing sets of terms of the form 2^i3^j, $i,j \in \mathbb{Z}$. We can represent signed floating-point numbers in a manner similar to the logarithmic number system described above if we allow the exponents in our 2^i3^j terms to take on positive and negative values. Specifically, we can approximate any number x by finding exponents b and t, which satisfy,

$$|x - s_x 2^b 3^t| < \epsilon, \; b,t \in \mathbb{Z}$$

where ϵ is a given tolerance and s_x is the sign of x. In practice, the exponents are represented to a specified precision (number of bits) which limits the set of possible values that can be represented well. Like the LNS system, this system, which we will call the double-base logarithmic number system (DBLNS), also called "2DLNS" in [6], easily supports multiplication and division but requires interpolation of a function for addition and subtraction. We will implement only multiplication and division here and leave addition and subtraction as projects for the reader.

We will fix the size of our exponents, at least for converting floating-point numbers to DBLNS, so that we can represent the input using a Python float data type. This is usually the same as a double-precision IEEE floating-point number. To quickly convert x to DBLNS, which means finding good integer values for b and t so that $|x - s_x 2^b 3^t|$ is as small as possible for the allowed range of exponents, we will create a sorted list of all representable values when $b \in [-600, 1023]$ and $t \in [-600, 646]$. We only need generate this list once so we do it outside of the DBLNS class when the module is imported,

```
from bisect import bisect_left

p = []
for i in range(-600,1023+1):
    for j in range(-600,646+1):
        p.append(((2.0**i)*(3.0**j),i,j))
p.sort()

gValues = []
gBexp = []
gTexp = []

```

```
13 for i in p:
14     v,b,t = i
15     gValues.append(v)
16     gBexp.append(b)
17     gTexp.append(t)
18 del p,v,b,t,i
19
20 class DBLNS:
21     def __init__(self, x):
22         idx = self.Closest(abs(x))
23         self.b = gBexp[idx]
24         self.t = gTexp[idx]
25         self.s = 0 if (x >= 0.0) else 1
```

Line 1 imports from the standard Python `bisect` module. This module allows for rapid searching of sorted lists which will let us quickly locate the best approximation to any input to the class. We will describe the `Closest()` method (line 22) below. For now, understand that it returns the index into the list of representable values, stored in the global `gValues`, of the closest value to `x`. The corresponding exponents are in `gBexp` and `gTexp` respectively so we store those (lines 23 and 24). Lastly, we set the sign (line 25). We now have a DBLNS representation for `x`.

The table is generated on import (lines 3 through 7) which computes all possible values, keeping the associated exponents, in line 6. This list is sorted (line 7, Python sorts on the first value of the tuple, which is what we want here). We then split the list into the global values `gValues`, `gBexp`, and `gTexp` and keep only these (lines 9 through 18).

We can now add the `Closest()` method and include a method to get the floating-point value and print it on the command line,

```
def Closest(self, x):
    pos = bisect_left(gValues, x)

    if (pos == 0) or (pos == len(gValues)):
        return pos

    if (gValues[pos]-x) < (x-gValues[pos-1]):
       return pos
    else:
       return pos-1

def Value(self):
    try:
        v = ((-1.0)**self.s)*(2.0**self.b)*(3.0**self.t)
    except:
        v = float('inf') if (self.s==0) else float('-inf')
    return v

```

```
19 def __str__(self):
20     return "%0.6f" % self.Value()
21
22 def __repr__(self):
23     return self.__str__()
```

Recall that `gValues` contains a sorted list of all possible floating-point values that we can represent for the range of exponents we are using. So, `bisect_left` returns the position in that list where `x` would fit (line 2). If we are outside the range, just return whatever `pos` is (line 5). Lines 7 through 10 decide if `x` is closer to the returned position or the one before it and returns the proper position. We use the returned position as an index into the b and t exponent lists (`gBexp` and `gTexp`) in order to define the DBLNS representation. Line 12 defines a simple function to return the floating-point value of the number. The `try` is necessary to capture the case when the exponents are beyond what can be represented as a Python float. In that case we simply return positive or negative infinity. Lastly, we overload the standard Python methods to allow us to print the floating-point value at the command line (lines 19 and 22).

With these preliminaries in place we can complete our DBLNS class by adding multiplication and division. These operations are particularly simple in this case, as they are in the LNS case, since,

$$xy = s_x 2^a 3^b \times s_y 2^c 3^d = ((s_x + s_y) \bmod 2)2^{a+c}3^{b+d}$$

$$x/y = s_x 2^a 3^b / s_y 2^c 3^d = ((s_x + s_y) \bmod 2)2^{a-c}3^{b-d}$$

which translates to,

```
1  def __mul__(self, y):
2      z = DBLNS(0)
3      z.s = (self.s + y.s) % 2
4      z.b = self.b + y.b
5      z.t = self.t + y.t
6      return z
7
8  def __div__(self, y):
9      z = DBLNS(0)
10     z.s = (self.s + y.s) % 2
11     z.b = self.b - y.b
12     z.t = self.t - y.t
13     return z
```

where for both multiplication and division we create an output object (lines 2 and 9) and update its sign and exponents appropriately with the proper sum or difference. Lastly, we return the new DBLNS number (lines 6 and 13).

A few example illustrate how to use the DBLNS class. Note that there will be a second or two pause when importing the class as the lookup table is initially generated. A hardware implementation would have this table in permanent memory. For example,

```
>>> from DBLNS import *
>>> x = DBLNS(3.141592)
>>> x
    3.140757
>>> x.b
    260
>>> x.t
    -163
>>> y = DBLNS(13)
>>> y
    13.002569
>>> p = x*y
>>> q = x/y
>>> p
    40.837911
>>> q
    0.241549
```

where we ask for $x = 3.141592$ and get 3.140757, the closest value calculated from the exponents `x.b` and `x.t`, $3.140757 = 2^{260}3^{-163}$. Similarly, $y = 13$ becomes $y = 13.002569$. The product (`p`) and quotient (`q`) are calculated giving 40.837911 (should be 40.840696) and 0.24159 (should be 0.24166). We will come back to these results in the Discussion below.

One thing left to point out about DBLNS numbers is that if the range of the exponents is fixed, as it would be in a true hardware implementation, we might quickly generate exponents that cannot be represented. One solution to this problem is to make use of approximations to unity. For example, all the following are within ±0.005 of 1,

Value	b	t
0.9958324437176391	168	−106
0.9968509430967009	653	−412
0.9968944606878651	−401	253
0.9979140462573112	84	−53
0.9989346746199259	569	−359
0.998978283176522	−485	306
1.0	0	0
1.0010227617964118	485	−306
1.001066461508586	−569	359
1.0020903140410862	−84	53
1.0031152137308417	401	−253
1.004141161648845	886	−559
1.0041849974949628	−168	106

which means that we could pre-multiply a value by one of these approximations to reduce the exponents to a range which fits the number of bits used to represent them. For example, consider this set of expressions,

```
>>> from DBLNS import *
>>> x = DBLNS(3.141592)
>>> x
    3.140757
>>> A = x*x*x*x
>>> A.b
    1040
>>> A.t
    -652
>>> A
    inf
>>> u = DBLNS(0)
>>> u.b = -401
>>> u.t = 253
>>> u
    0.996894
>>> z = x*u
>>> B = z*z*z*z
>>> B
    96.102380
>>> B.b
    -564
>>> B.t
    360
```

where we calculate x^4, first by simply multiplying x (A) and by scaling x by an approximation to unity (z=x*u) before multiplying (B). The final exponents for A are beyond the representable range so that we get infinity as the result. But, if we scale by a suitably chosen approximation to unity (u) we get a result of 96.102380 (actual is 97.305506). So, what just happened? We had $x = 2^{260}3^{-163}$ and we replaced it with $z = xu = 2^{260+(-401)}3^{-163+253} = 2^{-141}3^{90}$ which reduced the size of the exponents by nearly a factor of two. Because of this, z^4 resulted in exponents that were still within the allowed range. A full implementation of DBLNS would use this transformation to keep the exponents in range.

Discussion In this section we investigated two different ways to represent numbers using a double-base system. The first represented unsigned integers as a set of tuples, exponents of 2^i3^j, which sum to the number. We saw that an efficient mechanism exists for generating this representation. We also saw how to easily implement addition and multiplication of numbers in this format along with tricks for keeping the representation sparse. As interesting as double-base numbers are, the lack of subtraction and division operations for the unsigned form greatly limits its use. As with all the number systems in this chapter, DBNS numbers are intended for implementation in hardware for very specific calculations that require high speed. For example, Dimitrov and Jullien [6] gives an algorithm for computing $y = A^E \pmod{p}$ using unsigned DBNS numbers. This is a frequently used operation for cryptographic systems and so when the numbers involved are very large unsigned DBNS can be seen as an advantage. The second way we investigated involves double-base logarithmic numbers (DBLNS). These are numbers of the form $x = s_x 2^b 3^t$ where b and t are signed integer exponents and s_x is the sign of x. For a fixed range of exponents we saw how to implement basic multiplication and division operations and how to take advantage of unity approximations to keep the exponents in a set range. Multiplication and division of DBLNS numbers is a very fast operation. However, it is not hard to see that unless the range of integer exponents is large there are significant losses in precision when converting floating-point values to DBLNS. These errors are then further compounded when operations are performed on the numbers and when unity approximations are used. This limits the utility of DBLNS numbers. Again, Dimitrov and Jullien [6] offers examples of using DBLNS numbers for filter calculations It also offers block diagrams for DBLNS hardware implementations. The first section of this chapter examined logarithmic numbers (LNS) which is similar to the DBLNS but uses a floating-point exponent, represented in fixed-point format, to represent numbers as a power of a single base (2). In that section we saw that efficient multiplication and division of LNS numbers as well as addition and subtraction with look-up table interpolation of a 1-D function. Similar addition and subtraction operations for DBLNS numbers requires interpolating a 2-D function. On the whole, the advantages of LNS, which is a viable alternative to IEEE floating-point, seem to outweigh those of DBLNS.

10.4 Residue Number System

The residue number system (RNS) was described by Garner [7]. It is particularly useful for rapid multiplication and addition and, unlike the double-base number system above, also supports subtraction. However, division is difficult, as are other operations like magnitude comparison. Conversions to/from standard number formats are also less than straightforward. Our implementation will focus on representation, conversion, and the operations of unsigned addition, multiplication, and subtraction. Like the other number systems in this chapter, the advantages of RNS are best seen in hardware. Keeping space considerations in mind, we will not consider signed RNS numbers but will challenge the reader to do so in the exercises.

Representing Numbers in RNS The word "residue" in "residue number system" refers to the fact that numbers are represented as collections of remainders for a chosen set of bases. The remainder is found by the modulo operation which in C is the `%` operator so that `5 % 3 = 2` since the remainder of 5 divided by 3 is 2. Python uses the same operator as C.

In order for the representations to be unique the selected bases must be relatively prime, i.e., have no factors in common. Prime numbers naturally satisfy this requirement so we will use prime bases for our RNS implementation. How many bases and their values is up to the implementation. In Garner's original paper the bases for the examples are 2, 3, 5, and 7. The number of possible remainders for a number divided by 2 is, of course, two: 0 or 1. Similarly, for 3 it is three: 0, 1, 2. The pattern is clear, if the base is n then there are n possible remainders, $0, 1, \ldots, n-1$. So, if we select 2, 3, 5 and 7 as bases we can uniquely represent $2 \times 3 \times 5 \times 7 = 210$ possible numbers. So far so good but we haven't yet given concrete examples. Let's do so now with Garner's set of bases. In order to represent a number we use the remainders for the number divided by each of the bases like so,

	2	3	5	7
0	0	0	0	0
1	1	1	1	1
2	0	2	2	2
3	1	0	3	3
4	0	1	4	4
5	1	2	0	5
6	0	0	1	6
7	1	1	2	0
8	0	2	3	1
9	1	0	4	2
10	0	1	0	3

where 3 becomes 1 0 3 3 because `3 % 2 = 1`, `3 % 3 = 0`, `3 % 5 = 3`, and `3 % 7 = 3`.

The largest number we can represent with the bases above is 209 which becomes 1 2 4 6. What happens if we try to represent a larger number? Following the conversion rule we see that 210 becomes 0 0 0 0 and 211 is 1 1 1 1 which is the same as zero and one, respectively. So, after 209 the pattern of residues repeats which we would expect because 210 is the product of the bases and therefore is evenly divided by all of them.

Conversion to RNS Let's start our RNS implementation by deciding on the set of bases and how we are going to store them in a 32-bit unsigned C integer. If we use four bases we can allocate 8 bits to each residue. To maximize the range we should use bases that are close to $2^8 - 1 = 255$ so that we make maximum use of the bits available. A table of prime numbers reveals that the four largest primes less than 255 are: 233, 239, 241, and 251. Therefore, we will use these primes as the bases and allot 8 bits of the 32-bit unsigned integer to each of them. What sort of range will this give us compared to a standard 32-bit unsigned integer? Comparing gives,

$$\begin{aligned} 2^{32} - 1 &= 4{,}294{,}967{,}295 \\ 233 \times 239 \times 241 \times 251 - 1 &= 3{,}368{,}562{,}317 \end{aligned}$$

meaning we will retain 78.4% of the range of a standard 32-bit unsigned integer.

In order to store the residues we need to divide the 32 bits into four blocks of 8 bits. It will be handy to define bit masks and shift values so that we can quickly set and retrieve the value for any base. Once we have these we can define the conversion to RNS by using standard C operators. Putting it all together gives is the beginning of our RNS library,

```
1  #include <stdint.h>
2
3  typedef uint32_t rns_t;
4
5  #define B0 233
6  #define B1 239
7  #define B2 241
8  #define B3 251
9
10 #define M0  0xFF
11 #define M1  0xFF00
12 #define M2  0xFF0000
13 #define M3  0xFF000000
14
15 #define S0  0
16 #define S1  8
17 #define S2  16
18 #define S3  24
19
20 void pp(rns_t n) {
21     uint16_t n0,n1,n2,n3;
22     n0 = (n & M0) >> S0;
23     n1 = (n & M1) >> S1;
24     n2 = (n & M2) >> S2;
25     n3 = (n & M3) >> S3;
26     printf("%02x %02x %02x %02x", n0,n1,n2,n3);
27 }
28
29 rns_t to_rns(uint32_t n) {
30     rns_t ans = 0;
31     ans |= (n % B0) << S0;
32     ans |= (n % B1) << S1;
33     ans |= (n % B2) << S2;
34     ans |= (n % B3) << S3;
35     return ans;
36 }
```

where we include `stdint.h` to have access to the `uint32_t` data type (line 1). We define our RNS data type in line 3. Lines 5 through 8 define the prime numbers we are using for the bases. Lines 10 through 13 define bit masks on the 32-bit integer that, along with the shift values defined in lines 15 through 18, will let us select the value for any base. We store the bases so that the bits assigned to `B0` cover bits 0 through 7, `B1` cover bits 8 through 15, `B2` cover bits 16 through 23 and `B3` bits 24 through 31.

Lines 20 through 27 define a helper "pretty print" function to display the residues. It also serves as an example of how to extract the residues using the idiom of `(n & M1) >> S1` which selects the desired bits using the mask and then shifts down so that the bits are in the lowest position.

Lastly, lines 30 through 36 define our first library function, `to_rns`, which finds the residue for each base (`n % B0`) and shifts it into the proper place in the RNS representation (`(n % B0) << S0`). The bits are then logically OR'ed to set them without modifying other bits. The result is the RNS representation of the input number.

Let's use these functions to examine the conversion of 1234567890 to RNS. We call `to_rns(1234567890)` to get the RNS representation and then call `pp` to print the residues. Doing so gives us,

$$\begin{aligned} 1234567890 &= 94_{16},\ 06_{16},\ 52_{16},\ 2b_{16} \\ &= 148,\ 6,\ 82,\ 43 \end{aligned}$$

which is what we want to see since,

$$\begin{array}{rcrcr} 1234567890 & \% & 233 & = & 148 \\ 1234567890 & \% & 239 & = & 6 \\ 1234567890 & \% & 241 & = & 82 \\ 1234567890 & \% & 251 & = & 43 \end{array}$$

Now that we can represent RNS numbers, let's implement addition and multiplication.

Unsigned Addition and Multiplication The entire point behind considering the residue number system is that addition and multiplication are fast, highly parallel, operations. This especially true when RNS is implemented in hardware. It will not be true in terms of the number of operations performed by our RNS library but the library will illustrate the process.

Addition is performed base by base and there is no carrying of values between the bases. For each base we add the corresponding residues, modulo the base, and that is all. It is the lack of a carry that makes RNS potentially useful. It is not difficult to imagine a hardware implementation that performs this for each base simultaneously. Let's consider an example using the bases 2, 3, 5, and 7,

$$\begin{array}{rrrrrrr} & 104 & & 0 & 2 & 4 & 6 \\ + & 95 & + & 1 & 2 & 0 & 4 \\ \hline & & & 1 & \mathbf{4} & 4 & \mathbf{10} \\ & 199 & & 1 & 1 & 4 & 3 \end{array}$$

where the line with bold entries shows the result of simply adding each residue. The bold 4 and 10 exceed the base, 3 and 7, respectively, and so must be adjusted. While the modulo operation implies division it is simpler in this case because all that is needed is to subtract the base if the residue sum exceeds it. This produces the last line where it is claimed that 1 1 4 3 is an RNS representation of 199, which it is since: 199%2 = 1, 199%3 = 1, 199%5 = 4, and 199%7 = 3.

To add addition to our RNS library we will first define two helper functions, `rns_set` and `rns_get`, which set or get the value of a particular residue for a given RNS number. These are easily implemented,

```
1  void rns_set(rns_t *num, uint16_t val, uint8_t base) {
2      switch (base) {
3          case 0: *num |= (uint32_t)val << S0; break;
4          case 1: *num |= (uint32_t)val << S1; break;
5          case 2: *num |= (uint32_t)val << S2; break;
6          case 3: *num |= (uint32_t)val << S3; break;
7          default: break;
8      }
9  }
10
11 uint16_t rns_get(rns_t num, uint8_t base) {
12     uint16_t ans = 0;
13     switch (base) {
14         case 0: ans = (uint16_t)((num & M0) >> S0); break;
15         case 1: ans = (uint16_t)((num & M1) >> S1); break;
16         case 2: ans = (uint16_t)((num & M2) >> S2); break;
17         case 3: ans = (uint16_t)((num & M3) >> S3); break;
18         default: break;
19     }
20     return ans;
21 }
```

where to set a residue of an RNS number we supply the number, the residue value, and which base it belongs in (line 1). The rest of the function determines which base and shifts the residue value up accordingly before OR'ing it into the RNS number (e.g. line 3). To get a residue we supply the RNS number and which base (line 11) and, after masking the number appropriately, shift the residue down and return it (e.g. line 14).

With the helper functions above addition becomes particularly terse,

```
1 rns_t rns_add(rns_t x, rns_t y) {
2     rns_t ans = 0;
3     rns_set(&ans, (rns_get(x,0)+rns_get(y,0)) % B0, 0);
4     rns_set(&ans, (rns_get(x,1)+rns_get(y,1)) % B1, 1);
5     rns_set(&ans, (rns_get(x,2)+rns_get(y,2)) % B2, 2);
6     rns_set(&ans, (rns_get(x,3)+rns_get(y,3)) % B3, 3);
7     return ans;
8 }
```

where we set each residue to the sum of the corresponding residues of each input modulo the base. For example, line 3 sets the residue for base 0. Once all residues are set the new RNS number, with the proper sum value, is returned (line 7).

Multiplication of two RNS numbers is identical to addition in that it can be done in parallel and involves no carries. Instead of adding each residue, we multiply them instead (modulo the base). For example, consider,

	14		0	2	4	0
×	13	×	1	1	3	6
			0	2	**12**	0
	182		0	2	2	0

where the product of each residue is given followed by the residue product modulo the base. Note, the bold 12 implies that unlike addition, simple subtraction of the base value to find the residue is not sufficient for multiplication. The final answer is claimed to be 0 2 2 0 which is correct because $182\%2 = 0$, $182\%3 = 2$, $182\%5 = 2$, and $182\%7 = 0$. The implementation of multiplication, then, is a trivial modification of our addition routine,

```
rns_t rns_mul(rns_t x, rns_t y) {
    rns_t ans = 0;
    rns_set(&ans, (rns_get(x,0)*rns_get(y,0)) % B0, 0);
    rns_set(&ans, (rns_get(x,1)*rns_get(y,1)) % B1, 1);
    rns_set(&ans, (rns_get(x,2)*rns_get(y,2)) % B2, 2);
    rns_set(&ans, (rns_get(x,3)*rns_get(y,3)) % B3, 3);
    return ans;
}
```

where the only difference is to replace "+" with "*" for each residue.

Unsigned Subtraction If we recall that subtraction is defined by addition of the additive inverse we can see how to implement unsigned subtraction in RNS. The residues are all modulo the base which means that each residue is really a mathematical group and by definition every element of a group has an inverse so that the group operation performed on element a and its inverse, a^{-1}, will result in the identity element. The identity element is zero, so, for any of the residues, the inverse to a value, a, is what needs to be added to a to get to 0. This value is always the base minus a. So, if our bases are 2, 3, 5 and 7 and we have a set of residues, 1 2 0 4 (95), we calculate the inverse to be 1 1 5 3 because $1\,2\,0\,4 + 1\,1\,5\,3 = 0\,0\,0\,0$. Therefore, subtraction becomes addition of the inverse. For example, if we want to subtract 95 (1 2 0 4) from 134 (0 2 4 1) we should get $134 - 95 = 39$ or 1 0 4 4 because,

$$
\begin{aligned}
0\,2\,4\,1 - 1\,2\,0\,4 &= 0\,2\,4\,1 + 1\,1\,5\,3 \\
&= 1\,\mathbf{3}\,\mathbf{9}\,4 \\
&= 1\,0\,4\,4
\end{aligned}
$$

which is indeed 39, as expected.

The above means that we can add subtraction, $A - B$ assuming $A > B$, to our library by modifying addition to calculate the inverse of B first,

```
1 rns_t rns_inverse(rns_t y) {
2     rns_t ans = 0;
3     rns_set(&ans, B0 - rns_get(y,0), 0);
4     rns_set(&ans, B1 - rns_get(y,1), 1);
5     rns_set(&ans, B2 - rns_get(y,2), 2);
6     rns_set(&ans, B3 - rns_get(y,3), 3);
7     return ans;
8 }
9
10 rns_t rns_sub(rns_t x, rns_t y) {
11     rns_t ans = 0, iy = rns_inverse(y);
12     rns_set(&ans, (rns_get(x,0)+rns_get(iy,0)) % B0, 0);
13     rns_set(&ans, (rns_get(x,1)+rns_get(iy,1)) % B1, 1);
14     rns_set(&ans, (rns_get(x,2)+rns_get(iy,2)) % B2, 2);
15     rns_set(&ans, (rns_get(x,3)+rns_get(iy,3)) % B3, 3);
16     return ans;
17 }
```

where `rns_sub` is virtually identical to `rns_add` with the addition of first calculating the additive inverse of `y` (line 11). The inverse itself is calculated in `rns_inverse` (line 1) where each residue, r, of the argument is replaced with $B - r$, B the base of the residue.

Conversion from RNS Conversion to RNS is straightforward as shown above. Conversion from RNS back to a natural number is less obvious. We need to find values that map to special RNS representations. Once we have these values we can simply multiply the residues by them, sum the results, and get our answer (modulo the product of the bases). The special RNS representations are the ones where only one base has a value, of one, and all the others are zero. For example, our library has four bases so we need to find four numbers that in RNS are represented as 1 0 0 0, 0 1 0 0, 0 0 1 0, and 0 0 0 1. Why? Because, if we know the number that is 1 0 0 0, call it b_0, then an RNS number that has a residue of 3 for the first base will contribute $3b_0$ to the value of the number. Repeating for all the bases and summing, modulo the product of the bases, will give us our desired value.

An example is in order. Switching back to the simpler bases used above (2, 3, 5, and 7), we are looking for numbers that have RNS representations of 1 0 0 0, 0 1 0 0, 0 0 1 0, and 0 0 0 1. This is really saying that we are looking for an integer, A_0, which satisfies $A_0\%2 = 1, A_0\%3 = 0, A_0\%5 = 0$ and $A_0\%7 = 0$. This is just an application of the Chinese Remainder Theorem first presented by Sun Tzu centuries ago. There are different ways to locate the solution but since this is a book about computers, and computers are fast, we can simply use brute force searching. If we do this for the simple bases the computer responds immediately with 105, 70, 126, and 120. Therefore, we know that,

$$
\begin{array}{rcl}
105 & \rightarrow & 1\,0\,0\,0 \\
70 & \rightarrow & 0\,1\,0\,0 \\
126 & \rightarrow & 0\,0\,1\,0 \\
120 & \rightarrow & 0\,0\,0\,1
\end{array}
$$

which means that if we have $182 \rightarrow 0\ 2\ 2\ 0$ we can recover the original number as,

$$[105(0) + 70(2) + 126(2) + 120(0)]\%210 = 182$$

In order to add such a conversion to our library we need to find the four integers that map to the special RNS representations for the bases 233, 239, 241, and 251. Again, we use brute force searching, with time to get a cup of coffee, and learn that the special values are,

$$
\begin{array}{rcl}
3021585941 & \rightarrow & 1\,0\,0\,0 \\
1099363434 & \rightarrow & 0\,1\,0\,0 \\
1663315003 & \rightarrow & 0\,0\,1\,0 \\
952860257 & \rightarrow & 0\,0\,0\,1
\end{array}
$$

which we can encapsulate in the library as,

```
1  #define A0 3021585941u
2  #define A1 1099363434u
3  #define A2 1663315003u
4  #define A3  952860257u
5
6  uint32_t to_int(rns_t n) {
7      uint64_t ans;
8      ans = (uint64_t)A0*rns_get(n,0) +
9            (uint64_t)A1*rns_get(n,1) +
10           (uint64_t)A2*rns_get(n,2) +
11           (uint64_t)A3*rns_get(n,3);
12     return (uint32_t)(ans % (rns_maxint()+1));
13 }
```

where we multiply each special value by the residue for that base (e.g. line 8) and track the sum (`ans`) in a 64-bit integer to avoid overflow. The final return value is this sum modulo the maximum number of values in the RNS representation (line 12). The case in line 12 back to a 32-bit integer will not lose precision because of the modulo by a number that we know fits in a 32-bit value.

Discussion In this section we examined the essence of RNS, how to do basic arithmetic (ignoring difficult division), and how to convert numbers to and from RNS. In terms of practicality, at least for integer operations, RNS is a plausible option, especially for hardware cases which require low power and high performance.

Division remains more difficult. In [7] Garner speculates that division could be implemented as a trial multiplication and subtraction process, similar to doing long division by hand. Implementing such an algorithm in hardware might be difficult and worse yet, slow enough to undo the performance gains for addition, subtraction and multiplication.

Hopefully our brief introduction to the residue number system was enough to whet your appetite. RNS is still an active research area because of the promise of fast, carry-free, parallel operations. For example, consider [8, 9] and [10]. To dive deep into RNS, try [11].

10.5 Redundant Signed-Digit Number System

The final number system for this chapter is the redundant signed-digit number system (RSD). Like the residue and double-base number systems above, the redundant system allows a number to be represented in multiple ways in order to perform certain operations very quickly and efficiently. And, like the other number systems, this flexibility comes with a price when attempting to implement other operations.

The phrase "redundant signed-digit number system" refers to a positional number system where the digits are allowed to take on more values than are necessary. For example, a standard positional number, x, using radix (base) r requires digits $0, 1, \ldots, r-1$ and represents the number as a string of digits, $d_n d_{n-1} d_{n-2} \ldots d_0$, with the value,

$$x = \sum_{i=0}^{n-1} d_i r^i$$

The RSD system loosens the digit requirement and allows digits to take on values from $-\alpha, \ldots, -1, 0, 1, \ldots, \beta$ where α and β are $< r$. If $\alpha = \beta$ the system is symmetric. If $\alpha = \beta = r - 1$ the system is symmetric and maximally redundant. We will consider only symmetric, maximally redundant systems here. With these constraints, then, a radix-10 RSD number system has digits which range from $-9, -8, \ldots, 8, 9$ and therefore allows specific numbers to be represented in multiple ways depending upon the number of digits used. For example, this table shows select values, $0, \ldots, 99$, as they could be represented with two or three digits,

$0 =$	$00,$		$0 =$	000			
$1 =$	$01,$	$1\bar{9}$	$1 =$	$001,$	$01\bar{9},$	$1\bar{9}\bar{9}$	
$2 =$	$02,$	$1\bar{8}$	$2 =$	$002,$	$01\bar{8},$	$1\bar{9}\bar{8}$	
$3 =$	$03,$	$1\bar{7}$	$3 =$	$003,$	$01\bar{7},$	$1\bar{9}\bar{7}$	
$4 =$	$04,$	$1\bar{6}$	$4 =$	$004,$	$01\bar{6},$	$1\bar{9}\bar{6}$	
$17 =$	$17,$	$2\bar{3}$	$17 =$	$017,$	$02\bar{3},$	$1\bar{9}7,$	$1\bar{8}\bar{3}$
$18 =$	$18,$	$2\bar{2}$	$18 =$	$018,$	$02\bar{2},$	$1\bar{9}8,$	$1\bar{8}\bar{2}$
$19 =$	$19,$	$2\bar{1}$	$19 =$	$019,$	$02\bar{1},$	$1\bar{9}9,$	$1\bar{8}\bar{1}$
$20 =$	$20,$		$20 =$	$020,$	$1\bar{8}0,$		
$85 =$	$85,$	$9\bar{5}$	$85 =$	$085,$	$09\bar{5},$	$1\bar{2}5,$	$1\bar{1}\bar{5}$
$86 =$	$86,$	$9\bar{4}$	$86 =$	$086,$	$09\bar{4},$	$1\bar{2}6,$	$1\bar{1}\bar{4}$
$87 =$	$87,$	$9\bar{3}$	$87 =$	$087,$	$09\bar{3},$	$1\bar{2}7,$	$1\bar{1}\bar{3}$
$88 =$	$88,$	$9\bar{2}$	$88 =$	$088,$	$09\bar{2},$	$1\bar{2}8,$	$1\bar{1}\bar{2}$
$89 =$	$89,$	$9\bar{1}$	$89 =$	$089,$	$09\bar{1},$	$1\bar{2}9,$	$1\bar{1}\bar{1}$
$90 =$	$90,$		$90 =$	$090,$	$1\bar{1}0,$		
$99 =$	$99,$		$99 =$	$099,$	$1\bar{1}9,$	$10\bar{1}$	

where we use $\bar{9} = -9$. As the number of digits in the representation increases the number of possible representations also increases. Note that the range of an n-digit RSD number is twice that of an n-digit number for any radix (if symmetric). So, while a three digit decimal number can represent values from $[0, 999]$ a symmetric radix-10 RSD system can represent values from $[-999, 999]$.

Redundant Binary Representation (RBR) Even though RSD supports any radix the majority of the research and practical implementations of redundant signed-digit arithmetic have focused on radix-2 or binary. This is the "redundant binary representation" (RBR) and we will create a small library of RBR routines.

Following the definition above for the digits of a redundant representation the RBR digits are drawn from the set $\{-1, 0, 1\}$. Further, a practical implementation of RBR will encode the digits using two bits per digit according to this mapping,

Value	r	s
-1	0	0
0	0	1
0	1	0
+1	1	1

where r and s are the two bits stored for each digit value. For example, the RBR number $01\bar{1}011\bar{1}0_2$ would be stored as,

$$01\bar{1}011\bar{1}0_2 = 01\ 11\ 00\ 01\ 11\ 11\ 00\ 01_2$$

where it is immediately evident that RBR uses twice as much memory per value than standard binary representation. Conversion from RBR to the numerical value is straightforward using the table above. If there are n RBR digits, then memory is accessed in pairs of bits and converted according to,

$$x = \sum_{k=0}^{n-1} d_k 2^k = \sum_{k=0}^{n-1} (r_k + s_k - 1)2^k$$

where d_k is the RBR digit, $\{-1, 0, 1\}$, and r_k and s_k are the actual bits stored in memory using the table above. Let's verify that this indeed true,

$$\begin{aligned} 01\bar{1}011\bar{1}0_2 &= 0 \times 2^7 + 1 \times 2^6 - 1 \times 2^5 + 0 \times 2^4 + 1 \times 2^3 + 1 \times 2^2 - 1 \times 2^1 + 0 \times 2^0 \\ &= 2^6 - 2^5 + 2^3 + 2^2 - 2^1 \\ &= 64 - 32 + 8 + 4 - 2 \\ &= 42 \end{aligned}$$

and,

$$\begin{aligned} &01\ 11\ 00\ 01\ 11\ 11\ 00\ 01_2 \\ &= (0+1-1)2^7 + (1+1-1)2^6 + (0+0-1)2^5 + (0+1-1)2^4 \\ &\quad + (1+1-1)2^3 + (1+1-1)2^2 + (0+0-1)2^1 + (0+1-1)2^0 \\ &= 2^6 - 2^5 + 2^3 + 2^2 - 2^1 \\ &= 64 - 32 + 8 + 4 - 2 \\ &= 42 \end{aligned}$$

Conversion to RBR from standard binary is equally straightforward, simply take the binary value and encode it. There is no need to search for any representation (as was done for DBNS above). Therefore, we start our simple library with conversion routines to map an signed 32-bit integer to an encoded RBR integer stored in an unsigned 64-bit integer,

```
#include <stdint.h>
typedef uint64_t rbr_t;

rbr_t to_rbr(int32_t n) {
    rbr_t ans = 0;
    uint8_t i, s=0;

    if (n < 0) {
        s = 1;
        n = abs(n);
    }

```

```
13     for(i=0; i < 32; i++) {
14         ans += ((n & 1) ? 3 : 1) * (1llu << (2*i));
15         n >>= 1;
16     }
17     return (s) ? ~ans : ans;
18 }
19
20
21 int8_t rbr_digit(uint8_t b) {
22     switch (b) {
23         case 0: return -1;
24         case 1: return  0;
25         case 2: return  0;
26         case 3: return  1;
27     }
28 }
29
30 int32_t to_int(rbr_t n) {
31     int64_t ans = 0;
32     uint8_t i;
33
34     for(i=0; i < 32; i++) {
35         ans += rbr_digit(n & 3) * (1 << i);
36         n >>= 2;
37     }
38     return (int32_t)ans;
39 }
```

where we use `stdint.h` (line 1) to access the `uint32_t` data type (and similar). Line 2 defines the RBR type as an unsigned 64-bit integer. Conversion from binary to RBR starts in line 4. We track the sign of the argument, keeping a flag if it is negative and work with the absolute value (lines 8–11). We then loop over every bit of the argument (line 13) and encode that bit as 11_2 if it is set and 01_2 if it is not set (line 14) where `n & 1` returns the value of the lowest order bit of the argument. Once we have the encoded value we need to move it to the proper place in the 64-bit output and add it in. The phrase `* (1llu << (2*i))` shifts the encoded value to the proper place in the output, `ans`, where it is added to the existing output bits. Recall that each encoded bit takes two output bits hence shifting by `2*i`. Line 15 shifts the argument down one bit so that the next highest bit is in the lowest position. The loop then continues until all the bits of the argument have been encoded and stored. Line 17 returns the encoded RBR representation where if the argument is negative we flip all the bits. Why? Because, if we look at our encoding table we will see that flipping the bits of the pairs will change $01_2 \leftrightarrow 10_2$ and $00_2 \leftrightarrow 11_2$ which maps zero to zero and negative one to positive one and vice versa. This is equivalent to making the number negative because every positive power added (every 1 bit in the argument) is now added with a negative one.

The conversion from RBR to a standard binary integer starts at line 30. While the result will fit in a signed 32-bit integer, we use a signed 64-bit integer as an accumulator. Again, we loop over all 32 bits that will be in the output (line 34) and look at the encoded bits as pairs. The phrase `n & 3` extracts the two lowest bits, which is an encoded digit, and passes them to the `rbr_digit` helper function which implements our conversion table to return the proper bit value, $\{-1, 0, 1\}$. Once we know the digit value for the current bit we multiply it by `1 << i` to determine the numerical value we should add to the result. Line 30 shifts down by two bits to get to the next encoded digit. Line 32 returns the final answer cast to a 32-bit signed integer.

Let's look at the RBR encoding/decoding process,

$$123 \rightarrow 5555555555557\text{FDF}_{16}$$

$$-123 \rightarrow \text{AAAAAAAAAAAA}8020_{16}$$

where the right value is the RBR encoded version of 123 or -123. The number 123 fits in eight bits which becomes 16 bits in the encoded version. All other encoded bits will be zero. We encode zero as 01_2 or 10_2 so each pair of zero digits becomes either $0101_2 = 5_{16}$ or $1010_2 = A_{16}$, hence all the leading 5's and A's in the encoded versions. Let's look more closely at $7FDF_{16}$, which by pairs is,

$$01\ 11\ 11\ 11\ 11\ 01\ 11\ 11_{RBR} = 01111011_2 = 123$$

and for -123, 8020_{16}, which by pairs is,

$$10\ 00\ 00\ 00\ 00\ 10\ 00\ 00_{RBR} = 0\bar{1}\bar{1}\bar{1}\bar{1}0\bar{1}\bar{1}_2 = -123$$

showing that the RBR representation does indeed map back to 123 or -123 as we would expect.

RBR Addition and Subtraction Now that we can convert to and from RBR we can implement addition and subtraction. In hardware the addition process is highly parallelizable but, of course, here we will use standard C code. The key to addition in RBR, which is slightly different than RSD addition in bases greater than 2, is a translation table which converts the sum of two bits in $\{-1, 0, 1\}$ to a partial sum and transfer bit which is added to the partial sum of the next highest bit. So, instead of bit by bit addition with carry we generate the partial sum and transfer bits simultaneously, for each digit in the operands, and then shift the transfer bits and add. The selection of the partial sum and transfer bit values ensures that the sum will never require a carry. Let's look at an example which should clarify things.

Let's add $x = 123$ and $y = 35$ to see if we get 158. The RBR representations for 123 and 35 are, respectively, 01111011_{RBR} and 00100011_{RBR}. We will add these values by using a look up table which converts the bits of x and y to a new representation, the partial sum and the transfer bits. The table is,

x_i	y_i	$\rightarrow$	t_i	s_i
−1	−1		−1	0
−1	0		−1	1
0	−1		0	−1
−1	1		0	0
1	−1		0	0
0	0		0	0
0	1		0	1
1	0		1	−1
1	1		1	0

where the i-th bit of x and y are used to look up the i-th bit of s and t. Once we have s and t we need to shift t up one bit and then add to get the final result,

		0	1	1	1	1	0	1	1	$\leftarrow$	x
+		0	0	1	0	0	0	1	1	$\leftarrow$	y
		0	$\bar{1}$	0	$\bar{1}$	$\bar{1}$	0	0	0	$\leftarrow$	s
+	0	1	1	1	1	0	1	1		$\leftarrow$	t (shifted left)
	0	1	0	1	0	$\bar{1}$	1	1	0	$\rightarrow$	158

Subtraction is implemented by adding the inverse, $x - y \rightarrow x + \bar{y}$, which is the same as adding the negation of y. As we saw above for our conversion routines, flipping the bits of an RBR number changes the sign, so subtraction is simply adding the flipped bits of y.

Let's add addition and subtraction to our set of RBR routines,

```
int8_t rbr_enc(int8_t s) {
    switch (s) {
        case -1: return 0;
        case  0: return 1;
        case  1: return 3;
    }
}

int8_t rbr_add_partial(rbr_t s, rbr_t t) {
    int8_t z = rbr_digit(s & 3) + rbr_digit(t & 3);
    return rbr_enc(z);
}

int8_t rbr_add_st(rbr_t x, rbr_t y, int8_t *s, int8_t *t) {
    int8_t dx = rbr_digit(x & 3);
    int8_t dy = rbr_digit(y & 3);
    if ((dx == -1) && (dy == -1)) { *t = -1; *s =  0; }
    if ((dx == -1) && (dy ==  0)) { *t = -1; *s =  1; }
```

```
19     if ((dx ==  0) && (dy == -1)) { *t =  0; *s = -1; }
20     if ((dx == -1) && (dy ==  1)) { *t =  0; *s =  0; }
21     if ((dx ==  1) && (dy == -1)) { *t =  0; *s =  0; }
22     if ((dx ==  0) && (dy ==  0)) { *t =  0; *s =  0; }
23     if ((dx ==  0) && (dy ==  1)) { *t =  0; *s =  1; }
24     if ((dx ==  1) && (dy ==  0)) { *t =  1; *s = -1; }
25     if ((dx ==  1) && (dy ==  1)) { *t =  1; *s =  0; }
26 }
27
28 rbr_t rbr_add(rbr_t x, rbr_t y) {
29     rbr_t ans=0, s=0, t=0;
30     int8_t ds,dt;
31     uint8_t i;
32
33     for(i=0; i < 32; i++) {
34         rbr_add_st(x, y, &ds, &dt);
35         s += ds * (1llu << (i*2));
36         t += dt * (1llu << (i*2));
37         x >>= 2;
38         y >>= 2;
39     }
40     t <<= 2;
41
42     for(i=0; i < 32; i++) {
43         ans += rbr_add_partial(s,t) * (1llu << (2*i));
44         s >>= 2;
45         t >>= 2;
46     }
47     return ans;
48 }
49
50 rbr_t rbr_sub(rbr_t x, rbr_t y) {
51     return rbr_add(x, ~y);
52 }
```

where the code looks a little daunting at first but it isn't so bad. Let's start with subtraction (line 50). As stated above, subtraction is adding the inverse of the second argument so we simply return the sum of x and $\neg y$ where $\neg$ means logical NOT. Addition itself starts on line 28. We process each bit of the operands (x and y) in the loop at line 33. We take the current lowest order two bits of each (an encoded RBR digit) and use them to decide what values we will put in the corresponding bit positions of s and t. We use the scaling trick described above for our conversion routines to put encoded bit pairs in the proper place (lines 35 and 36). The call to `rbr_add_st` accomplishes this by converting the lowest digit of x and y to their actual digit values (lines 15 and 16) and then searches the lookup table, here implemented with a set of `if` statements (lines 17 through 25). These statements will set s and t properly. With the current lowest order digit processed x and y are shifted to the next digit (lines 37 and 38). When all digits have been processed line

40 shifts the transfer digits (t) up one position before adding them. The addition happens in the loop of lines 42 through 46. The design of the lookup table ensures that these additions will never produce a carry. At this point, `ans` contains our final sum (line 47).

The RBR representation seems cumbersome when implemented in code. In practice, hardware implementations, particularly on FPGAs, are parallelized which means that addition is a very rapid operation, even with the transfer digits.

Discussion The RSD system (including RBR) is, as all the systems of this chapter are, a special-purpose system which can be used in custom hardware situations for specific algorithms where the advantage of rapid arithmetic is useful. Intellectually, it is entertaining to consider a system like RSD and to see how it operates. RBR remains an area of active research. For some recent examples, consider [12–14] and [15].

10.6 Chapter Summary

In this chapter we reviewed four special purpose number systems: logarithmic numbers, double-base numbers, residue numbers, and redundant numbers. Each of these representations offers advantages in certain situations, most often realized best in hardware, but all, except logarithmic numbers, would be unlikely alternatives to standard representations. Logarithmic numbers offer a viable alternative to IEEE 754 in terms of working with floating-point numbers. Multiplication and division are extremely rapid in that case and while addition and subtraction are less precise, requiring function interpolation, modern hardware can store detailed information on the function to be interpolated so that even addition and subtraction can be rapid at the expense of some memory use. This makes LNS an attractive alternative for floating-point operations [16].

Double-base numbers are interesting, not least for their ability to be represented in two dimensions. Addition is very rapid and multiplication is as well, but subtraction and division are likely not possible in the unsigned form. The double-base logarithmic form has the same sort of possible advantages as LNS but use of integer exponents causes a serious lack of precision.

Residue numbers are attractive for their rapid implementation of addition, subtraction, and multiplication. The set of routines developed in this chapter was restricted to unsigned integers for space considerations but there is no reason signed numbers cannot also be represented as residues. Fixed-point floating-point operations are also possible. Again, special-purpose hardware can take advantage of the representation for specific algorithms.

Lastly, we considered redundant signed-digit numbers, specifically radix-2 redundant binary numbers. We saw how this system allows for rapid, parallel, implementation of addition and subtraction. The RSD representation is curious in that since each digit is signed there is no need for an overall sign for the number. This also makes comparison between numbers more difficult. The constant-time addition speed of RBR, in hardware, offers a distinct advantage in performance compared to two's complement addition, especially for larger bit widths (≥ 32), since the time for two's complement addition increases as the log of the bit width [17].

Exercises

10.1 Add a new function, `lns_lg2(x)`, which computes $\log_2(x)$, to the LNS number library of Sect. 10.2.

10.2 Add a new function, `lns_pow(x,n)`, which computes x^n, $n \in \mathbb{Z}$, to the LNS number library of Sect. 10.2.

10.3 Add a new function, `lns_sqrt(x)`, which computes $\sqrt{x}$, to the LNS number library of Sect. 10.2.

10.4 Add a new method to the Python DBNS class which computes x^y where x and y are both DBNS instances.

10.5 Add a new method to the Python DBNS class which computes $x!$, the factorial of x.

10.6 Addition and subtraction of DBLNS numbers involves interpolation of a 2-D function in a manner similar to the 1-D interpolation of the logarithmic number system. Specifically, addition (subtraction) is $2^a3^b \pm 2^c3^d = 2^a3^b(1 \pm 2^{c-a}3^{d-b})$ where the trick is to quickly interpolate $1 \pm 2^{c-a}3^{d-b}$ in a reasonable way. For this exercise skip interpolation and implement the function exactly using Python.

10.7 Add a new function to the RNS library, `rns_pow`, that takes an RNS number and an unsigned integer (`uint16_t`) as input and returns the RNS number raised to that power.

10.8 Add a new function to the RNS library, `rns_fact`, that takes an RNS number, x, and returns $x!$, the factorial of x.

10.9 Our library for RNS numbers focused on unsigned numbers only. Signed numbers can be represented in RNS in a couple of ways. One way is to split the range of the residues in two and use two's complement representation for negative residues. Another is to store the sign separately and use the `to_int` function to convert the arguments to integers, compare the magnitudes, and perform the proper operation, addition or subtraction, to get the proper magnitude result. The sign then can be set based on this knowledge. (Hint: Where should the sign bit be stored? One option: change from using `uint32_t` to `uint64_t` and use one of the high bits.)

***

10.10 Add multiplication to the RBR library. A naïve approach to $x \times y$ would add x to itself y times or add y to itself x times. Looping over the smaller of x and y would help, a little. Implement this (far) less than ideal approach.

10.11 Now add proper multiplication to the RBR library. Just as multiplication of two integers in binary is accomplished by adding shifted versions of multiplicand depending upon the value of the bits of the multiplier, RBR multiplication works the same way with the added condition that a bit value might be negative in which case the negation of the multiplicand is added. See Sect. 2.3.5 for an explanation of binary multiplication. Implement the corresponding RBR version. **

References

1. Coleman, J. Nicholas, et al. "The European logarithmic microprocessor." Computers, IEEE Transactions on 57.4 (2008): 532–546.
2. Haselman, Michael, et al. "A comparison of floating point and logarithmic number systems for FPGAs." Field-Programmable Custom Computing Machines, 2005. FCCM 2005. 13th Annual IEEE Symposium on. IEEE, 2005.
3. Dimitrov, Vassil S., Graham A. Jullien, and William C. Miller. "Theory and applications of the double-base number system." Computers, IEEE Transactions on 48.10 (1999): 1098–1106.
4. Dimitrov, Vassil S., Graham A. Jullien, and William C. Miller. "An algorithm for modular exponentiation." Information Processing Letters 66.3 (1998): 155–159.
5. Berthe, Valerie, and Laurent Imbert. "On converting numbers to the double-base number system." Optical Science and Technology, the SPIE 49th Annual Meeting. International Society for Optics and Photonics, 2004.
6. Dimitrov, Vassil S., and Graham A. Jullien. "Loading the bases: A new number representation with applications." IEEE Circuits and Systems Magazine 3.2 (2003): 6–23.
7. Garner, Harvey L. "The residue number system." IRE Transactions on Electronic Computers 2 (1959): 140–147.
8. Molahosseini, Amir Sabbagh, Saeid Sorouri, and Azadeh Alsadat Emrani Zarandi. "Research challenges in next-generation residue number system architectures." Computer Science and Education (ICCSE), 2012 7th International Conference on. IEEE, 2012.
9. Chen, Jienan, and Jianhao Hu. "Energy-efficient digital signal processing via voltage-overscaling-based residue number system." IEEE Transactions on Very Large Scale Integration (VLSI) Systems 21.7 (2013): 1322–1332.
10. Zalekian, Azin, Mohammad Esmaeildoust, and Amer Kaabi. "Efficient Implementation of NTRU Cryptography using Residue Number System." International Journal of Computer Applications 124.7 (2015).
11. Mohan, PV Ananda. Residue number systems: algorithms and architectures. Vol. 677. Springer Science and Business Media, 2012.
12. Ganesh, K. V., et al. "Constructing a low power multiplier using Modified Booth Encoding Algorithm in redundant binary number system." International Journal of Engineering Research and Applications Vol 2 (2012): 2734–2740.
13. Shukla, Vandana, et al. "A novel approach to design a redundant binary signed digit adder cell using reversible logic gates." 2015 IEEE UP Section Conference on Electrical Computer and Electronics (UPCON). IEEE, 2015.
14. El-Slehdar, A. A., A. H. Fouad, and A. G. Radwan. "Memristor based N-bits redundant binary adder." Microelectronics Journal 46.3 (2015): 207–213.

15. Cui, Xiaoping, et al. "A Modified Partial Product Generator for Redundant Binary Multipliers." IEEE Transactions on Computers 65.4 (2016): 1165–1171.
16. Coleman, J. N., and R. Che Ismail. "LNS with Co-Transformation Competes with Floating-Point." IEEE Transactions on Computers 65.1 (2016): 136–146.
17. Pai, Yu-Ting, and Yu-Kumg Chen. "The Fastest Carry Lookahead Adder." DELTA. 2004.

Index

A

AND, 25
Arbitrary precision floating-point
  Arithmetic, 270
  Comparing, 273
  Definition, 265
  Representing, 265
  Software libraries, 278
  Trigonometric and transcendental functions, 274
Ariane 5 rocket explosion, 118

B

Big integers, 139
  Addition and subtraction, 146
  Comba multiplication, 159
  Comparing, 144
  Divide-and-conquer, 165
  Division, 155
  in Python, 170
  Input and output, 142
  Karatsuba multiplication, 160
  Knuth division, 164
  Libraries, 167
  Representation, 140
  Schönhage and Strassen multiplication, 161
  School method multiplication, 152
Bits, 19
  Clearing, 32
  Masking, 30
  Rotations, 35
  Setting, 31
  Shifting, 33
  Testing, 31
  Toggling, 32
Boole, George, 25
Boolean algebra, 25
Bytes, 20

C

Checksum, 29
Compare instructions, 37
Converting numbers
  Binary to binary-coded decimal, 70
  Binary to decimal, 13
  Binary-coded decimal to binary, 72
  Decimal to binary, 11
  Fixed-point to floating-point, 187
  Floating-point to fixed-point, 187
  Hexadecimal and binary, 10
  Octal and binary, 11
  Others to decimal, 14
Cryptography, 177

D

Decimal floating-point, 215
  Biased exponent continuation field, 216
  Combination field, 216
  Continuation field, 216
  Declet, 218
  Densely packed decimal (DPD), 218
  in software (C), 225
  in software (Python), 231
  Infinity, 220
  Not-a-number (NaN), 220
  Rounding modes, 220
  Storage formats, 216
  Storage order, 221
Diffie-Hellman key exchange, 178

R.T. Kneusel, *Numbers and Computers*, DOI 10.1007/978-3-319-50508-4

Digital comparator, 37
Distribution of floating-point numbers, 82
Double-base number system
  Addition, 312
  Definition, 307
  Discussion, 323
  Graphical representation, 308
  Improving addition, 314
  Multiplication, 318
  Python implementation, 309
  Real number multiplication and division, 321
  Real number unity approximations, 322
  Representing integers, 308
  Representing real numbers, 319

E
Encryption, 28
Endianness
  Big-endian, 23
  Little-endian, 23
Experiments
  Comparison of fixed-point trigonometric functions, 199
  Decimal floating-point logistic map, 226
  Decimal floating-point to ASCII string, 221
  Floating-point uncertainty in comparing numbers, 125
  Floating-point uncertainty in repeated subtractions, 122
  Floating-point uncertainty in summing an array of numbers, 123
  Floating-point uncertainty in the Logistic Map, 119
  Illustrating uncertainty in floating-point calculations, 119
  Using rational numbers, 176

F
Fixed-point numbers
  Addition, 188
  Cosine (polynomial), 199
  Cosine (table), 196
  Cosine (Taylor series), 197
  Division, 193
  DOOM (case study), 211
  Exponential (Taylor series), 201
  Machine learning, 204
  Multiplication, 189
  Natural logarithm (Newton's method), 203
  Neural networks, 204
  Q notation, 183
  Sine (polynomial), 198
  Sine (table), 194
  Sine (Taylor series), 196
  Square root (Newton's method), 200
  Subtraction, 188
  Trigonometric functions, 194
  When to use, 211
Floating-point
  Addition/subtraction algorithm, 103
  Avoiding the pitfalls, 130
  Binary coded decimal floating-point, 110
  Comparison, 100
  Floating-point number, 81
  IEEE 754 rounding modes (binary), 97
  IEEE addition and subtraction, 102
  IEEE exceptions, 105
  IEEE floating-point in hardware, 108
  IEEE infinity (binary), 91
  IEEE multiplication, 103
  IEEE NaN (binary), 92
  Mantissa, 82
  Multiplication algorithm, 105
  Pitfalls, 117
  Real number, 81
  Rules of thumb for floating-point numbers, 130
  Significand, 82
  Subnormal numbers, 95
  Trapping exceptions, 106
  Using a tool to improve floating-point calculations, 131

H
Herbie, 131

I
IBM S/360 floating-point, 85
IEEE 754 number formats, 89
IEEE 754-2008, 81
Integers
  Binary addition, 41
  Binary subtraction, 43
  Binary-code decimal subtraction, 69
  Binary-coded decimal, 67
  Binary-coded decimal addition, 69
  Densely Packed Decimal, 73
  One's complement numbers, 55
  Power of two test, 46
  Sign extension, 63
  Sign-magnitude numbers, 54
  Signed addition, 58
  Signed comparison, 56

Signed division, 64
Signed multiplication, 60
Signed subtraction, 58
Two's complement numbers, 55
Unsigned, 21
Unsigned addition, 42
Unsigned division, 49
Unsigned multiplication, 47
Unsigned square root, 52
Unsigned subtraction, 44
Zoned Decimal, 74
Interval arithmetic, 236
Absolute value, 246
Addition and subtraction, 239
Comparisons, 247
Dependency problem, 256
in Python, 248
Monotonic functions, 253
MPFI library, 258
Multiplication, 240
Negation, 245
Powers, 243
Properties of intervals, 252
Reciprocal and division, 241
Sine and cosine, 254

L
Least-squares curve fitting, 198
Logarithmic number system, 293
Addition and subtraction, 302
Comparing numbers, 299
Discussion, 306
Multiplication and division, 298
Representation, 294
Logical operators, 25

M
Memory
Addresses, 22
Bit order, 23
Byte order, 23

N
Nibbles, 19
NOT, 26
Numbers
Babylonian numbers, 5
Binary numbers, 8
Decimal numbers, 8
Egyptian numbers, 3
Hexadecimal numbers, 9
Mayan numbers, 7
Octal numbers, 9
Roman numerals, 3

O
OR, 26

P
Parity, 29
Patriot missile failure, 118
Place notation, 5
Propagation of errors, 236

R
Radix point, 5
Rational arithmetic, 171
Addition, 174
Division, 175
GCD, 172
in Python, 171
Multiplication, 175
Subtraction, 174
Receiver operating characteristics curve, 208
Redundant signed-digit number system, 332
Addition and subtraction, 336
Discussion, 339
Redundant binary representation, 333
Residue number system, 324
Addition and multiplication, 327
Conversion from RNS, 330
Conversion to RNS, 325
Discussion, 331
Representing, 324
Subtraction, 329
RSA encryption, 179

S
Schleswig-Holstein parliament election, 119
Scikit Learn library, 206
Shannon, Claude, 19
Swift, Jonathan, 24

T
Torres y Quevedo, Leonardo, 84
Truth table, 25

V
Vancouver stock exchange, 119

W
Words, 20

X
XOR, 26

Z
Zuse, Konrad, 85

CPSIA information can be obtained
at www.ICGtesting.com
Printed in the USA
LVOW04*0741230217
525098LV00011BD/187/P